# Quantum and Optoelectronic Devices, Circuits and Systems

# Quantum and Optoelectronic Devices, Circuits and Systems

Editor

**Lucas Lamata**

MDPI • Basel • Beijing • Wuhan • Barcelona • Belgrade • Manchester • Tokyo • Cluj • Tianjin

*Editor*
Lucas Lamata
Física Atómica, Molecular y
Nuclear
Universidad de Sevilla
Sevilla
Spain

*Editorial Office*
MDPI
St. Alban-Anlage 66
4052 Basel, Switzerland

This is a reprint of articles from the Special Issue published online in the open access journal *Electronics* (ISSN 2079-9292) (available at: www.mdpi.com/journal/electronics/special_issues/micro_optoelectronic).

For citation purposes, cite each article independently as indicated on the article page online and as indicated below:

LastName, A.A.; LastName, B.B.; LastName, C.C. Article Title. *Journal Name* **Year**, *Volume Number*, Page Range.

**ISBN 978-3-0365-7989-4 (Hbk)**
**ISBN 978-3-0365-7988-7 (PDF)**

# Contents

# About the Editor

**Lucas Lamata**

Lucas Lamata's research up until now has been focused on quantum optics and quantum information, including pioneering proposals for quantum simulations of relativistic quantum mechanics, fermionic systems, and spin models, with trapped ions and superconducting circuits. He is also interested in new approaches to quantum simulation, and in this respect, he created, with his collaborators, the concept of digital–analog quantum simulations, which takes advantage of the scalability of analog quantum devices and the versatility and universality of digital ones, and may be a promising method of achieving new knowledge about quantum systems via NISQ devices in the near future. He has also analyzed the possibility of combining artificial intelligence and machine learning protocols with quantum devices. He enjoys working with experimentalists and has made proposals and participated in 16 experiments in collaboration with up to 17 prominent experimental groups in quantum science, with focuses on trapped ions, electrons in Penning traps, superconducting circuits, cold atoms, quantum photonics, and nuclear magnetic resonance. A total of 19 of his theoretical proposals for implementations have been carried out in experiments by top flight groups.

# Preface to "Quantum and Optoelectronic Devices, Circuits and Systems"

The fields of quantum electronics (e.g., quantum dots and superconducting circuits) and optoelectronics have garnered much interest in the past few years. They deal with quantum and optical platforms in combination with electronic and condensed matter systems, which may serve as building blocks for a plethora of applications in quantum computing and optical technology.

In this reprint, we reprint the articles in the Special Issue "Quantum and Optoelectronic Devices, Circuits and Systems", published in *Electronics* (MDPI).

**Lucas Lamata**
*Editor*

*electronics*

*Editorial*

# Quantum and Optoelectronic Devices, Circuits and Systems

**Lucas Lamata** [1,2]

[1] Departamento de Física Atómica, Molecular y Nuclear, Universidad de Sevilla, 41080 Sevilla, Spain; llamata@us.es
[2] Instituto Carlos I de Física Teórica y Computacional, 18071 Granada, Spain

## 1. Introduction

The fields of quantum electronics (e.g., quantum dots, superconducting circuits) as well as optoelectronics have produced many important research achievements in the past few years. They consist of quantum and optical platforms in combination with electronic and condensed matter systems, which can enable building blocks for a plethora of applications in quantum computing, optical technology, and information processing, such as, e.g., machine learning.

In this Special Issue, we have selected a series of works in this area dealing with a wide variety of topics, ranging from quantum dot applications, to quantum information processing, as well as machine learning for industry operations. We hope that the research here described may induce further developments in this area, and produce significant applications for industry and society alike.

In Section 2 we summarize the articles included in the Special Issue. Subsequently, in Section 3 we give a scope for future developments.

## 2. Brief Description of the Published Articles

Juárez-Barojas et al. [1] introduced a maintenance system that relies on artificial intelligence to monitor online the support bed expansion in a 30-L pilot-scale inverse fluidized bed reactor (IFBR). Their aim was to obtain a condition-based maintenance strategy by employing a single-level sensor for a biofilm inverse fluidizing bed as a source for virtual sensors. They implemented an artificial neural network on an embedded electronic system (Raspberry Pi 4) to achieve this.

Moiseev et al. [2] studied aspects of the electroluminescence spectra of narrow-gap type II InAs/InSb/InAs heterostructures containing a single layer of InSb quantum dots placed into a p-n-InAs junction. They used a forward and reverse bias in the temperature range of 77–300 K to investigate these luminescence properties as a function of the surface density of nano-objects buried in the narrow-gap matrix. Under certain conditions, they observed the suppression of negative interband luminescence and the dominance of interface recombination transitions at the InSb/InAs type II heterojunction at room temperature.

Gijare et al. [3] studied the electrochemical measurements of reduced graphene oxide-titanium oxide (rGO)/$TiO_2$) electrodes for the application of a glucose sensor. The paper analyses the sensitivity, stability, and reproducibility of the sensor electrodes employed to evaluate the concentration of glucose in the serum. This sensor was used for glucose level detection in natural blood serum and showed a relative standard deviation of 1.88%, which was in good agreement with the commercial glucose sensor values.

Venkatesan et al. [4] fabricated a planar Si/PEDOT: PSS heterojunction solar cell using three different solvents—ethylene glycol, acetonitrile, and dimethyl sulfoxide—to elucidate the best one. The obtained samples were characterized by diverse techniques, including diffuse reflectance spectroscopy, scanning electron microscopy, X-ray diffraction, and the current–voltage technique.

**Citation:** Lamata, L. Quantum and Optoelectronic Devices, Circuits and Systems. *Electronics* **2023**, *12*, 1717. https://doi.org/10.3390/electronics12071717

Received: 30 March 2023
Accepted: 3 April 2023
Published: 4 April 2023

Balakirev et al. [5] explored the influence of arsenic pressure during low-temperature GaAs overgrowth of InAs quantum dots on their optical properties. They observed a single broad line corresponding to the unimodal size distribution of quantum dots in the spectrum of quantum dots overgrown at a high arsenic pressure. Meanwhile, they also observed two distinct peaks (~1080 and ~1150 nm) at larger wavelengths in the spectra of samples with quantum dots overgrown at a low arsenic pressure.

Cherckesova et al. [6] developed a quantum protocol modification to standard quantum cryptography algorithms, increasing the cryptographic strength of the quantum protocol BB84.

Roslan et al. [7] analysed the charge transport properties of organic vanadyl 3,10,17,24-tetra-tert-butyl-1,8,15,22-tetrakis(dimethylamino)-29H,31H phthalocyanine (VTP). They showed that the I-V profile, demonstrated by a single VTP, showed a rectifying behaviour, and Schottky diode parameters, including the ideality factor, barrier height, shunt, and series resistance, were computed, among other studies.

Kotb et al. [8] demonstrated all basic optical logic operations, including XOR, AND, OR, NOT, NOR, XNOR, and NAND, through simulations using K-shaped compact silicon waveguides at the 1.55 μm telecommunication wavelength. These waveguides comprised three waveguide strips made of silicon and printed on silica.

Mohanty et al. [9] explored the effect of the tungsten nitride (WNx) diffusion barrier layer on resistive switching operations of the aluminum nitride (AlN)-based conductive bridge random access memory.

Quapp et al. [10] studied the twist map, with an interest in its use for the finite Frenkel–Kontorova model, explaining the meaning of tensile forces in some proposed models.

Sun et al. [11] proposed a wide reflected angle $Ti_3O_5/SiO_2$ DBR (WRA-DBR) for AlGaInP-based red and GaN-based green/blue flip-chip micro LEDs (RGB flip-chip micro-LEDs) to solve the drawbacks of the double-stack distributed Bragg reflector in micro-LEDs.

Shabbir et al. [12] reviewed the advantages of quantum dots in solar cells and quantum dot lasers, followed by a more in-depth discussion of applications in photodetectors. The diverse types of metallic materials, such as lead sulfide and indium arsenide, as well as nonmetallic materials, such as graphene and carbon nanotubes, were discussed.

Ahmad et al. [13] reviewed the usefulness of hybrid quantum dots as a theranostic system in different cancers and analysed diverse biomolecules conjugated hybrid quantum dots investigated for diagnostic/therapeutic applications in cancer. The properties of different biomolecules, such as folic acid, PEG, etc., as well as hybrid quantum dots on their biopharmaceutical attributes, were also discussed.

Parvin et al. [14] reviewed the diverse protocols for producing mono- and few-/multi-layer graphene. The impact of mono-/few-/multi-layer graphene was then studied concerning its quality and properties. The unique electrical features of graphene were highlighted, such as good carrier mobility, typical ambipolar behaviour, and a specific energy band structure, which might be employed in field effect transistors (FETs) and utilized in radio frequency (RF) circuits, sensors, memory, and other applications.

## 3. Future Directions

We expect that the field of quantum and optoelectronics will only grow in the future; a research avenue that could have a significant impact is the connection of the former with machine learning and artificial intelligence. This could serve, on the one hand, to facilitate the better design and control of the quantum and optoelectronics devices using machine learning and employing the physical devices in these areas to carry out more efficient machine learning through, e.g., neuromorphic architectures, memristors, and similar. Even if it is often difficult to predict the future evolution of a scientific field, our impression is that many applications for industry and society in these areas are yet to be developed.

**Acknowledgments:** We would like to thank all the authors for the papers they submitted to this Special Issue. We would also like to acknowledge all the reviewers for their careful and timely reviews to help improve the quality of this Special Issue. Last but not least, we would like to thank the Editorial Team of the Electronics journal for all the support provided in the publication of this Special Issue.

**Conflicts of Interest:** The author declares no conflict of interest.

## References

1. Juárez-Barojas, I.; Posada-Gómez, R.; Alvarado-Lassman, A.; Rodríguez-Jarquín, J.P. Condition-Based Maintenance of an Anaerobic Reactor Using Artificial Intelligence. *Electronics* **2023**, *12*, 799. [CrossRef]
2. Moiseev, K.; Ivanov, E.; Parkhomenko, Y. Long-Wavelength Luminescence of InSb Quantum Dots in Type II Broken-Gap Heterostructure. *Electronics* **2023**, *12*, 609. [CrossRef]
3. Gijare, M.; Chaudhari, S.; Ekar, S.; Shaikh, S.F.; Mane, R.S.; Pandit, B.; Siddiqui, M.U.H.; Garje, A. Facile Green Preparation of Reduced Graphene Oxide Using Citrus Limetta-Decorated rGO/$TiO_2$ Nanostructures for Glucose Sensing. *Electronics* **2023**, *12*, 294. [CrossRef]
4. Venkatesan, R.; Sheik Kadar Maideen, S.M.T.; Chandhiran, S.; Kushvaha, S.S.; Sagadevan, S.; Venkatachalapathy, V.; Mayandi, J. Fabrication and Characterization of Si/PEDOT: PSS-Based Heterojunction Solar Cells. *Electronics* **2022**, *11*, 4145. [CrossRef]
5. Balakirev, S.; Chernenko, N.; Kryzhanovskaya, N.; Shandyba, N.; Kirichenko, D.; Dragunova, A.; Komarov, S.; Zhukov, A.; Solodovnik, M. Photoluminescence Properties of InAs Quantum Dots Overgrown by a Low-Temperature GaAs Layer under Different Arsenic Pressures. *Electronics* **2022**, *11*, 4062. [CrossRef]
6. Cherckesova, L.V.; Safaryan, O.A.; Beskopylny, A.N.; Revyakina, E. Development of Quantum Protocol Modification CSLOE–2022, Increasing the Cryptographic Strength of Classical Quantum Protocol BB84. *Electronics* **2022**, *11*, 3954. [CrossRef]
7. Roslan, N.A.; Supangat, A.; Sagadevan, S. Investigation of Charge Transport Properties in VTP: PC71BM Organic Schottky Diode. *Electronics* **2022**, *11*, 3777. [CrossRef]
8. Kotb, A.; Zoiros, K.E. K-Shaped Silicon Waveguides for Logic Operations at 1.55 μm. *Electronics* **2022**, *11*, 3748. [CrossRef]
9. Mohanty, S.K.; Reddy, K.P.K.; Wu, C.-H.; Lee, P.-T.; Chang, K.-M.; Busa, P.; Kuthati, Y. Investigation of Barrier Layer Effect on Switching Uniformity and Synaptic Plasticity of AlN Based Conductive Bridge Random Access Memory. *Electronics* **2022**, *11*, 3432. [CrossRef]
10. Quapp, W.; Bofill, J.M. An Analysis of Some Properties and the Use of the Twist Map for the Finite Frenkel–Kontorova Model. *Electronics* **2022**, *11*, 3295. [CrossRef]
11. Sun, Y.; Shi, L.; Du, P.; Zhao, X.; Zhou, S. Rational Distributed Bragg Reflector Design for Improving Performance of Flip-Chip Micro-LEDs. *Electronics* **2022**, *11*, 3030. [CrossRef]
12. Shabbir, H.; Wojnicki, M. Recent Progress of Non-Cadmium and Organic Quantum Dots for Optoelectronic Applications with a Focus on Photodetector Devices. *Electronics* **2023**, *12*, 1327. [CrossRef]
13. Ahmad, J.; Garg, A.; Mustafa, G.; Ahmad, M.Z.; Aslam, M.; Mishra, A. Hybrid Quantum Dot as Promising Tools for Theranostic Application in Cancer. *Electronics* **2023**, *12*, 972. [CrossRef]
14. Parvin, N.; Kumar, V.; Joo, S.W.; Park, S.-S.; Mandal, T.K. Recent Advances in the Characterized Identification of Mono-to-Multi-Layer Graphene and Its Biomedical Applications: A Review. *Electronics* **2022**, *11*, 3345. [CrossRef]

*electronics*

*Article*

# Condition-Based Maintenance of an Anaerobic Reactor Using Artificial Intelligence

**Isaias Juárez-Barojas, Rubén Posada-Gómez, Alejandro Alvarado-Lassman * and José Pastor Rodríguez-Jarquín**

División de Estudios de Posgrado e Investigación, Tecnológico Nacional de México/Instituto Tecnológico de Orizaba, Orizaba 94320, Mexico
* Correspondence: lassman@prodigy.net.mx

**Abstract:** This paper proposes a condition-based maintenance system based on artificial intelligence for an online monitoring system of the support bed expansion in a 30-liter pilot-scale inverse fluidized bed reactor (IFBR). The main scope is to achieve a condition-based maintenance strategy using a single-level sensor for a biofilm inverse fluidizing bed as source for virtual sensors. The implementation of an artificial neural network was performed on an embedded electronic system (Raspberry Pi 4), both working together in real time. The signals estimated by the neural network are compared against the signals measured by the hardware sensors and, in case of detecting a failure in the physical measurement system, the artificial intelligence-based system then uses the signal estimated by the artificial neural network to maintain the correct operation of the IFBR. This system uses an artificial neural network to estimate the COD concentration of the effluent and the biogas production flow of a bioreactor, from the measurement of pH, the COD concentration of the influent, the inflow to the bioreactor and the signal coming from each of the conductivity sensors installed inside the reactor, which provide information about support media expansion in a pilot-scale inverse fluidized bed reactor. In addition, a fuzzy PI controller is presented, which was implemented in a Raspberry Pi electronic card, to regulate the COD concentration in the effluent of the bioreactor used as a case study.

**Keywords:** condition-based maintenance; artificial neural network; inverse fluidized bed reactor

**Citation:** Juárez-Barojas, I.; Posada-Gómez, R.; Alvarado-Lassman, A.; Rodríguez-Jarquín, J.P. Condition-Based Maintenance of an Anaerobic Reactor Using Artificial Intelligence. *Electronics* **2023**, *12*, 799. https://doi.org/10.3390/electronics12040799

Academic Editors: Lucas Lamata and Jian-Qiang You

Received: 26 December 2022
Revised: 30 January 2023
Accepted: 2 February 2023
Published: 5 February 2023



## 1. Introduction

This paper presents the implementation of a neural network that serves as the basis for the generation of a condition-based maintenance system for an anaerobic reactor, which is a complex process due to its nonlinear nature and that some of the variables involved in the process do not currently have commercial/professional sensors to be measured online, so it requires the implementation of virtual sensors to estimate the value of the same, in this case relying on a level sensor implemented in an anaerobic IFBR presented in [1], where control, diagnosis and maintenance were not achieved.

Online fault detection and isolation (FDI) has become a major issue due to the increasing demand for reliability and safety in industrial processes, as well as economic and environmental constraints [2]. When a fault occurs, it can affect the efficiency of the process, and if such a fault is not located at an early stage, it could lead to a catastrophic scenario including injury to personnel, environmental contamination and destruction of assets and equipment [3].

Nowadays, industrial processes are becoming increasingly complex, which increases the difficulty of performing online fault diagnosis quickly and efficiently [4], due to the nonlinear and highly coupled nature of the variables involved in these processes.

This last point is especially present in the case of anaerobic reactors, in which it is practically impossible to perform online measurement of all the variables involved in the process, thus fully justifying the need to develop an adequate fault diagnosis tool for this type of reactor [5].

Anaerobic reactors present an interesting case study for fault detection methods because it is very difficult to detect what is happening inside the reactor at a given instant, as well as to know how the concentrations of the different bacterial groups and of the organic pollutant load are evolving as a function of time [6]. In addition, given a set of indirect signals or measurements taken from a process, it can be difficult to identify whether the information coming from the process is reliable or not [7]. Furthermore, in the event that the process is not behaving as it should, it can be very difficult to know which component is responsible for the irregularity [8]. To solve this problem, the fault detection and isolation study presented below has been developed, which is based on a differential geometry approach for a class of nonlinear systems.

The observer-based approach to FDI [9] is among the most important contributions to the area of analytical redundancy-based FDI. Basically, it consists of defining a function of the system outputs, for each fault signal or set of fault signals, that can only be affected by that particular fault (or set of faults). The redundancy of the system is provided by the comparison of the output function in question with its estimate [10]. An observer-based FDI study can be found in [11]. Most of the proposed contributions were developed for linear systems; however, many processes are nonlinear. It should be noted that, even if the process operates at a certain operating point, the occurrence of a fault can move it away from the operating point, thus generating errors in the model that leads to the generation of false alarms [12].

A linear system is said to be observable at time $t_0$, if with the system in state $x(t_0)$ it is possible to determine that state from the observation of the system output, during a finite time interval. Consider the linear system in continuous time, represented by:

$$\dot{x}(t) = A(t)Bu(t), \qquad t \leq t_0 \qquad x(t_0) = x_0 y(t) = Cx(t) \tag{1}$$

where $x(t) \in \Re^n$ is the state vector, $A \in \Re^{nxn}$ is the state matrix, $y(t) \in \Re^n$ is the system matrix and $C \in \Re^{mxn}$. The observability of the system (Equation (1)) is established according to the observability matrix $\vartheta$:

$$\vartheta = \begin{bmatrix} C \\ CA \\ CA^2 \\ \vdots \\ CA^{(n-1)} \end{bmatrix} \tag{2}$$

The system is observable if and only if the observability matrix has full rank, that is, rank $(\vartheta) = n$.

For nonlinear systems, a concept related to observability is that of detectability or asymptotic observability, which establishes that a system is detectable if the difference between two trajectories with the same output history converges asymptotically to zero [13].

The observability analysis for nonlinear systems can be performed, in an approximate way, by linearizing the equations representing the model around an equilibrium point and determining the observability matrix in order to establish the local observability of the system in the vicinity of the selected equilibrium point. Another alternative to evaluate the observability of nonlinear systems is based on the use of differential geometry tools, through the construction of the observation space assuming that the output depends only on the state variables [14].

Biological processes, in most cases, are not completely observable, due to their nonlinear nature, so it is more appropriate to estimate some relationships between parameters, instead of estimating them individually [15].

- The estimation of states in biological reactors presents a critical drawback, due to the high uncertainty associated with the following factors [16].
- The variety of biological species present in the reactor makes it difficult to determine the role played by each of them in the process.

- The measurement of the initial total biomass does not correspond to the concentration of the effective biomass.
- The complexity of the system makes it difficult to describe the parameters associated with the kinetics of the biological reactions.
- The growth rates of the bacterial groups are time-varying.

In the work presented by authors of [17], a methodology for determining the global observability and detectability of biological reactors is explored. The method proposed in the article allows formalizing the design of the so-called asymptotic observers.

Recently, many efforts have been focused on the implementation of programmed sensors based on state observers, due to the advantages that the use of these sensors represents, compared to hardware sensors [18]. Some of these advantages are mentioned below:

The maintenance cost of a programmed sensor is much less expensive than that of a hardware sensor.

- Once a programmed sensor has been designed and tuned for a particular application, no periodic maintenance is required to recalibrate it (e.g., pH analyzers, oxygen analyzers, etc.).
- Programmed sensors can be used to predict information that is difficult to obtain using traditional methods, either due to inherent measurement delays or due to process operating conditions.
- Installation costs are cheaper in most cases compared to traditional methods of monitoring systems.

Programmed sensors are used in order to solve problems associated with the lack of sensors for real-time measurement of certain process variables. For example, authors of [19] used a support vector regression algorithm for COD (chemical oxygen demand) estimation in a perfectly stirred anaerobic reactor. One of the main difficulties in implementing programmed sensors is that the design contemplates previously obtaining a predictive model for a given process. Consequently, the success of a programmed sensor strongly depends on the accuracy of the mathematical model obtained for that process.

Many empirical equations exist in the literature to model the porosity or expansion of an IFBR. Campos-Díaz et al., 2012 [20], proposed the following model to estimate the porosity of the inverse fluidized bed, as a function of the Reynolds particle and Archimedes numbers:

$$1.753e^{-3.807} = Ar\left[0.75\left(24Re_p + C_1 Re_p^2\right)\right]^{-1} \tag{3}$$

where:

$$C_1 = 2.25(5.5/Re_p)^{0.34} \tag{4}$$

Authors of [20] developed an empirical correlation for predicting bed expansion as a function of physical and dynamic variables of the system, calculated using the following equation:

$$\frac{H}{H_0} = 0.988Ar^{-0.2124}Re^{0.4049}\phi_s^{-1.804}(D_c/d_p)^{0.4613} \tag{5}$$

where $\phi$ is the particle sphericity.

It is evident that at $Re_p < 0.2$, the correlation of the result of [20] adequately predicts the porosity of the bed, while at $Re_p > 0.7$, the correlation of the result of [21] is more in agreement with experimental data. However, both empirical correlations fail to provide a reasonable overall description at all evaluated operating conditions, and this might be associated with the experimental conditions in which they were fitted.

Therefore, taking into account that there are conductivity sensors placed inside the IFBR, which provide information on the expansion of the fluidized bed, it was decided to use the correlation of the result of [21] as a starting point to train an artificial neural network capable of finding a relationship between pH, influent COD concentration, inflow to the bioreactor and the signal coming from each of the conductivity sensors installed inside the reactor, to estimate the effluent COD concentration, the bioreactor biogas production flow rate (the sum of methane and carbon dioxide production), the methane production flow rate, the carbon dioxide production flow rate and the concentration of volatile fatty acids in the reactor effluent.

The importance of the work presented resides in the considerations made and how the process is treated to ensure the correct functioning of a control strategy and fault diagnosis in the same, to maximize the production of biogas and serve as an aid system to a human operator for the maintenance of the same, through the use of artificial intelligence techniques.

## 2. Materials and Methods

### *2.1. Support Expansion Monitoring System*

Figure 1 shows the block diagram of the bed expansion monitoring system. The expansion of the bed is produced by the flow generated by opening the control valve connected to the distribution tank. The flow is proportional to the valve opening and also dependent on the force of gravity. By circulating the fluid downward, the colonized bed, which has a lower conductivity than the effluent, begins to gradually expand.

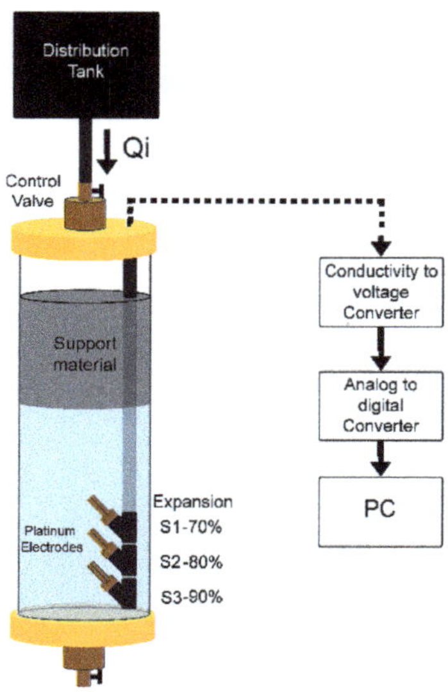


**Figure 1.** Block diagram of the support expansion monitoring system.

Predevelopment measurements were carried out with a commercial conductivity sensor to validate the above. The values obtained from the measurements using a conductivity electrode, model DFR0300, are shown below in Table 1. All conductivity measurements were performed at 25 °C.

**Table 1.** Conductivity comparison of different components involved in biodigestion.

| Component | Conductivity (mS/cm) |
|---|---|
| Citric wastewater | 2.18 |
| Expanding colonized support | 2.08–2.11 |
| Colonized support | 0–1.68 |

The electrode array allows for detecting the substrate change, thus identifying the level of substrate expansion. A conditioning card was developed to obtain the necessary voltage signal, and an acquisition card was used to capture the measured values.

The electrode model used for the array is CON2136-13 with a constant K = 10 and construction material of polysulphone and platinum. It was chosen because the measurement of the variable of interest must be continuous, which represents a problem for many of the metallic materials available, such as copper or tin, since, over time, the environment inside a biodigester is corrosive and can be degraded. Likewise, it is advisable in wastewater treatment to use electrodes with a counter of 10, considering the conductivity levels of the substrates in wastewater treatment. Moreover, the electrode performed adequately in a previous investigation using synthetic wastewater [1], where level measure was achieved, but a control and diagnosis systems were not achieved, conditions that are needed for a condition-based maintenance strategy.

A platinum conductivity probe is proposed at each point in the column. The material was chosen because the measurement of the variable of interest must be continuous, which represents a problem for many of the metallic materials available such as copper or tin, since, over time, the environment inside a biodigester becomes corrosive and can degrade them. Platinum, on the other hand, is immune to oxidation and to the biological activity of bacteria, making it one of the most suitable materials for this application.

The position of the electrodes is set at the bottom of the reactor column with a separation distance of 10 cm from each other and in diagonal position to avoid clogging. The voltage delivered by the conditioning circuit is calculated using Equation (6).

$$V_{out} = (V_{in})(G)(R_f) \tag{6}$$

where:

$V_{out}$: is the output voltage of the conditioning circuit.
$V_{in}$: is the alternating excitation voltage applied to the electrode.
G: solution conductivity.
Rf: feedback resistance of the conditioning circuit.

The value of the proposed feedback resistance in the conditioning circuit is 8.2 KΩ due to the conductivity probe constant and the conductivity values present in the reactor solutions previously presented. Dynamic testing of the system was performed, replicating the expansion of the colonized support. The support material used was colonized (bacteria biofilm around the support media) and noncolonized.

Figure 2 shows the results obtained from the dynamic expansion test of the support bed. This test consisted of introducing colonized support media recirculating in the reactor at a downward velocity of 1.41 m/h. It can be observed how after 650 s (10 min), sensor 1 detects a voltage decrease, which means that it has come into contact with the colonized support media; then, at 950 s (16 min), sensor 2 also comes into contact with the substrate, so its voltage decreases. Finally, sensor 3 comes into contact with the expanded support at 1230 s (20 min). After half an hour of operation, the recirculation is turned off, so the voltage values of the sensors begin to increase, but in reverse order, i.e., the voltage first increases in sensor 3, then in sensor 2, and finally, sensor 1 returns to its initial values.

### 2.2. Description of the Fuzzy Logic PI Type Controller

The fuzzy logic PI-type controller is commonly used to design a robust controller for ensuring suppression of disturbances. In a fuzzy PI-type controller (Figure 3), its nonlinear mapping between the control rules and the fuzzy reasoning can be represented as a function of the error e and the first-time derivative of the error Δe as follows:

$$\Delta u = FLC(e, \Delta e) \tag{7}$$

where Δu is the first-time derivative of the output control action. Δu can be expressed approximately as follows:

$$\frac{\Delta u(t)}{S\Delta u} = \frac{\Delta e(t)}{S\Delta e} + \frac{e(t)}{Se} \quad (8)$$

where $-Se < e(t) < Se$, $-S\Delta e < \Delta e(t) < S\Delta e$, $-S\Delta u < \Delta u(t) < S\Delta u$, or

$$u(t) = k_p e(t) + k_i \int e(t)dt \quad (9)$$

where $kp = S\Delta u / S\Delta e$ and $ki = S\Delta u/Se$.

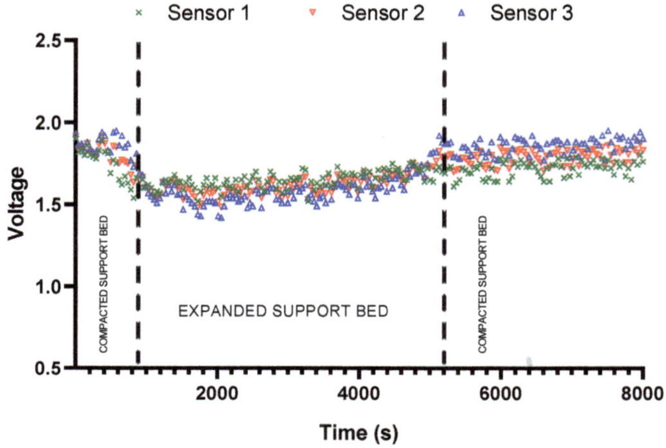


**Figure 2.** Dynamic test of support bed expansion.

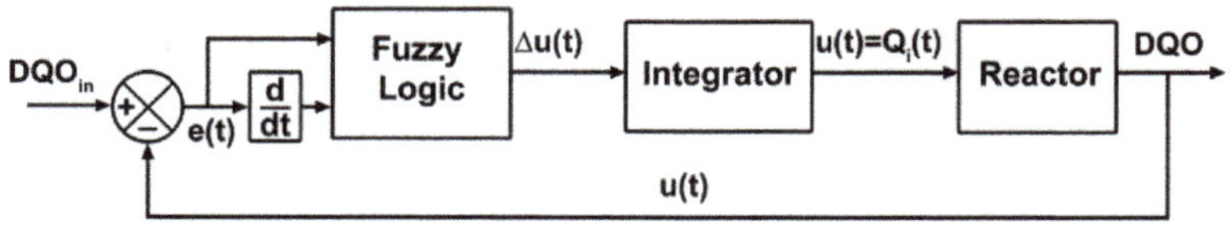


**Figure 3.** PI control system with fuzzy logic controller.

The membership functions for the inputs e and Δe are defined in the range (−1, 1), see Figure 4a,b, respectively. The fuzzy subsets are defined as follows: NL: negative large; N: negative; Z: zero; P: positive; and PL: positive large, for the input e. N: negative; Z: zero; and P: positive; for the input Δe.

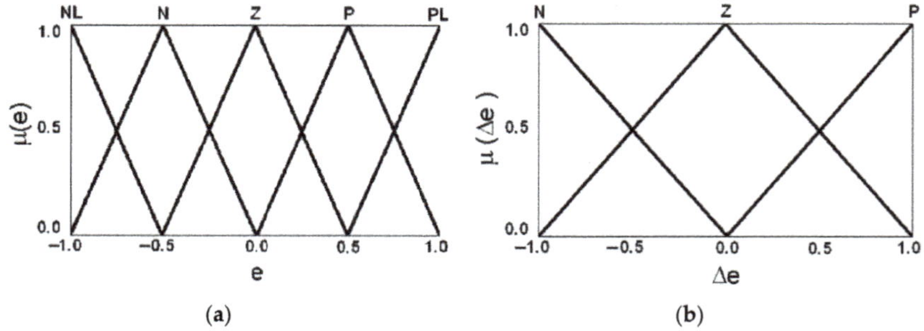


**Figure 4.** (**a**) Membership functions for the input e. (**b**) Membership functions for the input Δe.

The membership functions for the output Δu are defined in the range (−1, 1), see Figure 5. The fuzzy subsets for the output Δu are defined as follows: NL: negative large; NM: negative medium; NS: negative small; Z: zero; PS: positive small; PM: positive medium; PL: positive large; for the output Δu.

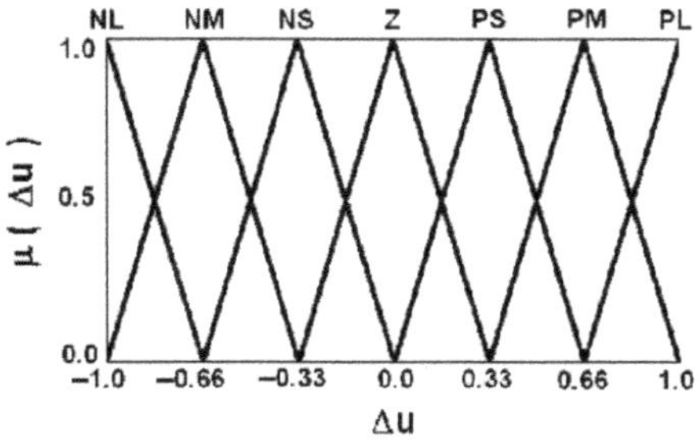


**Figure 5.** Membership functions for the output Δu.

*2.3. Description of Software Sensor Based on Artificial Neural Networks*

An artificial neural network was developed to find a relationship between pH, influent COD concentration, bioreactor inflow and the signal coming from each of the conductivity sensors installed inside the reactor to estimate the effluent COD concentration, the biogas production flow rate of the bioreactor (the sum of methane and carbon dioxide production), the methane production flow rate, the carbon dioxide production flow rate and the concentration of volatile fatty acids in the reactor effluent. The neural network was trained from hundreds of experimental tests performed on the bioreactor to relate biogas production to fluidized bed expansion, based on signals from each of the three bioreactor conductivity sensors.

The implementation of the artificial neural network was performed on an embedded electronic system (Raspberry Pi 4), both working together in real-time. This device was selected because it is a low-cost embedded system that does not require a computer to perform its work.

Figure 6 shows how the neural network was trained from the results obtained from the experimental tests performed with the bioreactor. The values of pH, COD concentration of the influent, inflow to the bioreactor and the signal coming from each of the conductivity sensors installed inside the reactor, during each of the experiments performed with the bioreactor, correspond to the input vector (A) of the neural network. This input vector contains 180 data points corresponding to the pH values, COD concentration of the influent, inflow to the bioreactor and the signal coming from each of the conductivity sensors installed inside the reactor during 6 months of experimental tests. There are 180 input neurons (B), which are connected to the neurons of the intermediate layer (C), and these neurons of the intermediate layer are connected to the output neurons (D), which generate the output signals estimated by the neural network, corresponding to the COD concentration of the effluent, the biogas production flow rate of the bioreactor (the sum of methane and carbon dioxide production), the methane production flow rate, the carbon dioxide production flow rate and the concentration of volatile fatty acids in the reactor effluent.

The objective of training the neural network is to make it capable of learning the relationship between the outputs and the input, so that when a new experiment is performed with the bioreactor, a new sample of pH, COD concentration of the effluent, inflow to the bioreactor and the signal coming from each of the conductivity sensors installed inside the reactor will be introduced into the neural network and it will be able, from its learning, to estimate the concentration values of the COD concentration of the effluent, the flow of biogas production from the bioreactor and the signal coming from each of the conductivity sensors installed inside the reactor that is introduced into the neural network and the neural network is able, based on its learning, to estimate the concentration values of the effluent

COD concentration, the bioreactor biogas production flow (the sum of methane and carbon dioxide production), the methane production flow, the carbon dioxide production flow and the concentration of volatile fatty acids in the reactor effluent.

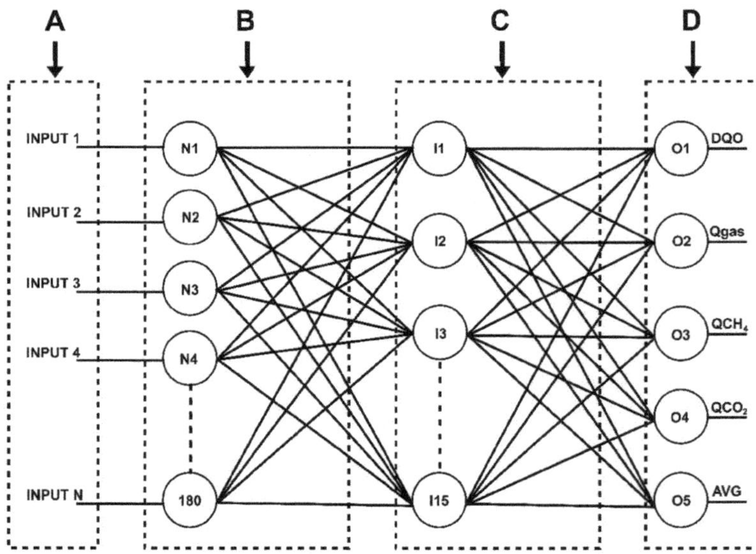


**Figure 6.** Neural network arrangement used for the determination of pH, Qgas, S1, S2 and S3.

One of the primary objectives of this section is to statistically validate the neural networks used to determine the output signals. One of the big questions that always has to be solved when working with neural networks is to know what the number of neurons in the intermediate layer or layers required for its error in the output to be optimal is. To achieve this objective, an automatic algorithm had to be developed to analyze the behavior of the network under different configurations of neurons in hidden layers.

In total, there are 180 data points from each of the 6 input signals to the neural network, which provides us a matrix of 1080 data points, which should generate a correlation with the data obtained in the experimental tests of the anaerobic reactor. Using the 180 data points from each of the 6 input signals to train an artificial neural network can result in an incorrect training of the network, because many of these variables may not have significant relevance. For this reason, a preprocessing of these data was performed to know the variance and correlation of the data in order to reduce the dimensionality of the input vector to the neural network. A PCA is a statistical methodology that performs a linear transformation through the creation of a new coordinate system for the input data, in which, on its axes, it places the data ordered from the highest variance to the lowest variance. In general terms, what a PCA does is create a new system in which it orders the data from most relevant to least relevant, thus reducing the dimensionality of the data.

A PCA was programmed using MATLAB software, using the 1080 data points of the input matrix, which corresponds to 180 dimensions for 6 input signals. According to the results of the PCA, statistically speaking, and looking at its percentage of re-relevance, the first 8 columns of the principal components were selected since they are the most important data. With this dimensional reduction, the training input vector used in the neural network was 8 dimensions.

A neural network was modeled with the 8 input data coming from the dimensional data reduction obtained by the PCA and experimented on with different configurations related to the number of hidden layers and number of neurons in the hidden layer. This neural network delivers 5 outputs as a result, which are the estimation of the COD concentration values of the effluent, the biogas production flow rate of the bioreactor (the sum of methane and carbon dioxide production), the methane production flow rate, the carbon dioxide production flow rate and the concentration of volatile fatty acids in the reactor effluent. Figures 7 and 8 show the structure of the neural network, composed of an input layer and an intermediate layer of neurons.

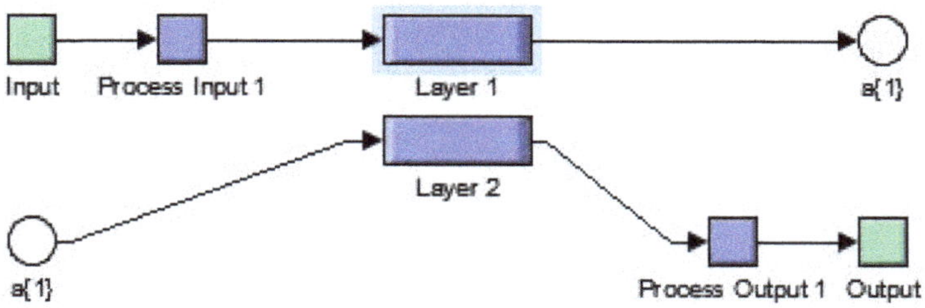


**Figure 7.** Neural network structure, input layer of neurons and intermediate layer of neurons.

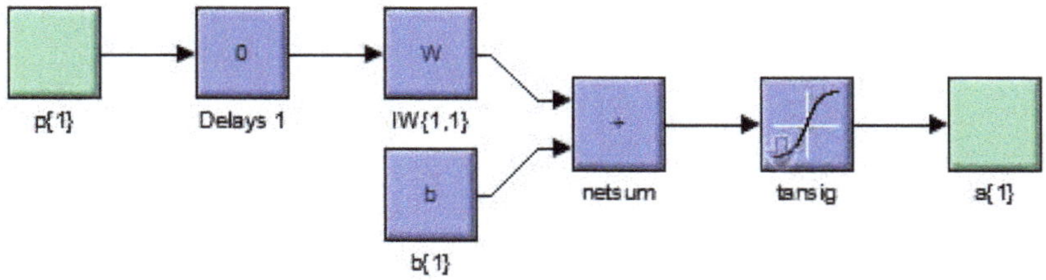


**Figure 8.** Input layer, composed of the normalization algorithm, the weights and the tansig function.

Figure 9 shows the behavior of a neural network using 5 neurons in the hidden layer of the neural network, obtaining a recognition error of 4.28%. Figure 10 shows the behavior of the neural network using 15 neurons in the hidden layer with an error of 1.45%. The graphs show the target (the blue line) representing the ideal recognition value, and the circles in the graph are the actual recognition values obtained for each target value. It can be seen that the neural network with 15 neurons in the hidden layer has a lower recognition error (Figure 10).

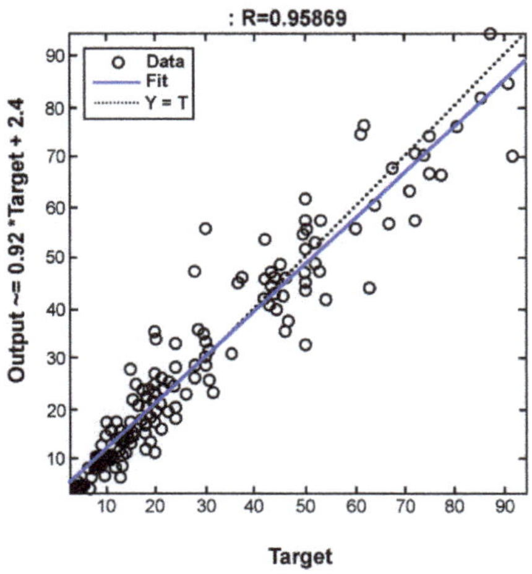


**Figure 9.** ANN using 5 neurons in the hidden layer.

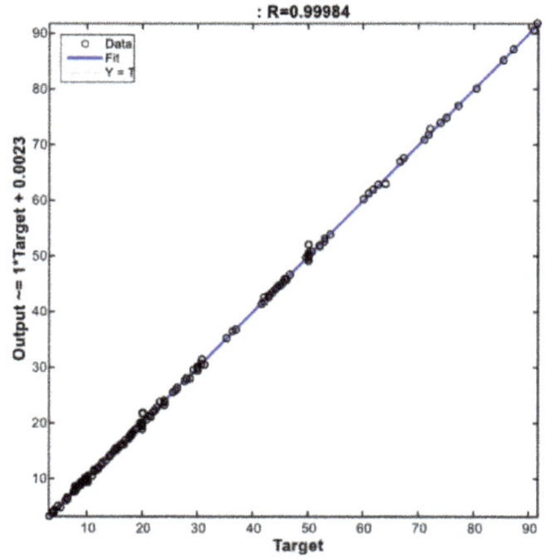


**Figure 10.** ANN using 15 neurons in the hidden layer.

Figure 11 shows the behavior of the neural network focused on estimating the values of the conductivity sensors, S1, S2 and S3, from the values of pH, COD concentration of the influent and inflow to the bioreactor. In this regard, after evaluating multiple configurations of neurons in the intermediate layer, it was determined that the best performance was obtained with a neural network with 10 neurons in the intermediate layer.

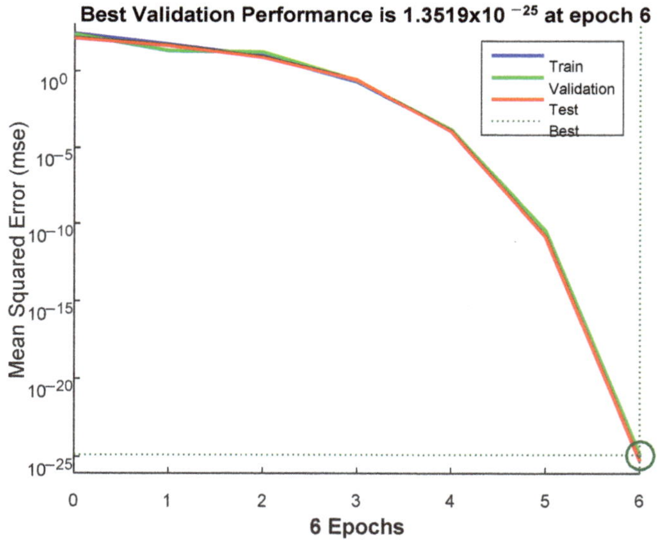


**Figure 11.** ANN with 10 neurons in the intermediate layer.

After training the neural network and performing the behavioral tests, validation experiments were conducted to determine the margin of error of the neural network in comparison with the hardware sensors. In this regard, the margin of error was less than 2%, which allowed establishing the threshold values for the fault diagnosis scheme developed in this project.

The microcontroller was programmed for real-time execution of the neural networks. Blocks were used to map the input signals from −1 to 1, and a Math Script node was used in which all the weights of the network as well as all the mathematical formulation to run the network were found. This block is connected directly to the outputs of the hardware sensors, which delivers an input vector to the network with the values of pH, COD concentration of the influent, inflow to the bioreactor and the signal coming from each of the conductivity sensors installed inside the reactor.

## 3. Results

### 3.1. Experimental Results of Condition-Based Maintenance Scheme

In this work, a fuzzy PI controller was used to regulate the COD concentration in the effluent of the fluidized bed reactor, having the inflow to the Qi bioreactor as a control variable. To validate the performance of the controller, experimental tests were carried out by feeding the reactor with citrus wastewater, which was diluted and conditioned to 6 g COD/L and pH = 7, during the first twenty days of experimentation. Subsequently, the wastewater was diluted to 3 gCOD/L, maintaining the original pH = 7. During the four weeks prior to the start of the experiment, the controller was operating at a set-point value of 1 gCOD/L. On the first day of experimentation, the set-point was changed from 1 gCOD/L to 0.5 gCOD/L, and was maintained unchanged for the forty days of experimentation. The reactor temperature was kept constant at 35 degrees Celsius during the experimentation.

### 3.2. Operation of the Condition-Based Maintenance Scheme

As mentioned above, the COD sensor is responsible for sending the feedback signal to the fuzzy PI controller that is responsible for regulating the COD concentration in the reactor effluent. Therefore, if the COD sensor fails, then the regulation of COD in the reactor effluent will also fail. To overcome this problem, the COD sensor condition-based maintenance scheme presented in Figure 12 was developed. This maintenance scheme receives the information from the fault diagnosis system (FDI), and when a COD sensor failure occurs, the COD sensor is then disabled and its signal is switched to the COD signal estimated by neural network 1, as long as this neural network is free of failure. However, in case that neural network 1 presents a failure, the feedback signal is then taken from neural network 2, as long as it is fault-free. In the extreme case that both neural network 1 and neural network 2 are faulted, the feedback signal is then taken from neural network 3, as long as it is fault-free.

Figure 13 shows the graphs of the COD concentration in the influent, both the experimentally obtained graph and the graphs estimated by asymptotic observers 2 and 3. In this figure, it can be seen that the three graphs are practically identical until day 60 of the experiment, when the sudden and permanent failure of sensor s1 is observed. In Figure 14, a transient is observed on day 60 of the experiment. This transient corresponds to the action of the condition-based maintenance scheme, whereby the signal from sensor s1 is switched by its estimate generated by observer 2, so that the controller continues to operate normally and continues to regulate the COD concentration in the reactor effluent to 0.5 g COD/L.

As can be seen, the condition-based maintenance scheme fulfills its function, allowing the system to continue to operate the control strategy, using the output obtained from an observer that was not affected by the fault, which, when diagnosed, allows a human operator to perform corrective maintenance without stopping the process; once this has been corrected, the system detects the correct operation of the sensor and the control strategy returns to its nominal behavior.

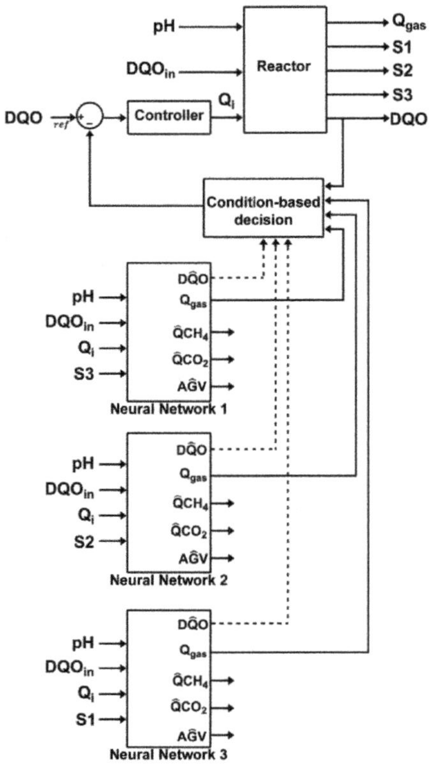


**Figure 12.** Condition-based decision scheme.

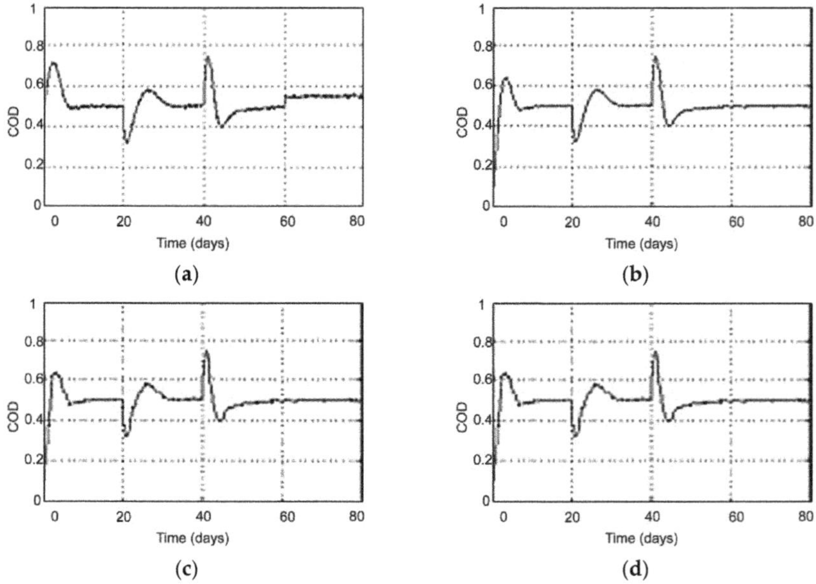


**Figure 13.** (**a**) Experimental measurement of COD of the reactor effluent. (**b**) Estimation of COD of neural network 1. (**c**) COD estimation of neural network 2. (**d**) COD estimation of neural network 3.

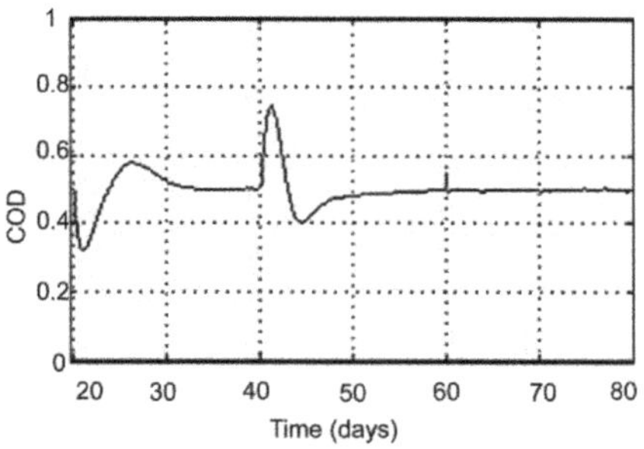


**Figure 14.** Graph of COD switched by the condition-based maintenance scheme due to a sudden and permanent failure of +50 mgCOD/L on day 60 of experimentation.

## 4. Conclusions

A real application of a fuzzy controller to a nonlinear multivariable anaerobic reactor process was investigated; this controller provides a useful fault detection strategy when the mathematical model is not available. In addition, an artificial neural network capable of finding a relationship between pH, influent COD concentration, inflow to the bioreactor and the signal coming from each of the conductivity sensors installed inside the reactor was developed to estimate the effluent COD concentration, the biogas production flow rate of the bioreactor (the sum of methane and carbon dioxide production), the methane production flow rate, the carbon dioxide production flow rate and the concentration of volatile fatty acids in the reactor effluent. The neural network was trained from hundreds of experimental tests performed on the bioreactor to relate biogas production to fluidized bed expansion, based on signals from each of the three bioreactor conductivity sensors.

In addition, the information obtained by this system allows not only to detect and isolate failures that may occur in the sensors during the anaerobic digestion process, but the information obtained from them can also be used as virtual sensors, as there are observers that are not affected by certain faults, which allows the controller to continue operating and leads to corrections in the physical sensor. This is vital in the case of bioreactors, due to their nature and the diverse conditions necessary for their operation, since stopping the process could reduce the population of biomass responsible for the processes of acidogenesis and methanogenesis that allow the production of biogas to an unacceptable minimum, and in an extreme case would require restarting the bioreactor, which would mean months of work.

In this work, the problem regarding the lack of online sensors to measure variables in anaerobic reactors was presented; in this case, we considered a sensor to measure the level of the reverse fluidized bed [1]; for a hybrid reactor reverse fluidized fixed bed, an original design of the Tecnológico Nacional de México Campus Orizaba—using this sensor as a basis—a control system and fault diagnosis based on artificial intelligence was made for the generation of condition-based maintenance.

An artificial intelligence strategy approach was used as an estimator, predictor and controller, as shown in this research.

**Author Contributions:** Conceptualization, A.A.-L. and J.P.R.-J.; Methodology, R.P.-G. and J.P.R.-J.; Software, I.J.-B. and J.P.R.-J.; Validation, R.P.-G. and J.P.R.-J.; Formal analysis, J.P.R.-J.; Investigation, I.J.-B.; Resources, A.A.-L.; Writing—original draft, I.J.-B.; Writing—review and editing, A.A.-L. and J.P.R.-J.; Supervision, R.P.-G. All authors have read and agreed to the published version of the manuscript.

**Funding:** This research received no external funding.

**Data Availability Statement:** Please contact the corresponding author for data.

**Conflicts of Interest:** The authors declare no conflict of interest.

## References


1. Marín-Peña, O.; Alvarado-Lassman, A.; Vallejo-Cantú, N.A.; Juárez-Barojas, I.; Rodríguez-Jarquín, J.P.; Martínez-Sibaja, A. Electrical conductivity for monitoring the expansion of the support material in an anaerobic biofilm reactor. *Processes* **2020**, *8*, 77. [CrossRef]
2. Phoungthong, K. An effective evaluation on fault detection in solar panels. *Energies* **2021**, *14*, 7770.
3. Safavi, S.; Safavi, M.A.; Hamid, H.; Fallah, S. Multi-sensor fault detection, identification, isolation and health forecasting for autonomous vehicles. *Sensors* **2021**, *21*, 2547. [CrossRef]
4. Ji, C.; Ma, F.; Wang, J.; Wang, J.; Sun, W. Real-time industrial process fault diagnosis based on time delayed mutual information analysis. *Processes* **2021**, *9*, 1027. [CrossRef]
5. Kazemi, P.; Bengoa, C.; Steyer, J.P.; Giralt, J. Data-driven techniques for fault detection in anaerobic digestion process. *Process Saf. Environ. Prot.* **2021**, *1*, 905–915. [CrossRef]
6. Avilés, J.D.; Torres-Zúñiga, I.; Villa-Leyva, A.; Vargas, A.; Buitrón, G. Experimental validation of an interval observer-based sensor fault detection strategy applied to a biohydrogen production dark fermenter. *J. Process Control* **2022**, *1*, 131–142. [CrossRef]
7. Anand, K.; Mittal, A.P.; Kumar, B. ANN-based sensorless adaptive temperature control system to improve methane yield in an anaerobic digester. *Biomass Convers. Biorefinery* **2022**, 1–21. [CrossRef]
8. Gupta, R.; Zhang, L.; Hou, J.; Zhang, Z.; Liu, H.; You, S.; Ok, Y.S.; Li, W. Review of Explainable Machine Learning for Anaerobic Digestion. *Bioresour. Technol.* **2022**, *369*, 128468. [CrossRef]
9. Wang, J.; Yang, C.; Xia, J.; Wu, Z.G.; Shen, H. Observer-based sliding mode control for networked fuzzy singularly perturbed systems under weighted try-once-discard protocol. *IEEE Trans. Fuzzy Syst.* **2021**, *6*, 1889–1899. [CrossRef]
10. Youn, W.; Lim, H.; Choi, H.S.; Rhudy, M.B.; Ryu, H.; Kim, S.; Myung, H. State estimation for HALE UAVs with deep-learning-aided virtual AOA/SSA sensors for analytical redundancy. *IEEE Robot. Autom. Lett.* **2021**, *6*, 5276–5283. [CrossRef]
11. Deng, C.; Wen, C.; Huang, J.; Zhang, X.M.; Zou, Y. Distributed observer-based cooperative control approach for uncertain nonlinear MASs under event-triggered communication. *IEEE Trans. Autom. Control* **2021**, *67*, 2669–2676. [CrossRef]
12. Lee, D.; Lai, C.W.; Liao, K.K.; Chang, J.W. Artificial intelligence assisted false alarm detection and diagnosis system development for reducing maintenance cost of chillers at the data center. *J. Build. Eng.* **2021**, *1*, 102110. [CrossRef]
13. Chen, Z.; Niu, B.; Zhao, X.; Zhang, L.; Xu, N. Model-based adaptive event-triggered control of nonlinear continuous-time systems. *Appl. Math. Comput.* **2021**, *8*, 126330. [CrossRef]
14. Touzé, C.; Vizzaccaro, A.; Thomas, O. Model order reduction methods for geometrically nonlinear structures: A review of nonlinear techniques. *Nonlinear Dyn.* **2021**, *105*, 1141–1190. [CrossRef]
15. Iida, Y.; Takatani, Y.; Kojima, A.; Ishii, M. Global trends of ocean CO2 sink and ocean acidification: An observation-based reconstruction of surface ocean inorganic carbon variables. *J. Oceanogr.* **2021**, *77*, 323–358. [CrossRef]
16. Ortega, R.; Bobtsov, A.; Nikolaev, N.; Schiffer, J.; Dochain, D. Generalized parameter estimation-based observers: Application to power systems and chemical–biological reactors. *Automatica* **2021**, *129*, 109635. [CrossRef]
17. Jerono, P.; Schaum, A.; Meurer, T. Observability analysis and robust observer design for a continuous yeast culture. *J. Process Control.* **2021**, *104*, 62–73. [CrossRef]
18. Li, C.; Wang, Y.; Liu, F.; Liu, Q.; Buss, M. Model-free incremental adaptive dynamic programming based approximate robust optimal regulation. *Int. J. Robust Nonlinear Control* **2022**, *32*, 2662–2682. [CrossRef]
19. Najafzadeh, M.; Niazmardi, S. A novel multiple-kernel support vector regression algorithm for estimation of water quality parameters. *Nat. Resour. Res.* **2021**, *30*, 3761–3775. [CrossRef]
20. Campos-Díaz, K.; Bandala-Gonzalez, E.; Limas-Ballesteros, R. Fluid bed porosity mathematical model for an inverse fluidized bed bioreactor with particles growing biofilm. *J. Environ. Manag.* **2012**, *104*, 62–66. [CrossRef]
21. Das, B.; Ganguly, U.P.; Bar, N.; Das, S.K. Holdup prediction in inverse fluidization using non-Newtonian pseudoplastic liquids: Empirical correlation and ANN modeling. *Powder Technol.* **2015**, *273*, 83–90. [CrossRef]

electronics

MDPI

Article

# Long-Wavelength Luminescence of InSb Quantum Dots in Type II Broken-Gap Heterostructure

**Konstantin Moiseev *, Eduard Ivanov and Yana Parkhomenko**

Ioffe Institute, 26 Politekhnicheskaya st., St. Petersburg 194021, Russia
* Correspondence: mkd@iropt2.ioffe.ru; Tel.: +7-812-292-7929

**Abstract:** The features of the electroluminescence spectra of narrow-gap type II InAs/InSb/InAs heterostructures containing a single layer of InSb quantum dots placed into the p-n-InAs junction were studied. The luminescent properties of the heterostructures under a forward and reverse bias in the temperature range of 77–300 K were investigated as a function of the surface density of nano-objects buried in the narrow-gap matrix. When applying the reverse bias to the heterostructures under study, the suppression of negative interband luminescence and the dominance of interface recombination transitions at the InSb/InAs type II heterojunction were observed at room temperature. The radiation, which corresponded to recombination transitions involving localized electron-hole states of the InSb quantum dots, was revealed and recorded at low temperatures.

**Keywords:** quantum dots; electroluminescence; InAs; InSb; type II heterojunction

## 1. Introduction

The mid-infrared spectral range (2.8–5.4 μm) is important for many applications of optical gas sensing [1] such as medical diagnosis by breath analysis and for environmental protection by localizing leaks of toxic and harmful gases [2]. Semiconductor heterostructures based on the narrow-gap InSb–InAs solid solutions system are very promising for the creation of light-emitting diodes operating in the atmospheric transparency window of 4.4–5.4 μm [3,4]. This spectral region contains the intense absorption bands of a number of nitrogenous chemical compounds ($N_2O$, NO, $(CN)_2$, HCN, etc.), which are of great interest for industrial gas analysis and the detection of some explosive and dangerous substances [5]. It is known that the use of quantum dots (QDs) in the active region of laser diodes increases the quantum efficiency and the operating temperature of these devices [6]. This opens up great opportunities for improving the design of mid-infrared optoelectronic devices. In contrast to well-studied wide-gap systems such as InAs/GaAs and InSb/InP, the energy band diagram of the narrow-gap InSb/InAs heterostructure exhibits a type II broken-gap heterojunction at each QD/matrix layer interface [7]. This specific feature of the InSb/InAs nanoheterostructures raises the question of identifying radiative transitions assisted by quantum dots. Recently, the photoluminescent (PL) properties of heterostructures based on the InSb QD layers inserted into the n-InAs matrix were studied [8]. A long-wave radiation was detected in the spectral range of 4–5 μm (0.25–0.3 eV) at low temperatures due to recombination transitions involving quantum dot eigenstates. The electroluminescence (EL) study was performed for heterostructures based on a single layer of the InSb QDs buried in the p-InAs matrix [9]. The EL spectra displayed positive luminescence when forward bias was applied and negative luminescence when reverse external bias was applied to the structure under study in the temperature range of $T$ = 77–300 K. Interband radiative recombination caused by the lightly doped p-InAs bulk was observed at both biases. Only a small contribution of interface-induced luminescence indicated the presence of the InSb QDs. In our current work, we present the results of the study of electroluminescence owing to radiative transitions involving quantum dot eigenstates. To fulfill this purpose, type II

**Citation:** Moiseev, K.; Ivanov, E.; Parkhomenko, Y. Long-Wavelength Luminescence of InSb Quantum Dots in Type II Broken-Gap Heterostructure. *Electronics* **2023**, *12*, 609. https://doi.org/10.3390/electronics12030609

Academic Editor: Lucas Lamata

Received: 31 December 2022
Revised: 20 January 2023
Accepted: 22 January 2023
Published: 26 January 2023

broken-gap heterostructures with a single layer of the InSb QDs located directly on the interface of the p-InAs/n-InAs junction were manufactured.

## 2. Materials and Methods

Narrow-gap heterostructures containing one layer of InSb QDs embedded in the InAs matrix were obtained using a combined epitaxial growth technology. The samples under study were grown on p-InAs(001) substrates doped with an acceptor impurity (Mn) to a hole concentration of $p_{300}$~$10^{17}$ $cm^{-3}$ at $T = 300$ K. At the first stage of the heterostructure fabrication, the InSb QDs array was deposited on the p-InAs substrate by means of liquid-phase epitaxy (Figure 1a). The technique used for the QDs growth and study on their structural properties in the InAs/InSb system were reported elsewhere [10,11]. Arrays of QDs in the surface density range of 0.5–2 × $10^{10}$ $cm^{-2}$ with a step of 2 × $10^9$ $cm^{-2}$ were obtained. The layer of InSb QDs was further overgrown with an epitaxial layer of indium arsenide using metal-organic vapor phase epitaxy (MOVPE). The n-InAs capping layer was intentionally undoped. It exhibited n-type conductivity with an estimated residual electron concentration of $n_{300}$~$6 \cdot 10^{16}$ $cm^{-3}$ at $T = 300$ K. The presence of the InSb QD arrays buried in the InAs matrix formed was confirmed by transmission electron microscopy investigations (Figure 1b). None of the structures obtained exhibited luminescence from QDs, except for samples with a limiting density of 2 × $10^{10}$ $cm^{-2}$. Therefore, the two most typical representatives from the whole series of samples were selected for the EL study (see Table 1). Additionally, the p-InAs/n-InAs structure (without the QD layer) was chosen as a test sample (see sample A in Table 1). It was obtained together with samples B and C in a single MOVPE process. Because epitaxial overgrowth was carried out using undoped compounds, it is possible to almost exclude the diffusion of the acceptor impurity across the p-InAs/n-InAs heterointerface from the p-InAs part of the matrix into the n-InAs layer being overgrown. Thus, a metallurgical p-n junction was formed inside the InAs matrix, at the interface of which the InSb QDs layer was located (see sample B and C in Table 1).

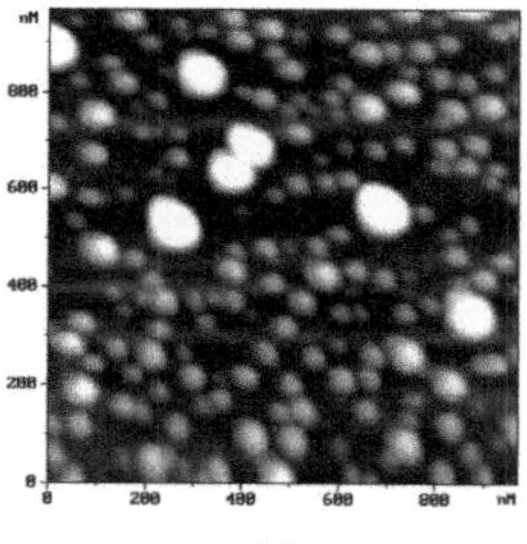

(a)

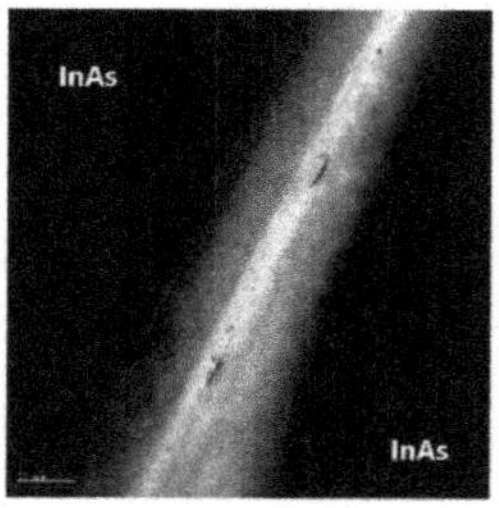


(b)

**Figure 1.** (**a**) AFM image of the InSb QD arrays on a surface of the InAs matrix with density $n_{QD} = 2 \times 10^{10}$ $cm^{-2}$. (**b**) TEM image of a cross-sectional view of the InAs/InSb QDs/InAs heterostructure. Light contrast is the position of the InSb QDs layer.

**Table 1.** Heterostructures with QDs obtained by combined epitaxial technology.

| Sample | Epitaxial Structure | QDs Surface Density($n_{QD}$) |
|---|---|---|
| A | p-InAs/n-InAs | - |
| B | p-InAs/InSb/n-InAs | 0.7 × $10^{10}$ $cm^{-2}$ |
| C | p-InAs/InSb/n-InAs | 2.0 × $10^{10}$ $cm^{-2}$ |

Samples for EL studies were performed as mesa diodes of 500 × 500 µm in size with a wide contact to the substrate side and a point contact to the upper epitaxial layer. A round mesa fabricated by standard photolithography and selective wet chemical etching had the diameter of $D$~300 µm; whereas, the point contact diameter was $d$~50 µm. The studied mesa diodes were mounted on the TO-18 package and located with a substrate side to the metal case. In this design, the radiation emerging from the structure was detected from the side of the capping epitaxial layer. The EL spectra were recorded using an automated setup based on a DK-480 monochromator (CVI Laser Corp., Albuquerque, NM, USA) with a diffraction grating of 150 lines/mm, an InSb photovoltaic detector cooled by liquid nitrogen (Judson Co, Montgomeryville, PA, USA), and an SR-810 lock-in amplifier (Stanford Research Systems Inc., Sunnyvale, CA, USA). The samples were supplied by rectangular current pulses with a duty cycle of $\eta$ = 50% and a frequency of $f$ = 512 Hz, both with forward and reverse external bias. The forward bias corresponds to the case where the positive potential was applied to the p-InAs substrate; whereas, the negative one was applied to the upper n-InAs layer. Therefore, reverse bias has the opposite configuration of polarities. The pulse amplitude of the driving current (i) was varied from 20 mA to 200 mA.

To analyze the obtained EL spectra, preliminary calculations of the band energy diagrams of the heterostructures studied in the experiment were carried out. Estimation of the energy spectrum of quantum-size structures was performed using available resources [12]. For instance, a single InSb QD can be represented as a narrow one-dimensional potential well with a set of localized levels (Figure 2a). This makes it possible to determine the energy levels for electrons and holes depending on the well width, material compositions and parameters of adjacent semiconductor compounds. It can be seen from the figure that the ground hole level of the InSb QD ($\mu_h$) is situated in energy above the bottom of the conduction band of the InAs matrix layer due to the peculiarity of the type II broken-gap heterojunction in the InSb-InAs system. As was shown in [13], structural crystalline defects in indium arsenide can create acceptor states (IS) on the surface of a semiconductor.

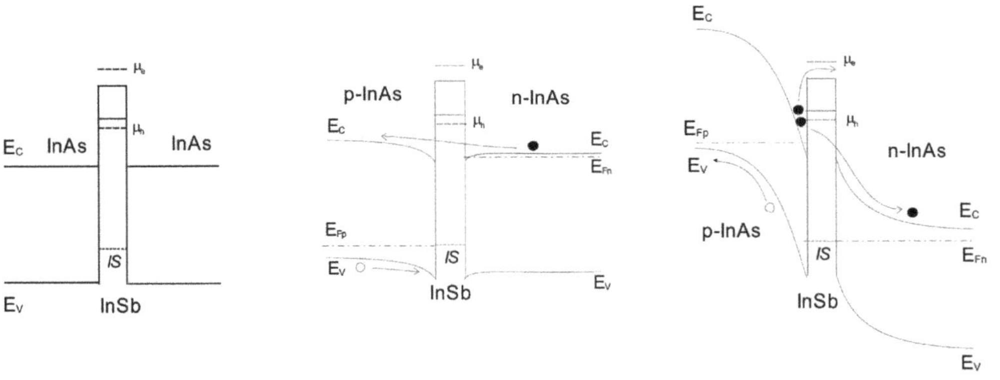


(**a**) (**b**) (**c**)

**Figure 2.** (**a**) Schematic energy band diagram of the type II InAs/InSb/InAs heterostructure with a layer of QDs. p-InAs/InSb QDs/n-InAs heterostructure under external forward (**b**) and reverse (**c**) bias. *IS*—the level of surface states at the interface InAs/InSb; μe and μh are the ground electron and hole states of the InSb QD, respectively.

### 3. Results

EL spectra at room temperature displayed similar features for all structures under the study on forward bias applied (Figure 3a). The typical EL spectrum contained one pronounced emission band with an intensity maximum near $h\nu_{300}(+)$ = 0.36 eV and FWHM = 48 meV, which corresponded to interband radiative transitions in the bulk of indium arsenide [14]. Additional emission bands can be distinguished near $h\nu_{IS}$ = 0.31 eV as a shoulder on the low-energy slope of the main EL band. At forward bias (Figure 2b), potential barriers near the InAs/InSb interface appear on the moving path of electrons in an external electric field from the populated conduction band of n-InAs to the unfilled conduction band of p-InAs. These barriers are induced by the QDs energy band structure and interface states at the heteroboundary between the matrix layers. It should be emphasized that no other radiative recombination transitions involving impurity states with an activation energy of less than 30 meV were detected, which can be explained by the depletion of shallow levels due to their thermal warming at $T = 300$ K.

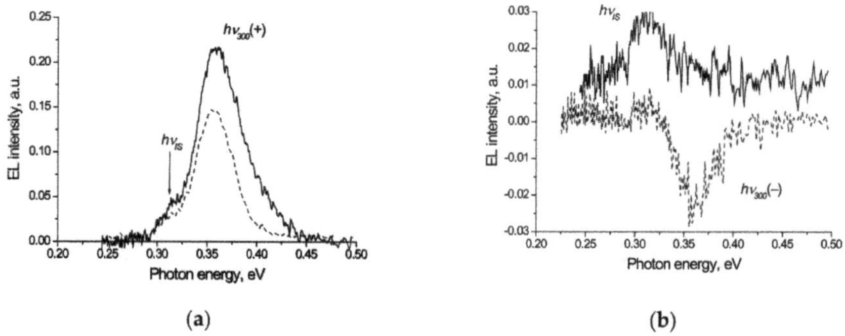


**Figure 3.** EL spectra obtained for *p*-InAs/*n*-InAs (dashed line) and *p*-InAs/InSb/*n*-InAs (solid line) heterostructures. The measurements were performed at forward (**a**) and reverse (**b**) biases, $T = 300$ K, and driving current $i = 50$ mA.

At reverse bias (Figure 2c), the shapes of the EL spectra at room temperature for samples with and without QDs were radically different from each other (Figure 3b). The EL spectrum for the p-InAs/n-InAs structure showed an emission band in the vicinity of $h\nu_{300}(-) = 0.36$ eV, reversed in phase by 180° relative to the zero-intensity level, i.e., negative luminescence was observed in the bulk of p-InAs [9,15]. Together with this band, a weak peak of positive EL was found at photon energy $h\nu_{IS} = 0.31$ eV. Contrary to sample A, the heterostructures with InSb QDs (p-InAs/InSb/n-InAs) demonstrated exclusively positive luminescence. Note that these EL spectra contain only one emission band and exactly with an intensity maximum in the vicinity of the photon energy of 0.31 eV. Radiation in the range of photon energy close to the InAs band gap was not detected for samples B and C at all. The space charge region is shunted by the ohmic contact of the type II broken-gap heterojunction due to the fact that the hole levels of the InSb QD are situated in energy above the conduction band of the InAs matrix layers. The main drop in external bias can be expressed as a linear slope in the region of the flat energy bands in the InAs layer that provides conditions for the free movement of electrons to the unfilled conduction band of the n-InAs and further to a metal contact without any potential barriers in their path. In this way, the process of extraction of charge carriers from the depleted regions can be significantly suppressed. A similar effect of negative luminescence suppression was recently observed for heterostructures with an ultrathin oxide layer separating the surfaces of two contacting semiconductors, which are forming a type II broken-gap heterojunction under conservation of the levels of localized states on the surface of indium arsenide [16]. Thus, radiative recombination transitions are realized with the participation of the interface level (IS) for any direction of the injection current (both forward and reverse biases) that is expressed in the presence of an emission band with unchanged polarity in the EL spectrum. Therefore, the $h\nu_{IS}$ emission band observed at room temperature can be ascribed to radiative transitions involving the surface states at the p-InAs/n-InAs heterointerface which are placed 50 meV above the top of the InAs valence band.

As the temperature is lowered to $T = 77$ K, the contribution of radiative recombination transitions with the participation of impurity states in p-InAs, which have an activation energy greater than the energy of thermal level smearing ($kT_{77}$~6 meV), becomes noticeable. The EL spectra for all the studied samples contained two distinct emission bands $h\nu_{77} = 0.408$ eV and $h\nu^{77}{}_{A} = 0.376$ eV upon application of forward external bias (Figure 4). The high-energy band ($h\nu_{77}$) can be attributed to interband radiative transitions in InAs [14].

The energy distance of the low-energy band $h\nu^{77}{}_A$ from the high-energy band was 32 meV, which points to radiative recombination transitions associated with impurity acceptor states of Mn situated in the forbidden gap of InAs [17]. Furthermore, the presence of an additional emission band is assumed, which can be revealed as a shoulder on the low-energy slope of the EL spectrum with an emission maximum near 0.35 eV. We suppose that it can be ascribed to radiative transitions involving interface states ($\Delta h\nu_{77}(IS)$~58 meV).

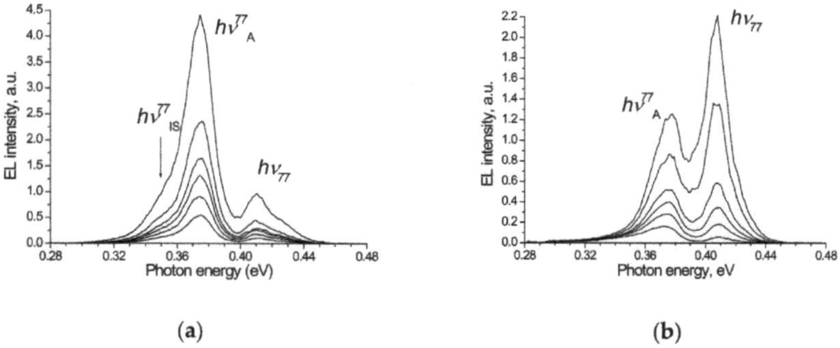


**Figure 4.** EL spectra obtained for different values of the driving current ($i$ = 20, 30, 40, 50, 75 and 100 mA) at forward bias applied at $T$ = 77 K: (**a**) for samples A and B; (**b**) for sample C.

The EL measurements revealed different behavior in intensity of the emission bands with increasing driving current for the samples under study depending on the surface density of the InSb QDs at the *p-n* junction interface. For sample A (without QDs) and sample B (with a low concentration of QDs), the dominance of the emission band $h\nu^{77}{}_A$ was retained over the entire range of driving currents (Figure 4a). For sample C (with a high concentration of QDs), a redistribution of intensity between the EL bands was observed, when the dominance of the low-energy band $h\nu^{77}{}_A$ at low injection levels ($i < 50$ mA) was replaced by the advantage of the high-energy band $h\nu_{77}$ at higher injection levels (Figure 4b). In this case, the current dependence of EL intensity for the $h\nu_{77}$ band that occurred by interband radiative transitions in indium arsenide was changing within equal range for all samples. In other words, an increase in the surface density of nano-objects at the interface led to a decrease in the contribution of radiative transitions involving impurity acceptor states of p-InAs to the total EL spectrum. According to the energy diagram shown in Figure 1b, the appearance of a space charge region on the p-InAs side near the heterointerface prevents electrons from penetrating into the region of flat bands, where radiative recombination with the participation of impurity levels takes place.

When reverse bias was applied to the heterostructures under study, the EL spectra at $T$ = 77 K showed no emission band corresponding to interband transitions in InAs (Figure 5), as well as at room temperature performance. This characteristic feature of EL behavior at low temperatures was previously observed for heterostructures with the InSb QDs layer at the type II broken-gap InAs/InSb heterointerface in the p-InAs matrix, where interband transitions began to manifest themselves as a negative luminescence at high pump levels only [9]. The main emission band ($h\nu_{IS+Channel}$) in the EL spectra recorded at reverse bias had an asymmetric shape with a sharp high-energy edge and exhibited a "blue" shift; i.e., the spectral position of the EL peak was shifting towards higher photon energies with an increase in driving current (Figure 6).

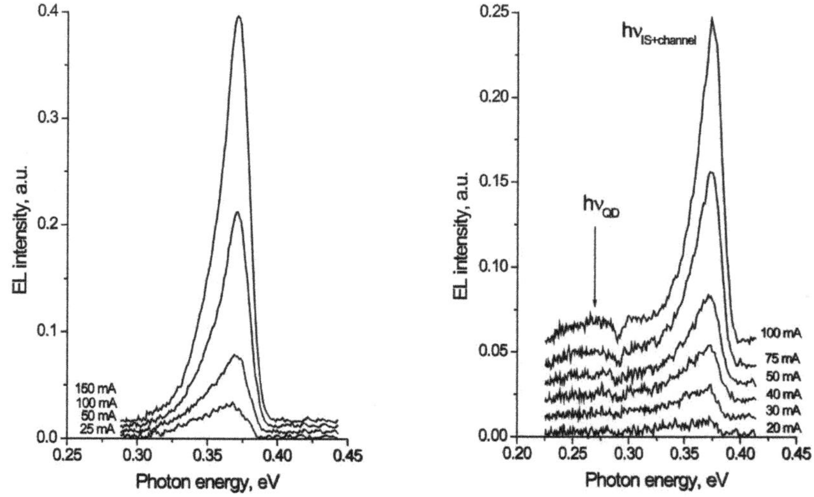


(a) (b)

**Figure 5.** EL spectra of *p*-InAs/InSb/*n*-InAs heterostructures with InSb QDs ((**a**) sample B and (**b**) sample C), measured at reverse bias at $T$ = 77 K.

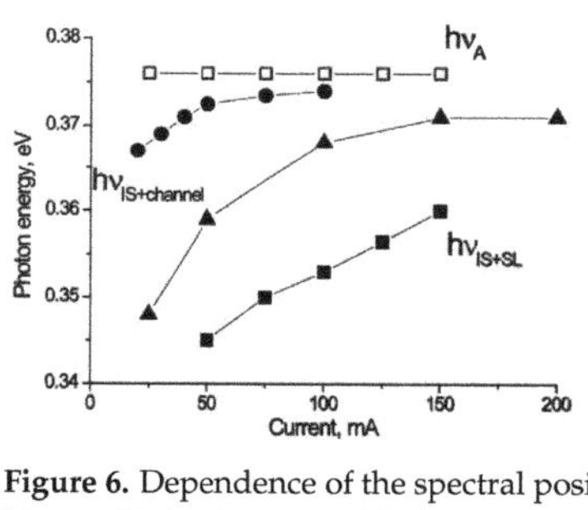


**Figure 6.** Dependence of the spectral position of the EL bands on the driving current at the reverse bias applied at $T$ = 77 K (filled symbols). For comparison, the dependence of the band $h\nu_A$ is shown when the forward bias has been applied (open squares).

Different behavior should be distinguished between the "blue" shift for each dependence of the EL emission band on the current. A linear dependence of the $h\nu_{IS+SL}$(i) band for sample A was observed in the photon energy range from 0.345 eV to 0.358 eV in the current interval $i$ = 50–150 mA (Figure 6, filled squares). Comparing the energy band diagram of sample A at reverse bias (not shown) and the photon energy range in which the shift of the band $h\nu_{IS+SL}$ was observed, one can estimate the energy position of the interface level IS at low temperature. It turns out that the linear "blue" shift observed in the EL spectra for sample A is the result of radiative transitions from shallow impurity levels (*SL*) formed by natural structural defects in the n-InAs capping layer to the interface level *IS*. Then, an increase in the reverse bias will cause a gradual depletion of the donor levels in n-InAs near the heterojunction. According to our estimates, the level *IS* was approximately 50 meV above the top of the InAs valence band at a low temperature ($T$ = 77 K), which is in good agreement with the experimental data obtained at room temperature (see above).

Note also that both dependences $h\nu_{IS+channel}$(i), which describe the "blue" shift for samples with QDs, demonstrated non-linear behavior and saturation at high pumping. The specific shape of these dependences was determined by the surface density of nano-objects at the InAs/InSb heterointerface. For sample C, with a high concentration of the InSb QDs at the interface ($n_{QD2} > 10^{10}$ cm$^{-2}$), luminescence began to appear at a photon energy $h_{IS+channel}$ = 0.366 eV (Figure 6, filled circles). In this case, the range of motion of the given EL band from the high-energy side was limited by the value of the radiative transition energy $h\nu_A$ = 0.376 eV (Figure 6, open squares). In contrast to samples A and C, the curve ascribed to sample B, with a low concentration of QDs at the interface ($n_{QD} < 10^{10}$ cm$^{-2}$), occupies an intermediate position, representing the result of a superposition of the two extreme cases considered (Figure 6, filled triangles). For this sample, the onset of luminescence was recorded at $h_{IS+channel}$ = 0.348 eV, and the "blue" shift was saturated at a photon energy of 0.37 eV.

## 4. Discussion

This section will consider the emergence of the band $h\nu_{IS\,+\,channel}$ and discuss its behavior. When reverse bias has been applied to the heterostructures with InSb QDs (samples B and C), a potential well on the p-InAs side near the InAs/InSb interface emerged as a result of bending in the conduction band (Figure 2c). A 2D-electron channel is formed at the p-InAs/InSb heterojunction under increasing the external electric field due to the filling of the potential well with electrons [18]. Then, there is the possibility of radiative recombination transitions from the quasi-Fermi level ($E_{Fp}$) states of the electron channel to the interface states *IS* [19]. The quantum energy of such transitions can be described by the following expression

$$h\nu_{IS+channel} = E_{IS} + E_{Fp}, \quad (1)$$

where $E_{IS}$ and $E_{Fp}$ are the energy of the interface level and Fermi level position relative to the bottom of the InAs conduction band, respectively. With a further increase in the external bias and, consequently, the filling of the electron channel, the $E_{Fp}$ level is shifted upward in energy that will be expressed in the "blue" shift of the EL band corresponding to these transitions. The subsequent saturation of the observed shift with increasing driving current is explained by the fact that the filling of the electron channel continues until electrons leak through the potential barrier formed by the band gap of the InSb QD. An additional current flow channel is formed already with the participation of the QD electron states. Accordingly, samples with various QD concentrations at the interface demonstrate different saturation levels of the "blue" shift of the EL bands (see Figure 6).

The appearance of the additional current flow channel upon application of a reverse bias enables filling of the eigenstates in the InSb QDs. Therefore, the EL spectra at low temperatures can contain the emission band corresponding to recombination transitions between localization levels in a quantum dot. The photon energy of such transitions can be estimated as

$$h\nu_{QD} = E_{QD} + \delta E_{QD} + \mu_h + \mu_e, \quad (2)$$

where $E_{QD}$ is the band gap of the InSb binary compound, $\delta E_{QD}$ is the deviation in the band gap of InSb due to internal compression strain in the InSb QD, $\mu_e$ and $\mu_h$ are the energies of the ground electron and hole levels of the InSb QD, respectively. The EL spectra of sample C exhibited the emission band in the region $h\nu_{QD}$ = 0.270 eV with a half-width of about 60 MeV (Figure 5b). Note that a sharp dip at the low-energy edge of the EL spectrum near a photon energy of 0.29 eV is associated with the absorption of optical radiation by $CO_2$ molecules ($\lambda$ = 4.27–4.29 μm) present in the laboratory atmosphere [3,5].

The spectral position of the $h\nu_{QD}$ band agrees well with the experimental data obtained by measuring the PL spectra at T = 77 K for structures with InSb QDs in an InAs matrix grown by molecular beam epitaxy [8]. It should be emphasized that the mentioned PL band was recorded only for structures with a surface density of nano-objects of $n_{QD} > 2{\cdot}10^{10}$ cm$^{-2}$. Because the intensity of EL radiation induced by transitions between localized levels in QDs also depends on the concentration of nano-islands, the band $h\nu_{QD}$ was not detected for samples with lower surface density of the InSb QDs (Figure 5a). In the EL spectra of sample C recorded at T = 300 K, the corresponding EL band was not observed either (Figure 3b). This was caused by the fact that the presented measurements were carried out using an InSb photovoltaic detector with a long-wavelength cutoff $\lambda$ = 5.3 μm that did not allow the recording of luminescence with a photon energy less than 0.23 eV. Attempts to obtain electroluminescence in heterostructures with InSb QDs in an InAs-based matrix were undertaken by other scientific groups; however, the low surface density of nano-objects prevents interpretation of the observed emission band with a photon energy near 0.37 eV as radiative recombination involving the states of InSb QDs [20]. It is highly likely that the claimed EL band is due to the participation of surface levels at the InAs/InSb heterointerface.

## 5. Conclusions

Thus, the EL spectra with participation of the eigenstates of InSb QDs placed in a narrow-gap matrix based on InAs compounds was observed for the first time. Narrow-gap heterostructures containing a single InSb QD layer embedded in the p-InAs/n-InAs junction were obtained using a combined epitaxial growth technology. Type II broken-gap p-InAs/InSb/n-InAs heterostructures exhibited radiative recombination in the bulk of indium arsenide and interface luminescence involving surface states at the InAs/InSb heterointerface at forward external bias in a wide temperature range. The observation of EL with the participation of the eigenstates of InSb QDs is possible under reverse bias applied when the QD states are filled from the 2D-electron channel at the type II broken-gap heterointerface.

**Author Contributions:** Conceptualization, K.M. and Y.P.; methodology, K.M.; software, E.I.; validation, K.M., E.I. and Y.P.; formal analysis, K.M.; investigation, E.I. and Y.P.; resources, E.I. and Y.P.; data curation, E.I. and Y.P.; writing—original draft preparation, K.M.; writing—review and editing, K.M. and E.I.; visualization, E.I. and Y.P.; supervision, K.M.; project administration, K.M.; funding acquisition, K.M. All authors have read and agreed to the published version of the manuscript.

**Funding:** This research received no external funding.

**Institutional Review Board Statement:** Not applicable.

**Informed Consent Statement:** Not applicable.

**Data Availability Statement:** Not applicable.

**Acknowledgments:** The authors are grateful to P.A. Dement'ev and V.N. Nevedomsky for performing AFM and TEM studies, respectively. The data were obtained using the equipment of the Joint Research Center "Material science and characterization in advanced technologies".

**Conflicts of Interest:** The authors declare no conflict of interest.

## References

1. Schiff, H.I.; Mackay, G.I.; Bechara, J. *Air Monitoring by Spectroscopy Techniques*; Sigrist, M.W., Ed.; John Wiley and Sons: New York, NY, USA, 1994; p. 560.
2. Wang, C.; Sahay, P. Breath Analysis Using Laser Spectroscopic Techniques: Breath Biomarkers, Spectral Fingerprints, and Detection Limits. *Sensors* **2009**, *9*, 8230–8262. [CrossRef] [PubMed]
3. Hudson, R.D. *Infrared System Engineering*; John Wiley and Sons: New York, NY, USA; London, UK; Sydney, Australia, 1969; p. 642.
4. Romanov, V.V.; Ivanov, E.V.; Pivovarova, A.A.; Moiseev, K.D.; Yakovlev, Y.P. Long-wavelength LEDs in the atmosphere transparency window of 4.6–5.3 μm. *Semicond.* **2020**, *54*, 253–257. [CrossRef]
5. Pierson, R.H.; Fletcher, A.N.; Gantz, E.C., St. Catalog of Infrared Spectra for Qualitative Analysis of Gases. *Analyt. Chem.* **1956**, *28*, 1218–1239. [CrossRef]
6. Zhukov, A.E.; Maksimov, M.V.; Kovsh, A.R. Device characteristics of long-wavelength lasers based on self-organizing quantum dots. *Semiconductors* **2012**, *46*, 1249–1273.
7. Wei, S.H.; Zunger, A. InAsSb/InAs: A type-I or a type-II band alignment. *Phys. Rev. B* **1995**, *52*, 12039–12044. [CrossRef] [PubMed]
8. Karim, A.; Gustafsson, O.; Hussain, L.; Wang, Q.; Noharet, B.; Hammar, M.; Anderson, J.; Song, J. Characterization of InSb QDs grown on InAs (100) substrate by MBE and MOVPE. *Proc. SPIE* **2012**, *8439*, 84391J.
9. Parkhomenko, Y.A.; Ivanov, E.V.; Moiseev, K.D. Specific features of electroluminescence in heterostructures with InSb quantum dots in an InAs matrix. *Semiconductors* **2013**, *47*, 1523–1527. [CrossRef]

10. Bert, N.A.; Nevedomsky, V.N.; Dement'ev, P.A.; Moiseev, K.D. TEM-study of free-standing self-assembled InSb quantum dots grown on InAs-rich surface. *Appl. Surf. Sci.* **2013**, *267*, 77–80. [CrossRef]
11. Moiseev, K.D.; Parkhomenko, Y.A.; Nevedomsky, V.N. Uniform InSb quantum dots buried in narrow-gap InAs(Sb,P) matrix. *Thin Solid Films* **2013**, *543*, 74–77. [CrossRef]
12. Available online: http://www.matprop.ru/levels (accessed on 22 November 2022).
13. Bunin, V.A.; Matveev, Y.A. Electronic structure of an anion vacancy in indium arsenide. *Sov. Phys. Semicond.* **1985**, *19*, 2018–2022.
14. Landolt-Börnstein. *Numerical Data and Functional Relationships in Science and Technology*; Madelung, O., Ed.; Springer: New York, NY, USA, 1982; Volume 17, p. 642.
15. Grigoryev, M.M.; Alekseev, P.A.; Ivanov, E.V.; Moiseev, K.D. Two-colour luminescence in a single type II InAsSbP/InAs heterostructure. *Semiconductors* **2013**, *47*, 28–32. [CrossRef]
16. Moiseev, K.; Ivanov, E.; Romanov, V.; Mikhailova, M.; Yakovlev, Y.; Hulicius, E.; Hospodková, A.; Pangrác, J.; Šimeček, T. Intense interface luminescence in type II narrow-gap InAs-based heterostructures at room temperature. *Phys. Procedia* **2010**, *3*, 1189–1193. [CrossRef]
17. Georghitse, E.I.; Postolaki, I.T.; Smirnov, V.A.; Untila, P.G. Photoluminescence of p-InAs<Mn>. *Sov. Phys. Semicond.* **1989**, *23*, 745–746.
18. Moiseev, K.D.; Berezovets, V.A.; Golenitskii, K.Y.; Averkiev, N.S. Band structure of a hybridized electron-hole system at a single broken-gap type II heterointerface. *Low Temp. Phys.* **2019**, *45*, 153–158. [CrossRef]
19. Mikhailova, M.P.; Moiseev, K.D.; Yakovlev, Y.P. Interface-induced optical and transport phenomena in type II broken-gap single heterostructures. *Semicond. Sci. Technol.* **2004**, *19*, R109–R128. [CrossRef]
20. Zhuang, Q.; Carrington, P.J.; Krier, A. Growth optimization of self-organized InSb/InAs quantum dots. *J. Phys. D Appl. Phys.* **2008**, *41*, 232003. [CrossRef]

*electronics*

MDPI

*Article*

# Facile Green Preparation of Reduced Graphene Oxide Using Citrus Limetta-Decorated rGO/$TiO_2$ Nanostructures for Glucose Sensing

**Medha Gijare [1], Sharmila Chaudhari [2], Satish Ekar [1], Shoyebmohamad F. Shaikh [3], Rajaram S. Mane [4,*], Bidhan Pandit [5], Muhammad Usman Hassan Siddiqui [6] and Anil Garje [7,*]**

1 Department of Physics, Baburaoji Gholap College, Pune 411027, India
2 Department of Physics, Anantrao Pawar College, Pirangut, Pune 412115, India
3 Department of Chemistry, College of Science, King Saud University, P.O. Box 2455, Riyadh 11451, Saudi Arabia
4 School of Physical Sciences, Swami Ramanand Teerth Marathwada University, Nanded 431606, India
5 Department of Materials Science and Engineering and Chemical Engineering, Universidad Carlos III de Madrid, Avenida de la Universidad 30, Leganés, 28911 Madrid, Spain
6 Unilever Lipton Jebal Ali Factory Engineering Department, Dubai P.O. Box 325, United Arab Emirates
7 Department of Physics, Sir Parashurambhau College, Pune 411030, India
* Correspondence: rajarammane70@gmail.com (R.S.M.); anil.garje@spcollegepune.ac.in (A.G.)

**Abstract:** The important electrochemical measurements of reduced graphene oxide-titanium oxide (rGO)/$TiO_2$) electrodes for the application of a glucose sensor are reported in the proposed work. Investigating the sensitivity, stability, and reproducibility of sensor electrodes that were made and used to evaluate the concentration of glucose in the serum is one of the novel aspects of this work. This study presents the use of citrus limetta (sweet lime) fruit peel waste to synthesize a green reduction of graphene oxide (rGO). The rGO/$TiO_2$ composite obtained using the microwave heating method is applied for measuring the structural and morphological properties by various means. A conducting fluorine-tin oxide substrate is used to modify the enzymeless glucose sensor electrode. The electrochemical measurements of rGO/$TiO_2$ sensor electrodes are carried out using the technique of cyclic voltammetry. The rGO/$TiO_2$ sensor electrode exhibits a high sensitivity of 1425 $\mu A/mM\ cm^2$ towards glucose concentration in the range of 0.1 to 12 mM. The sensor was found to be extremely stable and repeatable with a response time of 5 s along with a minimum detection limit of 0.32 $\mu M$ of glucose. The rGO/$TiO_2$ sensor shows relative standard deviation (RSD) of 1.14%, 1.34%, and 1.3% which reveals its excellent stability, repeatability, and reproducibility respectively. The sensor was used for glucose level detection in natural blood serum and shows an RSD of 1.88%. which is in good agreement with the commercial glucose sensor values.

**Keywords:** cyclic voltammetry; glucose sensor; graphene oxide; microwave heating; phytochemicals

**Citation:** Gijare, M.; Chaudhari, S.; Ekar, S.; Shaikh, S.F.; Mane, R.S.; Pandit, B.; Siddiqui, M.U.H.; Garje, A. Facile Green Preparation of Reduced Graphene Oxide Using Citrus Limetta-Decorated rGO/$TiO_2$ Nanostructures for Glucose Sensing. *Electronics* **2023**, *12*, 294. https://doi.org/10.3390/electronics12020294

Academic Editor: Wiesław Leonski

Received: 16 October 2022
Revised: 14 December 2022
Accepted: 19 December 2022
Published: 6 January 2023



## 1. Introduction

According to the International Diabetes Federation, the metabolic disorder known as diabetes has become a major issue for people around the world, affecting millions of people. This epidemic threatens human health and is one of the most significant contemporary barriers to long-term social development and economic growth. The ability to accurately monitor and control individuals' blood glucose levels is critical for diabetic patients to receive the proper diagnosis and treatment. To avoid diabetes emergencies like hypoglycemia (low blood sugar, $\leq$3 mM), the physiological glucose level must be checked on a routine basis [1]. Glucose detection has become increasingly important in the fight against diabetes and in preventing financial loss due to the significant financial cost caused by diabetes and its serious complications. As a result, for a few decades, developing reliable and low-cost glucose sensors with excellent sensitivity and selectivity has had enormous social and economic importance. Glucose biosensors have been significantly subsidized to monitor glucose levels in diabetic patients [2]. Many researchers are actively engaged in developing new biosensors for various applications in various disciplines since Clark and Lyons first proposed the concept of enzymatic electrodes in 1962. Enzymatic sensors confirmed good sensitivity and stability but demonstrated a few drawbacks like poor reproducibility, complex mobility, etc. [3,4]. Though enzymatic sensors have certain advantages, commercial applications are mitigated on account of their high manufacturing cost [5]. Non-enzymatic glucose sensors are a solution created in response to these limitations [6].

All candidate materials for the electrocatalyst in glucose detection, including metal oxide, carbon materials, polymers, graphene metal, graphene oxide materials, etc., have distinguished themselves [7]. Graphene is a two-dimensional material made up of a single layer of carbon atoms in a hexagonal lattice. It has received considerable attention due to its unusual physical, optical and electrical properties. It is the world's thinnest and toughest material, with exceptional electrical and thermal conductivity. Graphene is synthesized using a variety of methods, including micromechanical cleavage, chemical vapor deposition, epitaxial growth on silicon carbide liquid-phase exfoliation, graphene oxide reduction by various reducing agents, and graphite intercalation [8]. It is astonishing mechanical [9], thermal [10] and electrical [11] properties have captivated tremendous attentiveness. Reducers such as hydrazine or sodium borohydride are typically envisaged during the chemical preparation of GO [12,13]. These reducing agents have some instinctive disadvantages such as expensive and highly toxic. Moreover, the Van der Waals forces among the graphene layers produce an irreversible aggregation of GO [14].

In the case of green reduction, a study of suitable phytochemicals as reducing agents is necessary. Additionally, a green reduction is a simple, eco-friendly, and cost-effective method. In light of this motivation, the work that is being proposed provides an environmentally friendly reduction of GO using Citrus limetta (sweet lime/Mosambi fruit waste) and $TiO_2$ electrodes for electrochemical glucose sensing that does not require enzymes. Green reduction and bio-synthesis routes are employed to overcome the adverse effects of hazardous substances, where phytoextracts or bacteria microorganisms play eco-friendly (for reduction) and capping agent roles [15–17]. A germ-free environment is necessary to grow micro-organisms, and also expertise is needed in this process. The reduction of GO can be done using citrus plant extract, which has a slower reaction rate to control the crystal growth. In the past, the green synthesis of reduced graphene oxide (rGO) has been studied using several plants and fruit extracts [18]. The green nature, low cost, and easy local availability of plant extracts and fruit wastes make them suitable for reduced graphene oxide. Citrus limetta is the primary natural source of vitamin C and an energy booster [19]. Fruit peels of sweet lime are a waste to the juice industry, an important source of phytochemicals like flavonoids, pectins, carotenoids, hesperidin, terpineol alkaloids, essential oils, *d*-limonene, and extraordinary antioxidants as well as antibacterial properties [20]. Graphene obtained via the green reduction process can be functionalized with metal oxides for various applications including electrochemical supercapacitors, batteries, fuel cells, magnetic data storage, water splitting, chemical sensors, water filters, dye removal, and dye-sensitized solar cells, etc. [21–24].

In the present study, citrus limetta was chosen as a reducing agent for the synthesis of rGO because it has a small effect on the environment and doesn't have any harmful biological reactions that could happen. The soxhlet extraction method is preferred due to its simplicity and effectiveness in collecting sweet lime peel extract. Functionalized graphene with some metallic oxide revealing extraordinary sensing properties is familiar for a long [25]. The electrochemical deposition method adopted by Luo et al. to produce graphene and CuO nanocubes [26]. The rGO obtained by Yang et al. using the chemical reduction method demonstrated excellent sensitivity and selectivity of 4 mM and 7.5 mM linear ranges y [27]. Komathi et al. presented the development of electrochemical biosensors by functionalizing titanium dioxide ($TiO_2$) nanowires/graphene as nano-stocks [28]. According to these findings, the graphene-based metal/metal oxide nanoparticles endow good sensitivity, a wide linear range, and a low detection limit with good selectivity. Hence, the rGO/$TiO_2$ composite has been selected as a sensor electrode in the proposed work and used as an enzyme-free biosensor to detect glucose concentration efficiently using a cyclic voltammetry (CV) test. Moreover, very little literature on green synthesis of the rGO/$TiO_2$-based biosensors is available is familiar. The as-fabricated glucose biosensor is also used to investigate glucose levels in the existence of obstructing substances in real blood serum.

## 2. Materials and Methods

### 2.1. Materials

All the chemicals used were of the analytical grade, meaning they were of the highest purity. Graphite powder (98%), $KMnO_4$ (99%), $NaNO_3$ (98%), $H_2O_2$ (30%), $H_2SO_4$ (99.99%), HCl (30%), $H_3PO_4$ (85%), KCl (99.99%) and NaCl (99%) $TiO_2$ (99.9%), etc., were obtained from Finar chemicals (India). Qualigens fine chemicals in India provided polyvinyl alcohol (PVA, 99%), ethanol, Fluorine-doped Tin Oxide (FTO) substrates, D (+) glucose, dopamine, L-ascorbic acid, D (-) fructose, and lactose, phosphate buffer solution (PBS) was prepared according to AAT Bioquest, Inc. [29] and sweet limes were bought from a local market.

### 2.2. Preparation of GO

A modified Hummer method was used to synthesize GO [30]. Briefly, the concentrated $H_2SO_4$ and $H_3PO_4$ were mixed in a 9:1 proportion in a round bottom flask, followed by graphite powder and $NaNO_3$ in a 2:1 proportion. To achieve the desired result, it was necessary to stir the mixture at 100 °C continuously for an hour before adding the $KMnO_4$. Adding $H_2O_2$ to the mixture served the purpose of oxidation, resulting in the formation of a yellow-colored product. $SO_4{}^{2-}$ is completely removed from the sample by repeatedly washing with 5% HCl and deionized (DI) water, followed by centrifugation (4000 rpm). After nearly 24 h of drying at 60 °C, the resultant product (GO) was obtained which was collected for the next processes.

### 2.3. Preparation of Aqueous Phytoextract

Fresh sweet limes were purchased from a local market and cleaned using DI water. The citrus limetta fruit waste, such as peel, was cut into small pieces with a domestic cutter. In soxhlet extraction, 15 gm of finely cut peels were placed in a thimble of soxhlet tube, and the DI water was used as the solvent. During condensation, solutes are moved to the reservoir for repeated cycles. The system was cooled for 10–12 h and then the collected extract was preserved at 4 °C for further use.

### 2.4. Mechanism of Reduction of GO by Citrus Limetta Phytoextract

Citrus Limetta phytoextract can reduce GO by utilizing the oxidized products found in sweet lime fruit peel. These oxidized products include flavonoids, polyphenols, tannis, terpenoids, and saponins, among others. These phytochemicals show high binding linkage towards $O_2$- containing functionalities to produce their respective oxides and water molecules. Flavonoid compounds are reduced to the phenolic hydroxyl groups. While reacting with the $O_2$- containing groups, the phenolic hydroxyl group may release water molecules on GO sheets. GO contains reactive, hydroxyl, epoxide, and reactive carbonyl species. As shown in Figure 1, acidic polyphenols in the aqueous GO and Citrus Limetta phytoextract were combined, ultra-sonicated, and refluxed at 95 °C. Through a nucleophilic reaction, epoxide fraction and phytoextracts could react with each other. This caused the carbon-oxygen bond to break. Carbonyl and hydroxyl groups in polyphenols could be attacked by nucleophiles in the same way, which would destroy water molecules. GO is changed into rGO in this process, which makes graphene, which is a stable black compound.

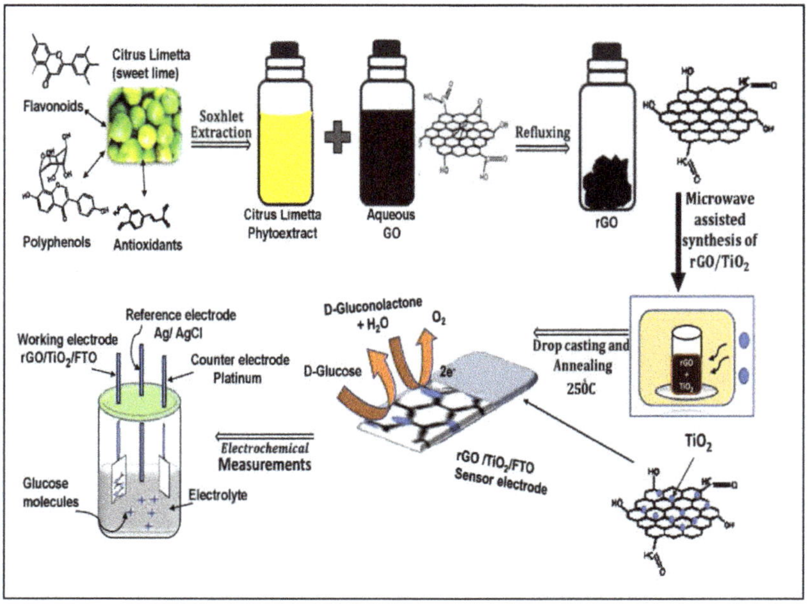


**Figure 1.** Mechanism of green reduction of GO and rGO/$TiO_2$/FTO for glucose sensing application.

*2.5. Reduction of GO Using Aqueous Citrus Limetta Peel Extract*

Briefly, 100 mg GO (0.1 mg/mL) was disseminated in ultrapure water by probe sonication for 40 min. The GO solution was mixed with 5 wt.% of 10 mL of aqueous sweet lime fruit extract. The solution's composition was refluxed at 98 °C for 8 h until it turned brown-black. The supernatant (rGO) was repetitively washed with DI water which was centrifuged at 4500 rpm, followed by drying at 100 °C in the oven.

*2.6. Synthesis of rGO/$TiO_2$ Nanocomposite*

The rGO/$TiO_2$ composite was prepared using microwave treatment. A fine powder of rGO and $TiO_2$ in a 2:1 ratio was dispersed in DI water. After sonication, the mixture was stirred. The compound was kept at 300 °C for 15 min in a microwave (500 W). Finally, the rGO/$TiO_2$ nanocomposite dry powder was collected. First, 10 mg of rGO/$TiO_2$ nanocomposites were mixed with 5 mL of DI water and 5 µL of PVA and sonicated in a bath for 30 min. Then, 10 µL of the mixture was drop-cast onto a clean conducting FTO substrate. The working area of the electrode was 1 $cm^2$. The as-prepared electrode was annealed to remove the binder at a temperature of 250 °C for 1 h. A three-electrode setup consisting of a working electrode made of rGO/$TiO_2$/FTO, a reference electrode made of Ag/AgCl, and a counter electrode made of platinum was used to conduct electrochemical measurements for glucose detection.

*2.7. Glucose Sensing through rGO/$TiO_2$*

For rGO/$TiO_2$ nanocomposite preparation, in-situ thermal reductions of rGO and $TiO_2$ were developed. The epitaxial growth process was used to prepare $TiO_2$ nanobelts. Using Phosphate-Buffered Saline (PBS) as an electrolyte, the glucose sensing performance of the rGO/$TiO_2$/FTO modified electrode was evaluated. When glucose dissolves in PBS, D-gluconolactone and hydrogen peroxide are produced ($H_2O_2$). Electro-oxidation of glucose at $TiO_2$ grain boundaries also produces D-gluconate with $H^+$ ions. As a result, gluconolactone is the primary product of the oxidation process that hydrolyzes gluconic acid. By releasing electrons, lowering the barrier potential between successive grains, and increasing electrical conductivity, $H^+$ ions reduce pre-absorbed oxygen [31].

$$\text{Glucose} + O_2 \rightarrow H_2O_2 + \text{gluconic acid} \quad (1)$$

### 2.8. Characterization Techniques

The as-synthesized GO, rGO, and rGO/$TiO_2$ sample products were analyzed using a Bruker D8-Advanced Diffractometer with Cu Kα radiation (λ = 0.154 nm) from 5° to 85° at a scanning rate of 2°/min for structural elucidation. Field Emission Scanning Electron Microscopes (FE-SEM): FEI Nova Nano SEM 450 was used to examine the surface morphology and elemental composition of GO, rGO, and rGO/$TiO_2$ samples. A Micro-Raman spectrometer (Jobin-Yuon HR 800 UV) with a He–Ne (633 nm) laser excitation source was used to conduct the Raman analysis. The multifunctional X-ray photoelectron spectroscopy (XPS) was used to perform XPS analysis on GO and rGO (PHI ulvac probe III Scanning Microprobe). The Fourier-transform infrared spectroscopy (FTIR) study was conducted using an FTIR-6100 spectrometer (JASCO) in transmission mode in the wavenumber range 4000–400 $cm^{-1}$.

## 3. Results and Discussion

### 3.1. Structure Elucidation Confirmation

The crystal structures of the GO, rGO, and rGO/$TiO_2$ were characterized and confirmed using X-ray diffraction spectra, respectively. As shown in Figure 2a, the reflection peak obtained at 11.8° indicates the presence of oxygen functionalities on the GO surface, which has an interlayer basal spacing of 7.4 Å, suggesting the presence of oxygen functionalities on the GO surface (0 0 2). The spacing between the graphite grains was 3.36 Å. There could be several oxygen-containing groups on the edges of each layer in the first one, which eventually would have increased the distance between them. A broad peak around 25.17° was assigned to the (0 0 2) basal plane of rGO with a separation of 0.37 nm. The fall in basal space and broad peak at 24.7° suggested that GO was transformed to rGO. The prior XRD patterns of rGO/$TiO_2$ confirmed peaks at 25.3°, 37.8°, 48.0°, 53.8°, 55.0°, 62.6°, and 68.8°, for the (1 0 1), (0 0 4), (2 0 0), (1, 0 5), (2 1 1), (2 0 4), and (1 1 6) reflection planes), respectively of anatase $TiO_2$ (JCPDS#21-1272 [32]. The narrow and sharp peaks confirmed a high degree of crystallinity of rGO/$TiO_2$, supporting an easy electron transfer in an electrochemical reaction. Nevertheless, there were no broad peaks corresponding to rGO in the rGO/$TiO_2$ nanocomposites. The rGO surface can be extensively coated on the $TiO_2$ nanostructures, which has the effect of suppressing XRD signals that are caused by the stacking of rGO layers [33]. The average crystallite size (D) calculated using the Debye-Scherrer formula is as follows,

$$D = (0.9 \times \lambda)/(\beta \times \text{Cos}\theta) \quad (2)$$

where, λ is a wavelength of Cu Kα radiation of 1.54 Å, and β is the full width of the diffraction peak at the half maximum at Bragg diffraction angle 2θ. Table 1 shows the XRD results of the rGO/$TiO_2$ nanocomposite. The average crystallite size for $TiO_2$ was $35 \pm 2$ nm.

The D-bands of GO (black color), rGO (red color), and rGO/$TiO_2$ (blue color) at 1345 $cm^{-1}$, 1349 $cm^{-1}$, and 1353 $cm^{-1}$ were respectively responsible for the obtained peaks (Figure 2b). They correspond to the structural defects developed by attaching various functional assemblies with carbon. The peaks at 1593 $cm^{-1}$, 1598 $cm^{-1}$, and 1604 $cm^{-1}$ reflect the G band, which appears due to the primitive scattering of the $E_{2g}$ phonon of $sp^2$ carbon atoms in the same experiment. The wide 2D band of GO with a higher wavenumber was found at 2917 $cm^{-1}$, confirming the multilayered nature of GO. The reduction of GO to rGO using citrus limetta phytoextract, which causes rGO sheets to stack, was predicted by the two D bands of rGO. Furthermore, the ID/IG ratio was used to measure structural disorder, which increased slightly from 0.85 to 0.87 as GO was reduced to rGO. The higher

ID/IG ratio could be assigned to higher structural defects. As a result, defects form in addition to removing oxide functional groups attached to the GO surface. Despite this, the ID/IG ratio for the rGO/$TiO_2$ nanocomposite was 0.6. This was assigned to the microwave heating process during the reduction of $sp^3$ carbon to $sp^2$ carbon, causing the $sp^2$ domain size of carbon atoms to become decreased [34]. In rGO/$TiO_2$ spectrum, the peaks below 1000 $cm^{-1}$ may be attributed to the $A_g$ and $B_g$ modes of $TiO_2$. The crystalline size of the rGO/$TiO_2$ nanocomposite was estimated using Tuinstra- Koening relationship as [35,36],

$$L_a = \left(2.4 \times 10^{-4}\right) \times \left(\frac{I_G}{I_D}\right) \times \lambda^4 \tag{3}$$

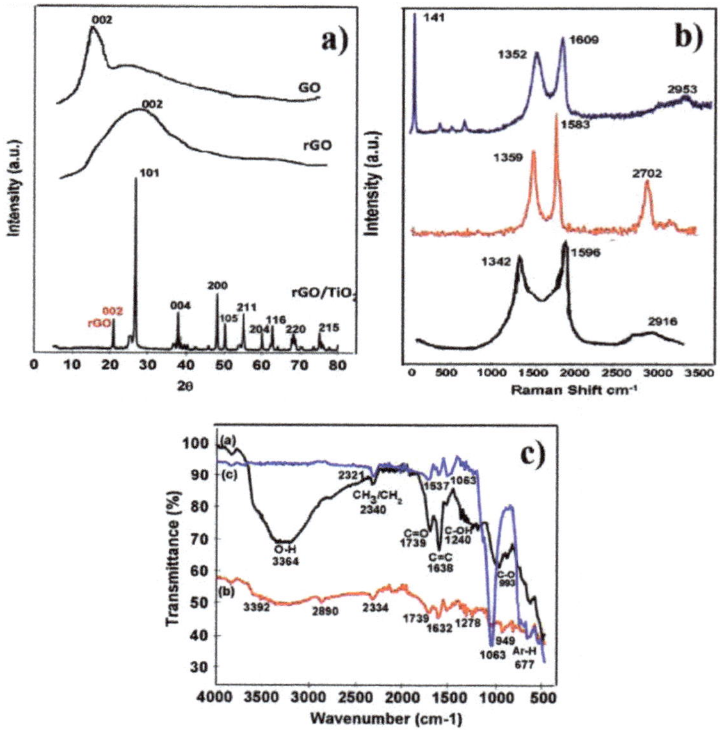


**Figure 2.** (**a**) X-ray diffraction, (**b**) Raman shift, and (**c**) FTIR analysis spectra of GO, rGO, and rGO/$TiO_2$.

**Table 1.** XRD analysis for rGO and $TiO_2$ products.

| No | Two Theta (deg.) | (h k l) | Phase | Crystallite Size (D) ± 2 (nm) |
|---|---|---|---|---|
| 1 | 20.9 | 002 | rGO | 19 |
| 2 | 26.7 | 101 | $TiO_2$ | 81 |
| 3 | 37.89 | 004 | $TiO_2$ | 14 |
| 4 | 48.14 | 200 | $TiO_2$ | 41 |
| 5 | 49.89 | 105 | $TiO_2$ | 46 |
| 6 | 55.13 | 211 | $TiO_2$ | 41 |
| 7 | 60.05 | 204 | $TiO_2$ | 22 |
| 8 | 62.77 | 116 | $TiO_2$ | 22 |
| 9 | 68.31 | 220 | $TiO_2$ | 23 |
| 10 | 75.15 | 215 | $TiO_2$ | 47 |

$L$ = crystallite size, $\lambda$ = wavelength of the laser source, and $I_D/I_G$ is the Raman peak intensity ratio of $D$ and $G$ bands, respectively. The nanocomposite's crystallite size was found to be 34 $\pm$ 2 nm, which agrees well with XRD analysis. The $O_2$ functional groups in GO were confirmed by the broad O-H stretching vibrational band, C=O, and epoxy group at 3364 $cm^{-1}$, 1739 $cm^{-1}$, and below 1000 $cm^{-1}$, respectively (Figure 2c). A few absorption peaks at 2340 $cm^{-1}$, 1638 $cm^{-1}$, and 993 $cm^{-1}$ of C-H stretching, bending vibrations of C=C, and aromatic =C-H around, respectively, were obtained. For rGO (curve b of Figure 2c), the peak at 3340 $cm^{-1}$ changed to broad and less intense than GO, revealing the significant removal of the hydroxyl group. Also, it was noticed that the other peaks became less intense compared to the GO at the same location. It could be due to the incomplete removal of $O_2$ functionalities while synthesizing GO using citrus limetta phytoextract. In short, the polar functional groups support hydrophilic properties and high dispersibility of synthesized rGO in DI [37]. The peak noted at 677 $cm^{-1}$ was credited to Ti-O-Ti and Ti-O-C in rGO/$TiO_2$ composite (curve c).

### 3.2. Morphological Study

The FE-SEM surface micrograph of GO (Figure 3a), synthesized using an improved Hummer method, clearly demonstrated a two-dimensional sheet-like structure. As GO has a multiple laminar layer structure the edges of individual sheets are distinguished from one another in the FE-SEM images. The films are stacked one on top of the other, with some wrinkles. As per reported literature [38], the thickness of GO sheet was significantly greater than the thickness of the single-layer graphene. The increase in thickness results from the oxygen-containing functional groups that were introduced. It can also be noted that the GO sheets were thicker at the edges [38]. Since the oxygen-containing functional groups could primarily be combined at the edges of the GO. It is clear from the FE-SEM images that the GO sheets are firmly suspended in their frames. The FE-SEM image of rGO (Figure 3b) obtained from GO demonstrates nanosheets' agglomeration and the irregular arrangement of stacked crumpled sheets. This could be attributed to the nucleophilic reaction of flavonoids and polyphenols, which occurred during the synthesis of rGO and involved the removal of water molecules from carbonyl and hydroxyl groups. The difference in morphological characteristics between GO's leafy, stacked structure and rGO wrinkled structure suggests that the green reduction process has prime importance in GO to rGO conversion.

The granular shape of $TiO_2$ nanoparticles displayed in Figure 3c is attributed to the aggregation of nanoparticles during the evaporation process due to microwave heating [39]. The elemental compositions of rGO and rGO/$TiO_2$ were studied using an energy-dispersive spectrum. Figure 3d depicts carbon (C) and oxygen (O) contents in rGO and titanium (Ti) in rGO/$TiO_2$. From the EDS analysis, it is evident that the rGO contains around 84.88 at. % C, 15.12 at. % O whereas rGO/ $TiO_2$ avails about 65.99 at.% C, 28.95 at.% O and 28.95 at.% Ti, confirming the formation of $TiO_2$ nanostructures on the rGO surface.

### 3.3. X-ray Photoelectron Spectroscopy Analysis

Figure 4a–c displays the XPS spectrum of the C1s spectrum of GO, rGO, and rGO/$TiO_2$. The peak in Figure 4a that corresponds to the C1s bands in deconvoluted GO into C-C, C=O, and C(O)OH bonds, respectively, has binding energies of 284.8, 286.5, and 288.5 eV [40]. When compared to GO, the C=O and C(O)OH bands of the rGO C1s spectrum show a marked decrease in intensity, indicating that rGO has undergone deoxygenation. However, as shown in Figure 4b, after being reduced, the peak oxygen intensity with functionalities is sharply reduced. These results have demonstrated the decrease in GO. The C-C, C=O, and C(O)OH bonds, respectively, are ascribed to the peaks with binding energies of 284.5, 285.4, and 283.7 eV that correspond to the C1s bands in deconvoluted rGO. The variation in the chemical surrounding of carbon atoms after reduction may cause a shift in binding energy [41]. Additionally, the surface C/O atomic ratios increased from 2.12 to 5.57, validating the successful removal of $O_2$ groups from the GO surface using aqueous peel extract of citrus limetta. In Figure 4c, three peaks found in the C1s deconvoluted spectrum of $rGO/TiO_2$ are for C-C/C-H $sp^3$ hybridized carbon bonds, C-O bonds, and the $sp^2$ carbon bonds of graphene (c), respectively.

**Figure 3.** FE-SEM images of; (**a**) GO, (**b**) rGO, (**c**) $rGO/TiO_2$, and (**d**) EDS spectrum of $rGO/TiO_2$.

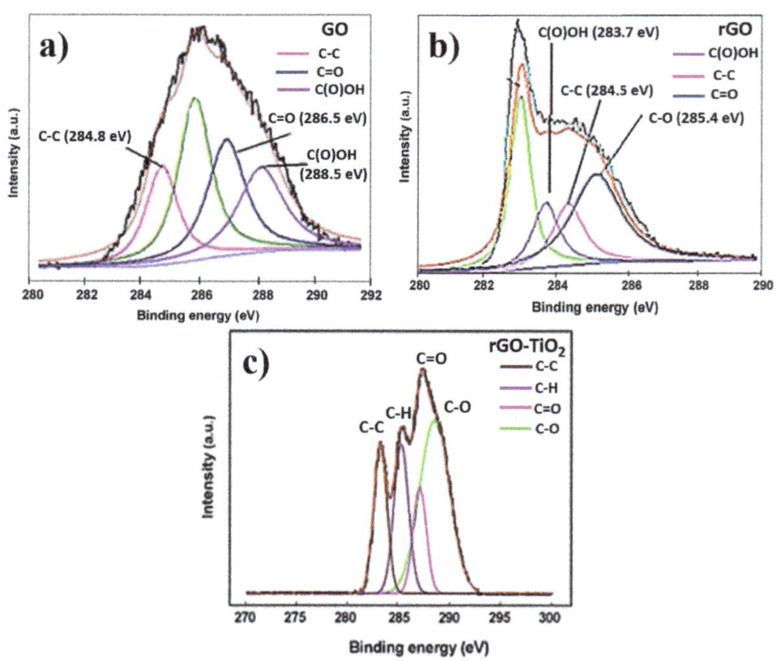


**Figure 4.** XPS binding energy spectra of C1s spectrum of (**a**) GO, (**b**) rGO, and (**c**) $rGO/TiO_2$.

### 3.4. Electrochemical Non-Enzymatic Detection

The electrochemical characteristics of pristine FTO, $TiO_2$/FTO, rGO/FTO, and rGO/$TiO_2$/FTO samples were investigated using CV scans in a PBS (7.4 pH) solution containing 2 mM D (+) glucose at a scan rate of 50 mV/s and a working area of 1.0 $cm^2$ (Figure 5a). For FTO, it was found that a faradic peak does not exist. It was noticed that rGO/FTO had a clearly defined peak for glucose oxidation that occurred at −0.25 V. At −0.25 V, the $TiO_2$/FTO and rGO/$TiO_2$/FTO oxidation peaks were significantly enhanced. The current rGO/$TiO_2$/FTO reaction was practically double in comparison to the rGO/FTO response when the voltage was set to −0.25 V.

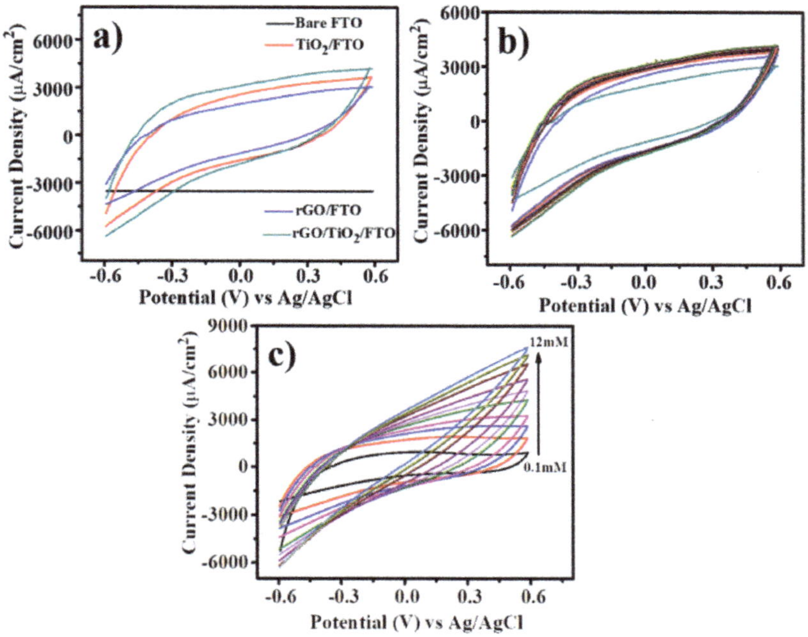


**Figure 5.** (**a**) CV response of various electrodes, (**b**) Effect of variable scan rate on rGO/$TiO_2$/FTO electrode, and (**c**) Effect of variable glucose concentration on rGO/$TiO_2$/FTO electrode.

Consequently, it is confirmed that the rGO/$TiO_2$/FTO composite is essential for obtaining increased sensitivity and electrocatalytic activity. A further investigation into the electrochemical effect on rGO/$TiO_2$/FTO with variable scan rate was carried out using CV which is illustrated in Figure 5b. The electrode was scanned between 10 and 300 mV/s in an electrolyte containing 2 mM D (+) glucose using a CV between −0.6 V and +0.6 V. Regarding current flow, the oxidation that took place at −0.25 V is the most notable. that the rGO/$TiO_2$ nanocomposite undergoes an electrochemical reaction, demonstrating the enzyme detection capability. Furthermore, electrochemical measurements were carried out at the optimal potential of −0.25 V vs. Ag/AgCl, which was optimal.

Figure 5c displays the I-V responses of the modified sensor electrode at a scan rate of 50 mv/s for a variable glucose concentration in the range of 0.1 to 12 mM in a PBS electrolyte solution (7.4 pH) (c). The sensor's current density increases linearly as the glucose concentrations increase. The rGO/$TiO_2$ composite's high electrical conductivity, electron mobility, and electrocatalytic activity are in charge of enhancing the electron transport process between the sensor electrode and the PBS solution during a redox reaction.

### 3.5. Calibration of the Sensor

The calibration curves for various glucose concentration rage are presented in Figure 6. A calibration curve for glucose concentrations in 0.1 to 12 mM is presented in Figure 6a. This curve can be divided into two linear glucose concentration ranges: 0.1–1 mM and 2–12 mM, as presented in Figure 6b,c. For the calibration curve (0.1–1 mM), the equivalent linear regression equations are Ip = 1488.1c + 930.4 and $R^2$ = 0.9862 (N = 4). The sensor's sensitivity determined from the slope of the graph is 1425 μA/cm$^2$ (S/N < 3). For the calibration curve (2–12 mM) the equations are Ip = 498.5c + 1503.7 and $R^2$ = 0.9878 (N = 6). The sensor's sensitivity determined from the slope of the graph is 491.75 μA/cm$^2$ (S/N < 3).

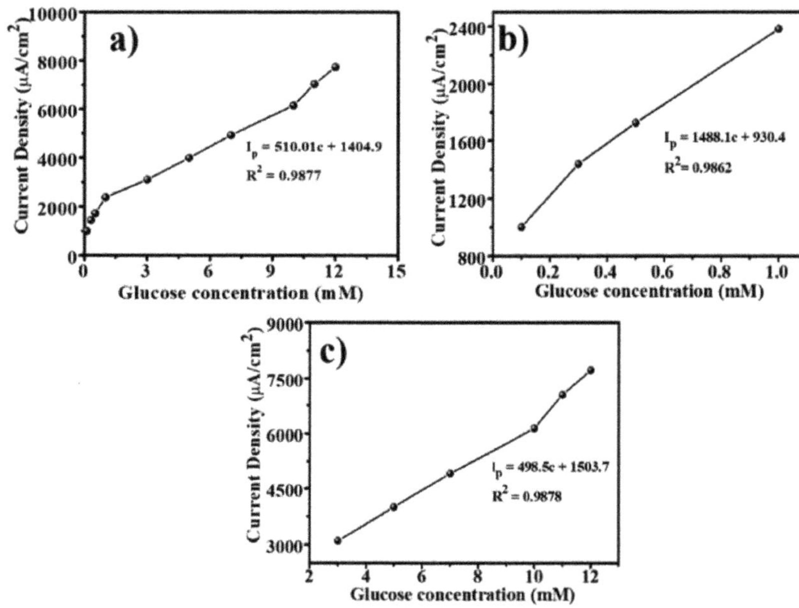


**Figure 6.** Calibration curve for glucose concentration determination using the Glucose liquid color kit (**a**) 0.1–12 mM (**b**) 0.1–1 mM (**c**) 3–12 mM for different glucose concentration.

### 3.6. Selectivity and Stability of the Sensor

The proposed biosensor was examined for interfering species in blood serum, including ascorbic acid, dopamine, fructose, and lactose. The selectivity of the rGO/$TiO_2$/FTO sensor electrode is shown in Figure 7a. The sample's stability (aging effect) tested periodically for 30 days with a 1 mM glucose concentration, yielding an RSD of 1.14%, is presented in Figure 7b. For ten cycles, the sensor electrode's reproducibility was investigated. To observe the current response of each sensor, 2 mM of glucose concentration was used. The RSD of 1.3% confirms the significant reproducibility.

### 3.7. Repeatability of the Sensor

The repeatability of the proposed sensors was examined by measuring the current response five times over a week. The proposed sensor's performance is compared to previously reported values, and the results are tabulated in Table 2. The sensor exhibits good recovery with an RSD of 1.34%.

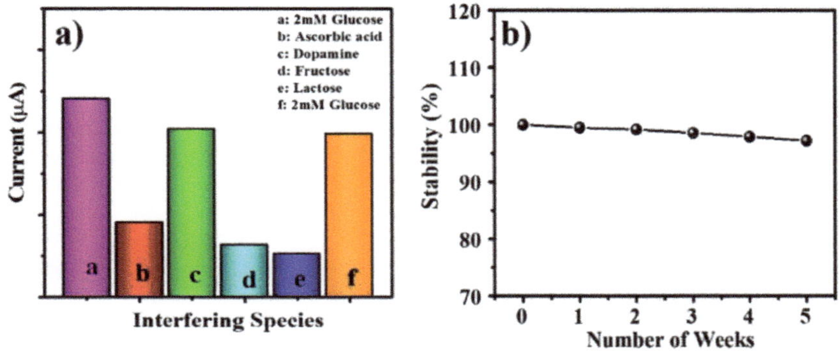


**Figure 7.** (**a**) Selectivity and (**b**) stability of $rGO/TiO_2/FTO$ sensor electrode.

**Table 2.** Comparative analysis on sensitivity, detection limit, and linear range of $rGO/TiO_2/FTO$ glucose sensor.

| Electrode | Detection Potential (V) | Sensitivity μA/mM1cm2 | Linear Range/mM | Detection Limit/μM | Reference |
|---|---|---|---|---|---|
| $rGO/TiO_2/FTO$ | −0.25 | 1425 | 0.1–12 | 0.32 | This work |
| Nafion/$GO_x$/rGO-6/ZnONRs/Au/PET | −0.8 to +0.8 | 2.26 | 11.5 | 37.5 | [42] |
| Ni Ps/ATP/r-GO | 0.0 to −1.5 | 1414.4 μA | 1–710 μM | 0.37 | [43] |
| $NiCo_2O_4$-rGO | – | 548.9 | 1–25 | 0.35 | [44] |
| rGO/NiO | −0.6 | 6.2 | 1–15 | 19.35 | [45] |
| $GO_x$/n-$TiO_2$/PANI/GCE | – | 6.31 | 0.02–6 | 18 | [46] |
| 3D Ag-GR-$TiO_2$ | – | 12 | – | – | [47] |
| $TiO_2$–r-GO–PANI | – | – | 10–180 μM | 7.46 | [48] |

### *3.8. Investigation of Glucose Level in Blood Serum*

The glucose concentration level in blood serum samples was determined using the prepared sensor electrode. The blood serum samples from 40 to 50 years of age were collected from the certified pathology laboratory. Using the standard dilution method, these samples were multiplied by 100 before being subjected to analysis. The goal of the study was to compare the results for the proposed sensor with the certified values obtained from the pathology laboratory. The experimental data for blood serum glucose concentration are shown in Table 3.

**Table 3.** Investigation of glucose in blood serum and its recovery.

| Serum Sample | Glucose Added (mM) | Glucose Obtained (mM) | Recovery (%) | RSD (%) |
|---|---|---|---|---|
| 1 | 2.00 | 1.97 | 98.5 | 1.5 |
| 2 | 4.00 | 3.92 | 98.8 | 1.2 |
| 3 | 6.00 | 5.87 | 97.5 | 2.2 |
| 4 | 8.00 | 7.89 | 98.6 | 1.4 |
| 5 | 10.00 | 9.71 | 97.1 | 2.9 |

The average recovery rate of 98.2% (RSD 1.88%) exhibits the reliability of the proposed sensor for serum glucose biosensor application.

## 4. Conclusions

The $rGO/TiO_2$ sensor electrode adduced a high sensitivity of 1425 μA/mM $cm^2$ in the linear range of 0.1 mM to 12 mM, which is quite good compared with the available literature data. It shows a quick response time of 5 s, a correlation coefficient ($R^2$) ~0.98, and a LOD of

0.32 μM. It also presents high selectivity towards the interfering species. The RSD of 1.14%, 1.34%, and 1.3% demonstrate excellent stability, repeatability, and reproducibility of the sensor respectively. This reveals the use of $rGO/TiO_2$ electrodes for applications involving glucose sensing. The sensor has an RSD of 1.88% when measuring glucose levels in natural blood serum. Using a commercially available glucose sensor, the measured values agree well with those obtained. Moreover, the process used to prepare rGO and $rGO/TiO_2$ composite is simple, rapid, cost-effective, and eco-friendly which has the potential for commercial/market viability.

**Author Contributions:** Conceptualization, M.G., S.C. and A.G.; Methodology, M.G.; Software, M.U.H.S.; Validation, S.F.S. and R.S.M.; Formal analysis, R.S.M.; Investigation, S.E., R.S.M. and A.G.; Resources, S.F.S.; Writing—original draft, M.G., S.C. and S.E.; Writing—review & editing, R.S.M. and A.G.; Visualization, B.P.; Supervision, S.E., R.S.M. and A.G.; Project administration, S.F.S.; Funding acquisition, S.F.S. All authors have read and agreed to the published version of the manuscript.

**Funding:** The authors sincerely thank the Researchers Supporting Project number (RSP2023R370), King Saud University, Riyadh, Saudi Arabia, for financial support.

**Data Availability Statement:** Not applicable.

**Acknowledgments:** The authors sincerely thank the Researchers Supporting Project number (RSP2023R370), King Saud University, Riyadh, Saudi Arabia, for financial support.

**Conflicts of Interest:** The authors declare no conflict of interest.

## References

1. Zhu, H.; Li, L.; Zhou, W.; Shao, Z.; Chen, X. Advances in non-enzymatic glucose sensors based on metal oxides. *J. Mater. Chem. B* **2016**, *4*, 7333–7349. [CrossRef]
2. Bruen, D.; Delaney, C.; Florea, L.; Diamond, D. Glucose Sensing for Diabetes Monitoring: Recent Developments. *Sensors* **2017**, *17*, 1866. [CrossRef]
3. Sai-Anand, G.; Gopalan, A.-I.; Kang, S.-W.; Komathi, S.; Lee, K.-P. One Pot Synthesis of New Gold Nanoparticles Dispersed Poly(2-aminophenyl boronic acid) Composites for Fabricating an Affinity Based Electrochemical Detection of Glucose. *Sci. Adv. Mater.* **2014**, *6*, 1356–1364. [CrossRef]
4. Gopalan, A.; Muthuchamy, N.; Lee, K. A novel bismuth oxychloride-graphene hybrid nanosheets based non-enzymatic photo-electrochemical glucose sensing platform for high performances. *Biosens. Bioelectron.* **2017**, *89*, 352–360. [CrossRef] [PubMed]
5. Kumar, G.G.; Amala, G.; Gowtham, S.M. Recent advancements, key challenges and solutions in non-enzymatic electrochemical glucose sensors based on graphene platforms. *RSC Adv.* **2017**, *7*, 36949–36976. [CrossRef]
6. Gopalan, A.; Muthuchamy, N.; Komathi, S.; Lee, K.-P. A novel multicomponent redox polymer nanobead based high performance non-enzymatic glucose sensor. *Biosens. Bioelectron.* **2016**, *84*, 53–63. [CrossRef] [PubMed]
7. Dong, Q.; Ryu, H.; Lei, Y. Metal oxide based non-enzymatic electrochemical sensors for glucose detection. *Electrochim. Acta* **2021**, *370*, 137744. [CrossRef]
8. Manthey, J.A.; Grohmann, K. Phenols in Citrus Peel Byproducts. Concentrations of Hydroxycinnamates and Polymethoxylated Flavones in Citrus Peel Molasses. *J. Agric. Food Chem.* **2001**, *49*, 3268–3273. [CrossRef]
9. Ghule, B.G.; Shinde, N.M.; Raut, S.D.; Gore, S.K.; Shaikh, S.F.; Ekar, S.U.; Ubaidullah, M.; Pak, J.J.; Mane, R.S. Self-assembled α-$Fe_2O_3$-GO nanocomposites: Studies on physical, magnetic and ammonia sensing properties. *Mater. Chem. Phys.* **2022**, *278*, 125617. [CrossRef]
10. Balandin, A.A. Thermal properties of graphene and nanostructured carbon materials. *Nat. Mater.* **2011**, *10*, 569–581. [CrossRef]
11. Wang, Z.; Chen, Y.; Li, P.; Zhou, J.; He, J.; Zhang, W.; Guo, Z.; Li, Y.; Dong, M. Modulation of N-bonding configurations and their influence on the electrical properties of nitrogen-doped graphene. *RSC Adv.* **2016**, *6*, 92682–92687. [CrossRef]
12. Shin, H.-J.; Kim, K.K.; Benayad, A.; Yoon, S.-M.; Park, H.K.; Jung, I.-S.; Jin, M.H.; Jeong, H.-K.; Kim, J.M.; Choi, J.-Y.; et al. Efficient Reduction of Graphite Oxide by Sodium Borohydride and Its Effect on Electrical Conductance. *Adv. Funct. Mater.* **2009**, *19*, 1987–1992. [CrossRef]
13. Ren, P.-G.; Yan, D.-X.; Ji, X.; Chen, T.; Li, Z.-M. Temperature dependence of graphene oxide reduced by hydrazine hydrate. *Nanotechnology* **2010**, *22*, 055705. [CrossRef]
14. Wang, Y.; Zhang, P.; Liu, C.F.; Zhan, L.; Li, Y.F.; Huang, C.Z. Green and easy synthesis of biocompatible graphene for use as an anticoagulant. *RSC Adv.* **2012**, *2*, 2322–2328. [CrossRef]
15. Mhamane, D.; Ramadan, W.; Fawzy, M.; Rana, A.; Dubey, M.; Rode, C.; Lefez, B.; Hannoyer, B.; Ogale, S. From graphite oxide to highly water dispersible functionalized graphene by single step plant extract-induced deoxygenation. *Green Chem.* **2011**, *13*, 1990–1996. [CrossRef]

16. Salas, E.C.; Sun, Z.; Lüttge, A.; Tour, J.M. Reduction of Graphene Oxide *via* Bacterial Respiration. *ACS Nano* **2010**, *4*, 4852–4856. [CrossRef] [PubMed]
17. Nasrollahzadeh, M.; Maham, M.; Rostami-Vartooni, A.; Bagherzadeh, M.; Sajadi, S.M. Barberry fruit extract assisted in situ green synthesis of Cu nanoparticles supported on a reduced graphene oxide–$Fe_3O_4$nanocomposite as a magnetically separable and reusable catalyst for the O-arylation of phenols with aryl halides under ligand-free conditions. *RSC Adv.* **2015**, *5*, 64769–64780. [CrossRef]
18. Arias, B.; Ramón-Laca, L. Pharmacological properties of citrus and their ancient and medieval uses in the Mediterranean region. *J. Ethnopharmacol.* **2005**, *97*, 89–95. [CrossRef] [PubMed]
19. Manthey, J.A.; Grohmann, K. Concentrations of Hesperidin and Other Orange Peel Flavonoids in Citrus Processing Byproducts. *J. Agric. Food Chem.* **1996**, *44*, 811–814. [CrossRef]
20. Majhi, S.; Mirzaei, A.; Kim, H.; Kim, S. Reduced Graphene Oxide (rGO)-Loaded Metal-Oxide Nanofiber Gas Sensors: An Overview. *Sensors* **2021**, *21*, 1352. [CrossRef]
21. Pradhan, S.; Konwar, K.; Ghosh, T.; Mondal, B.; Sarkar, S.; Deb, P. Multifunctional Iron oxide embedded reduced graphene oxide as a versatile adsorbent candidate for effectual arsenic and dye removal. *Colloid Interface Sci. Commun.* **2020**, *39*, 100319. [CrossRef]
22. Molak, A.; Paluch, M.; Pawlus, S.; Klimontko, J.; Ujma, Z.; Gruszka, I. Electric modulus approach to the analysis of electric relaxation in highly conducting $(Na_{0.75}Bi_{0.25})(Mn_{0.25}Nb_{0.75})O_3$ ceramics. *J. Phys. D Appl. Phys.* **2005**, *38*, 1450–1460. [CrossRef]
23. Pradhan, S.S.; Ghosh, T.N.; Marik, A.; Raul, K.K.; Sarkar, S.K. Magnetodielectric effects in three reduced graphene oxide–polymer nanocomposites. *Bull. Mater. Sci.* **2020**, *43*, 208. [CrossRef]
24. Ghule, B.G.; Shinde, N.M.; Raut, S.D.; Shaikh, S.F.; Al-Enizi, A.M.; Kim, K.H.; Mane, R.S. Porous metal-graphene oxide nanocomposite sensors with high ammonia detectability. *J. Colloid Interface Sci.* **2021**, *589*, 401–410. [CrossRef] [PubMed]
25. Liu, S.; Sun, H.; Liu, S.; Wang, S. Graphene facilitated visible light photodegradation of methylene blue over titanium dioxide photocatalysts. *Chem. Eng. J.* **2013**, *214*, 298–303. [CrossRef]
26. Luo, L.; Zhu, L.; Wang, Z. Nonenzymatic amperometric determination of glucose by CuO nanocubes–graphene nanocomposite modified electrode. *Bioelectrochemistry* **2012**, *88*, 156–163. [CrossRef] [PubMed]
27. Yang, Z.; Hao, X.; Chen, S.; Ma, Z.; Wang, W.; Wang, C.; Yue, L.; Sun, H.; Shao, Q.; Murugadoss, V.; et al. Long-term antibacterial stable reduced graphene oxide nanocomposites loaded with cuprous oxide nanoparticles. *J. Colloid Interface Sci.* **2019**, *533*, 13–23. [CrossRef] [PubMed]
28. AAT Bioquest, Inc. Quest Calculate™ PBS (Phosphate Buffered Saline) (1X, pH 7.4) Preparation and Recipe. AAT Bioquest. Available online: https://www.aatbio.com/resources/buffer-preparations-and-recipes/pbs-phosphate-buffered-saline (accessed on 7 December 2022).
29. Komathi, S.; Muthuchamy, N.; Lee, K.-P.; Gopalan, A.-I. Fabrication of a novel dual mode cholesterol biosensor using titanium dioxide nanowire bridged 3D graphene nanostacks. *Biosens. Bioelectron.* **2016**, *84*, 64–71. [CrossRef]
30. Zaaba, N.I.; Foo, K.L.; Hashim, U.; Tan, S.J.; Liu, W.W.; Voon, C.H. Synthesis of Graphene Oxide using Modified Hummers Method: Solvent Influence. *Procedia Eng.* **2017**, *184*, 469–477. [CrossRef]
31. Zhang, L.; Hai, X.; Xia, C.; Chen, X.-W.; Wang, J.-H. Growth of CuO nanoneedles on graphene quantum dots as peroxidase mimics for sensitive colorimetric detection of hydrogen peroxide and glucose. *Sens. Actuators B Chem.* **2017**, *248*, 374–384. [CrossRef]
32. Shaikh, S.F.; Mane, R.S.; Min, B.K.; Hwang, Y.J.; Joo, O.-S. D-sorbitol-induced phase control of $TiO_2$ nanoparticles and its application for dye-sensitized solar cells. *Sci. Rep.* **2016**, *6*, 20103. [CrossRef]
33. Nipane, S.V.; Lee, S.-W.; Gokavi, G.S.; Kadam, A.N. In situ one pot synthesis of nanoscale $TiO_2$-anchored reduced graphene oxide (RGO) for improved photodegradation of 5-fluorouracil drug. *J. Mater. Sci. Mater. Electron.* **2018**, *29*, 16553–16564. [CrossRef]
34. Atchudan, R.; Edison, T.N.J.I.; Perumal, S.; Karthikeyan, D.; Lee, Y.R. Effective photocatalytic degradation of anthropogenic dyes using graphene oxide grafting titanium dioxide nanoparticles under UV-light irradiation. *J. Photochem. Photobiol. A Chem.* **2017**, *333*, 92–104. [CrossRef]
35. Ghosh, T.N.; Pradhan, S.S.; Sarkar, S.K.; Bhunia, A.K. On the incorporation of the various reduced graphene oxide into poly(vinyl alcohol) nano-compositions: Comparative study of the optical, structural properties and magnetodielectric effect. *J. Mater. Sci. Mater. Electron.* **2021**, *32*, 19157–19178. [CrossRef]
36. Kołodziej, A.; Długoń, E.; Świętek, M.; Ziąbka, M.; Dawiec, E.; Gubernat, M.; Michalec, M.; Wesełucha-Birczyńska, A. A Raman Spectroscopic Analysis of Polymer Membranes with Graphene Oxide and Reduced Graphene Oxide. *J. Compos. Sci.* **2021**, *5*, 20. [CrossRef]
37. Zhu, J.; Zeng, G.; Nie, F.; Xu, X.; Chen, S.; Han, Q.; Wang, X. Decorating graphene oxide with CuO nanoparticles in a water–isopropanol system. *Nanoscale* **2010**, *2*, 988–994. [CrossRef]
38. Song, J.; Wang, X.; Chang, C.-T. Preparation and Characterization of Graphene Oxide. *J. Nanomater.* **2014**, *2014*, 276143. [CrossRef]
39. Qianqian, Z.; Tang, B.; Guoxin, H. High photoactive and visible-light responsive graphene/titanate nanotubes photocatalysts: Preparation and characterization. *J. Hazard. Mater.* **2011**, *198*, 78–86. [CrossRef]
40. Stankovich, S.; Dikin, D.A.; Piner, R.D.; Kohlhaas, K.A.; Kleinhammes, A.; Jia, Y.; Wu, Y.; Nguyen, S.T.; Ruoff, R.S. Synthesis of graphene-based nanosheets via chemical reduction of exfoliated graphite oxide. *Carbon* **2007**, *45*, 1558–1565. [CrossRef]
41. Jana, M.; Saha, S.; Khanra, P.; Murmu, N.C.; Srivastava, S.K.; Kuila, T.; Lee, J.H. Bio-reduction of graphene oxide using drained water from soaked mung beans (*Phaseolus aureus* L.) and its application as energy storage electrode material. *Mater. Sci. Eng. B* **2014**, *186*, 33–40. [CrossRef]

42. Mao, Q.; Jing, W.; Zhou, F.; Liu, S.; Gao, W.; Wei, Z.; Jiang, Z. Depositing reduced graphene oxide on ZnO nanorods to improve the performance of enzymatic glucose sensors. *Mater. Sci. Semicond. Process.* **2021**, *121*, 105391. [CrossRef]
43. Shen, Z.; Gao, W.; Li, P.; Wang, X.; Zheng, Q.; Wu, H.; Ma, Y.; Guan, W.; Wu, S.; Yu, Y.; et al. Highly sensitive nonenzymatic glucose sensor based on nickel nanoparticle–attapulgite-reduced graphene oxide-modified glassy carbon electrode. *Talanta* **2016**, *159*, 194–199. [CrossRef] [PubMed]
44. Ni, Y.; Xu, J.; Liu, H.; Shao, S. Fabrication of RGO-$NiCo_2O_4$ nanorods composite from deep eutectic solvents for nonenzymatic amperometric sensing of glucose. *Talanta* **2018**, *185*, 335–343. [CrossRef] [PubMed]
45. Alghazzawi, W.; Danish, E.; Alnahdi, H.; Salam, M.A. Rapid microwave-assisted hydrothermal green synthesis of rGO/NiO nanocomposite for glucose detection in diabetes. *Synth. Met.* **2020**, *267*, 116401. [CrossRef]
46. Tang, W.; Li, L.; Zeng, X. A glucose biosensor based on the synergistic action of nanometer-sized $TiO_2$ and polyaniline. *Talanta* **2015**, *131*, 417–423. [CrossRef]
47. Jang, H.D.; Kim, S.K.; Chang, H.; Jo, E.H.; Roh, K.M.; Choi, J.-H.; Choi, J.-W. Synthesis of 3D Silver-Graphene-Titanium Dioxide Composite via Aerosol Spray Pyrolysis for Sensitive Glucose Biosensor. *Aerosol Sci. Technol.* **2015**, *49*, 538–546. [CrossRef]
48. Du, J.; Tao, Y.; Xiong, Z.; Yu, X.; Xie, A.; Luo, S.; Li, X.; Yao, C. Titanium Dioxide–Graphene–Polyaniline Hybrid for Nonenzymatic Detection of Glucose. *Nano* **2019**, *14*, 1950093. [CrossRef]

electronics

MDPI

*Article*

# Fabrication and Characterization of Si/PEDOT: PSS-Based Heterojunction Solar Cells

**Ragavendran Venkatesan [1], Sheik Moideen Thaha Sheik Kadar Maideen [1,2], Saravanan Chandhiran [1], Sunil Singh Kushvaha [3,4], Suresh Sagadevan [5,*], Vishnukanthan Venkatachalapathy [6,7] and Jeyanthinath Mayandi [1,*]**

1 Department of Material Science, School of Chemistry, Madurai Kamaraj University, Madurai 625021, India
2 Department of Chemistry, School of Advanced Sciences, Vellore Institute of Technology, Vellore 632014, India
3 CSIR-National Physical Laboratory, Dr. K.S. Krishnan Road, New Delhi 110012, India
4 Academy of Scientific and Innovative Research (AcSIR), Ghaziabad 201002, India
5 Nanotechnology and Catalysis Research Centre, Universiti Malaya, Kuala Lumpur 50603, Malaysia
6 Department of Physics/Centre for Materials Science and Nanotechnology, University of Oslo, P.O. Box 1048 Blindern, NO-0316 Oslo, Norway
7 Department of Materials Science, National Research Nuclear University "MEPhI", 31 Kashirskoe sh, 115409 Moscow, Russia
* Correspondence: drsureshsagadevan@um.edu.my (S.S.); jeyanthinath.chem@mkuniversity.org (J.M.)

**Abstract:** In this study, we fabricated a planar Si/PEDOT: PSS heterojunction solar cell using three different solvents—ethylene glycol, acetonitrile, and dimethyl sulfoxide—to find the best one. The fabricated samples were characterized by diffuse reflectance spectroscopy, scanning electron microscopy, X-ray diffraction, and current–voltage. Diffused reflectance spectrum analysis showed reduced reflectance compared to the bare silicon wafers. The absorbance spectrum shows the change in absorption of the Si-coated PEDOT: PSS which was more than a 50% increase in the UV region, and for the EG sample, there was a 20% increase in the entire visible spectrum. This indicates that the solvent plays a major role in the bandgap between the Si and Si/PEDOT: PSS. Scanning electron microscope (SEM) was used to examine the surface morphology of Si/PEDOT: PSS as agglomerated, island-formed surfaces and carbon-layered Si-PEDOT: PSS. Cross-sectional images show the thickness of the PEDOT: PSS layer on the silicon wafer surface. The X-ray diffraction (XRD) pattern shows the characteristic peaks for silicon (69.5°), and Si/PEDOT: PSS shows a forbidden Si (200) peak at 32°. Current–voltage measurements have shown the characteristic diode curve for all fabricated cells. This characteristic diode curve indicated the presence of a heterojunction. Ethylene glycol-containing cells showed current output as 0.2 μA with $V_{oc}$ of 0.2 V.

**Keywords:** silicon; PEDOT: PSS; heterojunction; reflectance; solar cell; organic solvents

**Citation:** Venkatesan, R.; Sheik Kadar Maideen, S.M.T.; Chandhiran, S.; Kushvaha, S.S.; Sagadevan, S.; Venkatachalapathy, V.; Mayandi, J. Fabrication and Characterization of Si/PEDOT: PSS-Based Heterojunction Solar Cells. *Electronics* **2022**, *11*, 4145. https://doi.org/10.3390/electronics11244145

Academic Editor: Lucas Lamata

Received: 9 November 2022
Accepted: 8 December 2022
Published: 12 December 2022



## 1. Introduction

Solar energy is the only potential renewable energy source that can compete with existing low-cost thermal power. The solar energy market is dominated by silicon, accounting for almost 90% of the market share [1]. However, their high cost, high energy-intensive fabrication techniques, and low payback period still cause them to lag behind other energy sources. Research has been undertaken to reduce the cost and material usage while simultaneously increasing its efficiency. The low efficiency of silicon solar cells is due to their high reflectance over the surface, sub-bandgap spectrum losses, and high recombination rate [2]. Two different forms of Si—pure Si, and amorphous Si—were used to build the cells. However, the use of photovoltaic cells has been limited owing to their high processing costs. High-purity single-crystal materials and a lack of effective mass-production techniques have been used to produce thin silicon films. Although single-crystalline silicon solar cells are the most efficient and advanced of all the cells, they found difficult to implement owing to their high cost [3]. Thus, there are alternatives to silicon in the form of thin-film materials such as cadmium telluride [4], copper indium: diselenide (CIS), dye-sensitized solar cells (DSSC) [5,6], perovskite solar cells [7], and polymer solar cells. However, these technologies are immature and require further investigation.

Silicon processing has been considered as a technology development, and research has focused on the modification of existing conventional silicon solar cells using nanotechnology. Organic heterojunction solar cells are the promising alternatives to achieve high-efficiency and low-cost silicon solar cells. Energy-intensive emitter formation can be eliminated in the hybrid solar cells (HSC) based on the nanostructures. Emitter junctions can be processed in a solution and also eliminated the use of additional antireflection layers. First, planar bulk silicon organic solar cells were fabricated [8–11]. Silicon organic heterojunction (SOH) solar cells use organic materials such as poly-(3-hexythiophene) (P3HT) [12], Spiro-OmeTAD [13], poly-(Si-cyclooctatetrocene) [14], polyaniline [15], 4-Tricyanovinyl-N, N-diethylaniline [16], and PEDOT: PSS [17–19]. Lin et al. first demonstrated the possibility of using PEDOT: PSS as a hole-transport layer in a Si/organic hybrid solar cell and achieved a PCE of 0.008% in a planar configuration [20]. PEDOT: PSS was considered as the hole transport layer owing to its superior properties such as high transmittance (>90%), high conductivity, good absorption in the visible spectrum, and a favorable bandgap of ~1.6 eV [21]. The HOMO (highest occupied molecular orbital) of PEDOT: PSS and valence band (VB) of Silicon is nearly similar to ~5.1 eV [20]. He et al. investigated the effect of Si surface termination conditions on the device performance. Their work proved that oxide-terminated (SiOx-Si) cells exhibit superior performance compared to the hydrogen-terminated (H-Si) cells. Hydrogen-terminated cells exhibited 0.002% while oxide-terminated cells showed 10.6% [22] Zhao et al. improved the efficiency of planar Si/PEDOT: PSS cells by 12.70% by growing a $SiO_2$ layer at the interface of Si and PEDOT: PSS, resulting in a hydrophilic surface allowing PEDOT: PSS to adhere easily on Si [9]. Avasthi et al. demonstrated an 11.70% efficient planar Si/PEDOT: PSS HSC by spin-coating a PEDOT: PSS layer at the temperatures below 100 °C [11]. Leung et al. demonstrated the effect of a solvent on the conductivity of the PEDOT: PSS layer. Ethylene glycol (EG) proved to be more efficient in improving the conductivity [18].

Planar-Si cells with fluorosurfactant (FS) as the surfactant and EG as the co-solvent exhibited a record with the efficiency of 13.3%. The co-solvent-optimized PEDOT: PSS also acts as an antireflection coating [18]. Despite the simple fabrication, planar Si/PEDOT: PSS cells have a high surface recombination rate because the separation of the electron–hole pair occurs only at the junction, whereas electron–hole pairs in PEDOT: PSS and bulk silicon are recombined. Moreover, they have a very small PN-junction surface area. To reduce the material cost, Wang et al. fabricated a planar Si/PEDOT: PSS Solar cell on thin-film with Si deposited using a thermal chemical vapor deposition reactor (CVD) and achieved an efficiency of 8.2% [10]. Recently, nanostructures have been incorporated into Si wafers to enhance light trapping and reduce reflection. Theoretically, SiNW/PEDOT: PSS solar cells can surpass the efficiencies of commercial crystalline silicon solar cells when a radial p–n junction is associated between PEDOT: PSS and Si Nanowires [23].

Therefore, the present work aims to fabricate a hybrid solar cell using Si/PEDOT: PSS, and then analyze the various solvents for dissolving PEDOT: PSS to choose the best one. We optimized the formation of Si/PEDOT: PSS and analyzed the influence of the spin-coating speed and annealing temperature of PEDOT: PSS to achieve a radial p–n junction between Si and PEDOT: PSS to attain the enhanced efficiency.

## 2. Experimental Details

### *2.1. Materials*

3,4 Poly (3,4-ethylene dioxythiophene) poly (styrene sulfonate) (PEDOT: PSS) was purchased from Sigma Aldrich; acetone (AR grade) was purchased from MERCK; and hydrofluoric acid (51%), dimethyl sulphoxide (DMSO), ethylene glycol, acetonitrile, Triton X-100 was purchased from CDH. All the chemicals were purchased in India.

### 2.2. Sample Preparation

In this study, an n-type electronic-grade silicon wafer (100 μm) with a thickness of 525 μm was used. Before the etching process, the silicon wafers were made to cut to the desired size of 2 × 2 cm and cleaned using DI water under ultrasonication for 10 min. Furthermore, the oxide layer of the silicon wafers were extracted using $H_2O$: HF in a volume ratio of 10:1 for 3 min at room temperature. Three different polymeric solutions were prepared using various solvents: (i) PEDOT: PSS solution (1.3% in $H_2O$) was prepared using 7% ethylene glycol (EG) and 0.25% Triton-X. (ii) A PEDOT: PSS solution (1.3% in $H_2O$) was prepared using 5% acetonitrile and 0.25% Triton-X. (iii) PEDOT: PSS solution (1.3% in $H_2O$) was prepared using 5% DMSO and 0.25% Triton-X. The cleaned and oxide-removed wafers were dipped into the acetone solution for 5 min to induce adhesiveness. Subsequently, the wafers were dried and spin-coated with 50 μL PEDOT: PSS at 3000 rpm for 60 s. The coated wafers were then annealed at 150 °C for 30 min. To improve the adhesive nature of the silicon wafer, a carbon coating was applied to the wafer. The black camphor smoke was subjected to the falling onto silicon wafers. The silicon wafers were maintained far above the camphor smoke to maintain the thickness of the carbon layer. The contacts from the silicon wafers were made with silver paste using the screen-printing method.

### 2.3. Instrumentation

The prepared samples were characterized using scanning electron microscopy (SEM) using a VEGA TESCAN 3 model with a tungsten filament. The diffuse and specular reflectance properties of the silicon nanostructures were measured in the range from 200 nm to 800 nm using an integrating sphere with 8° incidence angle (JASCO V-650 UV–VIS spectrophotometer) to determine the reflection properties. X-ray diffraction was performed using a RIGAKU SmartLab X-ray diffractometer with a CuKα radiation source (λ = 1.5406 Å) operated at 30 kV. All the samples were scanned from 20° to 80° at a scan rate of 0.02°/s at the 2θ position at room temperature. The current-voltage characterization was performed using an AAA-grade PET solar simulator with a Keithley 4200 source meter with an air mass of 1.5 (AM 1.5, 100 mW/cm$^2$).

## 3. Results and Discussion

### 3.1. Optical Properties

The optical properties of the Si-coated PEDOT: PSS were studied using a JASCO (V 650) UV–VIS Spectrometer by DRS method to obtain the surface reflectance and absorbance. Figure 1a, b shows the reflectance and absorption spectra of PEDOT: PSS-coated N-type silicon wafers using different solvents. Figure 2a, b shows the reflectance and absorption spectra of carbon-coated Si-PEDOT: PSS using different solvents (DMSO, EG, acetonitrile) and compared with a standard N-type silicon wafer as a reference. The obtained surface absorption and reflectance (300–800 nm) of Si-PEDOT: PSS fabricated using acetonitrile, DMSO, and ethylene glycol are listed in Table 1. The results show that the surface reflectance was reduced significantly from a bare silicon wafer to a PEDOT: PSS-coated silicon wafer [18]. The absorbance spectrum shows the Si-coated PEDOT: PSS has significantly more than a 50% increase in the UV region and for the EG sample there is a 20% increase in the entire visible spectrum [24]. Similarly, the surface reflectance of the carbon-coated Si-PEDOT: PSS also decreased from that of the bare silicon wafer, as shown in Table 1.

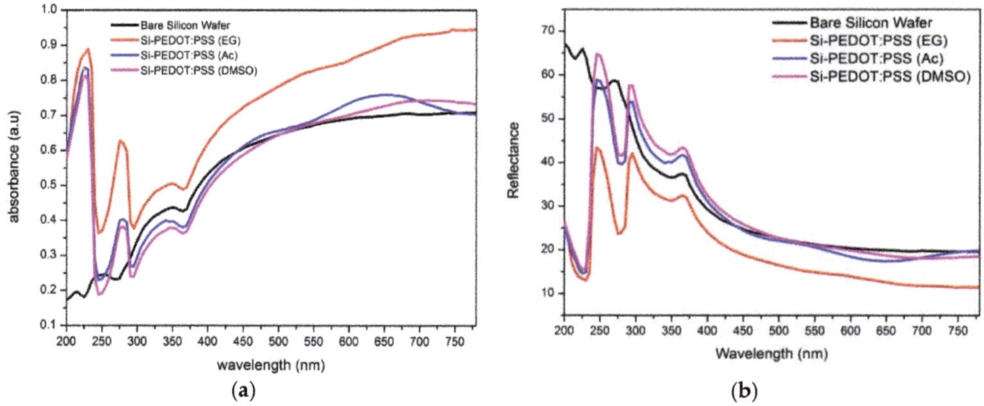


**Figure 1.** (**a**) Absorption and (**b**) reflectance spectra N-type Si wafer with PEDOT: PSS prepared in different solvents (EG) ethylene glycol, (AC) acetonitrile, and (DMSO) dimethyl sulphoxide with bare silicon as reference.

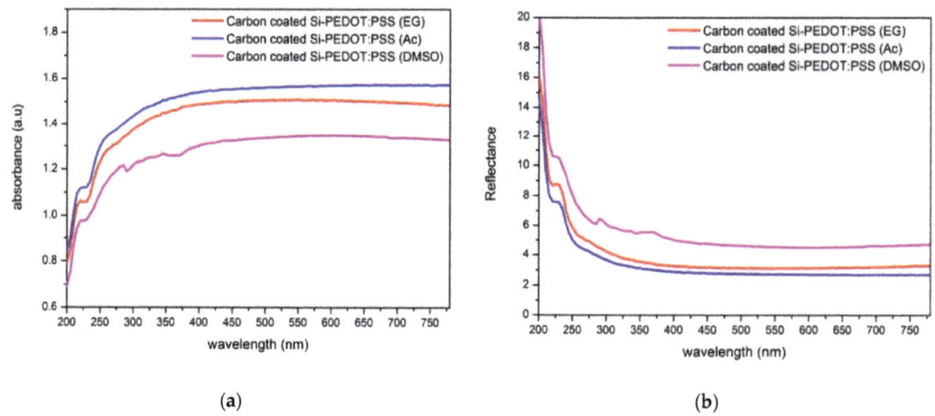


**Figure 2.** (**a**) Absorption and (**b**) reflectance spectra carbon coated N-type Si wafer with PEDOT: PSS prepared in different solvents (EG) ethylene glycol, (AC) acetonitrile, and (DMSO) dimethyl sulphoxide.

**Table 1.** Optical properties of Si-PEDOT: PSS and carbon coated Si-PEDOT: PSS.

| Samples | Optical Mode | Bare Silicon | EG | Acetonitrile | DMSO |
|---|---|---|---|---|---|
| Si/PEDOT: PSS | Reflectance | 67 | 43 | 58 | 62 |
| | Absorbance | 0.24 | 0.87 | 0.84 | 0.81 |
| Carbon coated—Si/PEDOT: PSS | Reflectance | 67 | 8.7 | 7.5 | 10.2 |
| | Absorbance | 0.24 | 1.06 | 1.11 | 0.99 |

### *3.2. Surface Morphology Analysis*

The morphologies of the bare and carbon-coated Si/PEDOT: PSS wafers were analyzed using a scanning electron microscope. The morphology of the PEDOT: PSS deposited on the Si wafer surface is crucial for device performance. The type of PN junction formed between Si and PEDOT: PSS and the lateral or radial PN junctions was investigated by SEM. Figure 3 displays the top view (a–c) and cross-sectional view (d) of PEDOT: PSS-coated Si. The SEM image reveals that PEDOT: PSS on the surface of Si results in the formation of a lateral PN junction with a low PN junction surface area. This improper coverage (agglomeration and island formation) of PEDOT: PSS might be due to the smaller spacing of the highly dense Si, surface tension, and hydrophobicity of the wafer. PEDOT: PSS was deposited and agglomerated on the surface of silicon [8]. Figure 3d shows a cross-sectional view of the PEDOT: PSS-coated silicon wafer.

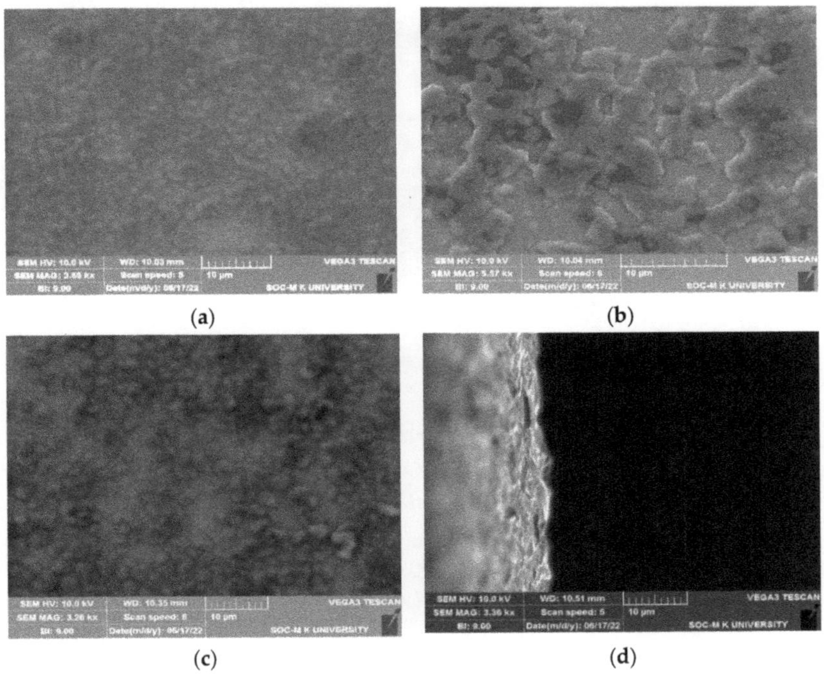


**Figure 3.** SEM images of Si/PEDOT: PSS fabricated using (**a**) ethylene glycol, (**b**) acetonitrile, (**c**) DMSO, respectively; (**d**) a cross section image of Si/PEDOT: PSS.

The surface morphologies of the carbon-coated samples were investigated using scanning electron microscopy (SEM). The morphology and thickness of the carbon layer play major roles in the morphology of the resulting surface. Therefore, we attempted to engineer a carbon layer using camphor smoke. Figure 4 shows the SEM image and cross-section image of the carbon-containing n-type silicon wafer coated with PEDOT: PSS using (a, d) ethylene glycol; (b, e) acetonitrile; and (c, f) DMSO solvent. The cross-section image shows a 1.5 to 2 μm range.

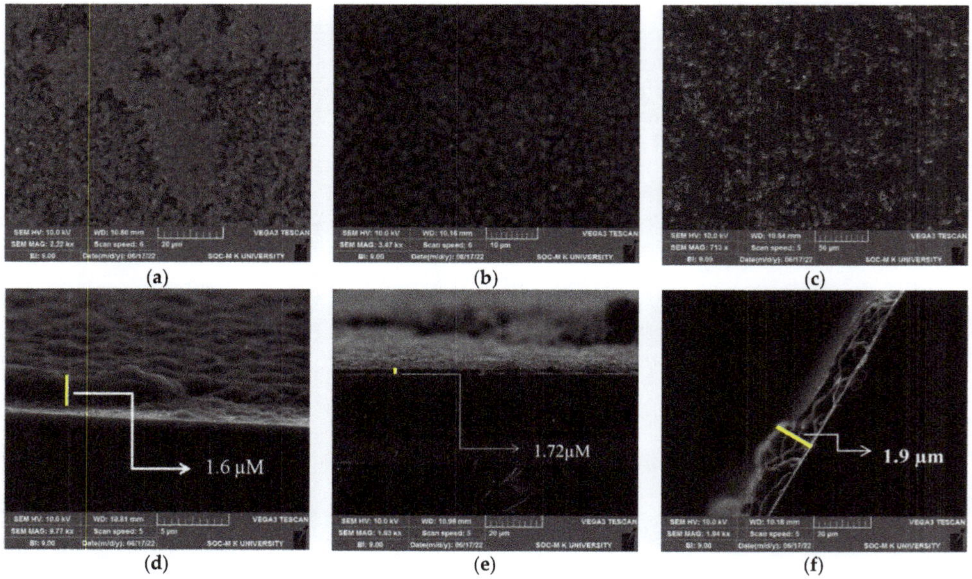


**Figure 4.** SEM images and cross-section image of the carbon-containing n-type silicon wafer coated with PEDOT: PSS using (**a**,**d**) ethylene glycol; (**b**,**e**) acetonitrile; and (**c**,**f**) DMSO solvent.

### *3.3. Structural Properties*

The structural characterization of Si-PEDOT: PSS was performed by X-ray diffraction (XRD), which gives an idea of the crystallite size, structure, and chemical composition, as shown in Figure 5a,b. All three PEDOT: PSS-coated Si wafers were subjected to XRD analysis. The characteristic peak of Si was obtained at 69.5°, and the sharp peak indicated the crystalline nature of Si. Because the Si peaks of the above four data points appear to be the same, the peaks are not identical. There were slight deviations in the Si peak (0.1°) due to the solvent effect. Owing to the presence of different solvents, the forbidden Si (200) peaks were not identical to those of the bare silicon wafer [25]. The peak appearing at 25°–30° corresponds to the $SiO_2$ layer, indicating the presence of $SiO_2$ in the wafers. The peak obtained at 32° corresponds to the forbidden Si interface peak obtained owing to the presence of PEDOT: PSS, as shown in Figure 5a. Table 2 presents the d-spacing and the two θ values for the different Si wafers.

**Table 2.** d-spacing and 2θ value calculated from XRD data.

| Wafer | d-Spacing (Å) | 2θ (degree) |
|---|---|---|
| Si-PEDOT: PSS (EG) | 1.35828 | 69.09817 |
| Si-PEDOT: PSS (Ac) | 1.35828 | 69.08817 |
| Si-PEDOT: PSS (DMSO) | 1.36027 | 68.98316 |

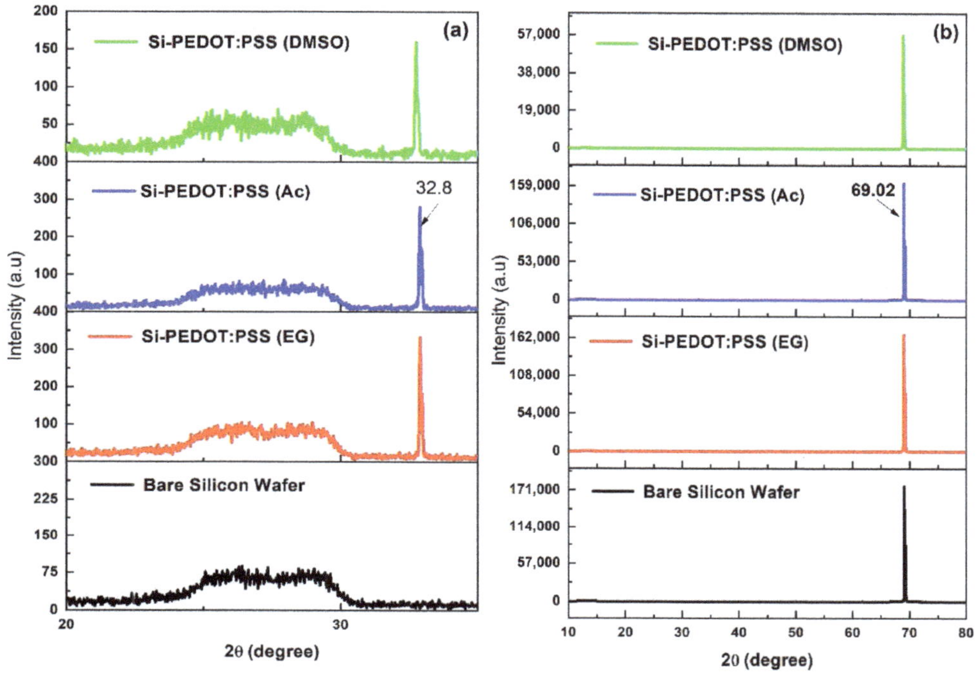


**Figure 5.** Powder X-ray diffraction of (**a**) $SiO_2$ and Si/PEDOT: PSS; (**b**) Si peak for all the samples.

### 3.4. I–V Analysis

Diode characteristics were measured using a solar simulator with a light equivalent to an air mass of 1.5 (AM 1.5, 100 mW/cm$^2$). The Voc, Isc, and diode characteristics were calculated using the projected area of the solar cell. Figure 6a shows the characteristic I–V curve for the diode formation, although the current obtained (EG) is very low of $I_{sc}$ = 0.2 µA and the $V_{oc}$ is 0.2 V. The characteristic I–V curve shows the formation of a heterojunction between PEDOT: PSS and silicon wafer. Figure 6b shows a characteristic I–V curve, which indicates the formation of a heterojunction between PEDOT: PSS and the carbon-coated silicon wafer [26]. A comparison of the performance of this study with those reported in the literature is presented in Table 3.

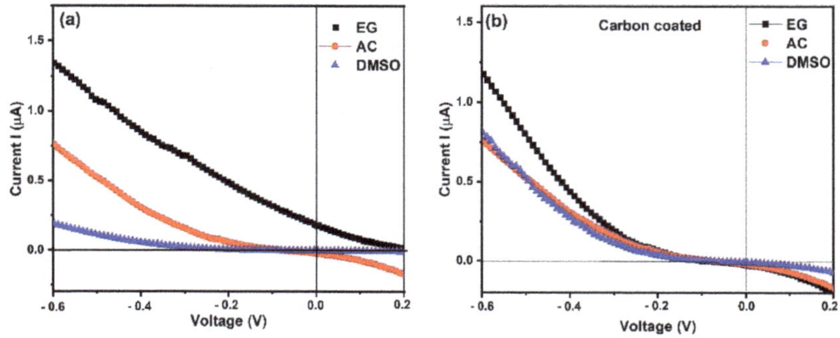


**Figure 6.** Current–voltage measurement of (**a**) Si-PEDOT: PSS wafers and (**b**) carbon coated Si-PEDOT: PSS wafers.

**Table 3.** Comparison of the performance of this work with others reported in the literature.

| S. No. | Findings | PCE | Reference |
|---|---|---|---|
| 1 | Proper surface termination and light trapping in Si/organic hybrid cells can potentially deliver very high PCE. | 10.6% | [8] |
| 2 | Effects of H-Si and SiOx-Si surface termination lead to PCE change from 0.04% to 11.3%. | 11.3% | [27] |
| 3 | Hole-conducting polymer PEDOT: PSS provides a high level of surface passivation on c-Si wafers leading to the efficiencies up to 12.3%. | 12.3% | [28] |
| 4 | Superior photovoltaic properties of solar cells could be realized the engineering electrode/semiconductor interface. | 12.2% | [29] |
| 5 | By adding 4% PFI into PEDOT: PSS, the device achieved an FF of 0.70, which improved by 20% compared to the non-treated one. | 9.90% | [30] |
| 6 | Planar Silicon with a hydrogen-terminated Si (H-Si) surface exhibits high power conversion efficiency (PCE) of 8.27~12.70%. | 12.70% | [9] |
| 7 | Cells with longer Si NWs lead to poor performance because of lower shunt resistance and higher recombination rate. | 12.70% | [22] |
| 8 | The minimal-defect interface between PEDOT: PSS and the planar Si substrate Planar-Si HSCs delivered 13.3% PCE. | 13.3% | [18] |
| 9 | Study on an effective method to improve the conductivity of PEDOT: PSS films with n-Si using EG and Triton. | 14.5% | [24] |
| 10 | The mixed PEDOT: PSS solution of DMSO and FS31 achieves higher conductivity and a smaller contact angle. | 11.07% | [31] |
| 11 | Exposure-oxidation treatment of an H-terminated Si substrate enhances the performance of planar hybrid Si/PEDOT: PSS solar cells. | 13.31% | [32] |
| 12 | 10% concentration of EG in PEDOT: PSS polymer device efficiency is the maximum. | 4.69% | [26] |
| 13 | The cell performance of the siloxane layer at both interfaces shows the best result of >17% efficiency. | 17.34% | [33] |
| 14 | Significant improvement in the performance due to the presence of nanostructures leads to the enhancement in light trapping and photo-carrier collection. | 8.15% | [34] |
| 15 | An attempt to find the best solvent for making PEDOT: PSS hybrid solar cell with silicon. Ethylene glycol shows better result. | $I_{sc}$ = 0.2 μA $V_{oc}$ = 0.2 V | Current work |

## 4. Conclusions

In summary, we fabricated the heterojunction solar cells using three solvents: ethylene glycol, acetonitrile, and dimethyl sulfoxide. The fabricated hybrid solar cell was investigated using UV-DRS, SEM, XRD, and IV. The interpreted results showed that ethylene glycol-containing wafers have lower reflectance and higher absorbance than other solvent-containing wafers. The higher absorbance of light leads to a broad spectrum of photovoltaic devices, which results in the higher cell efficiency. Current–voltage measurements have shown the characteristic diode curve for all the fabricated cells. This characteristic diode curve indicated the presence of a heterojunction. The EG-containing cells displayed current output on I–V characteristics. Further optimization of the concentration, annealing temperature, and spin-coating RPM may yield better characterization results for the fabricated material used in this work.

**Author Contributions:** R.V.: Conceptualization, methodology, Formal analysis, Data curation, Visualization, and Original draft preparation. S.M.T.S.K.M.: Conceptualization, methodology, Formal analysis, Data curation, Visualization, and Original draft preparation. S.C.: Formal analysis, Data curation, and Visualization. S.S.K.: Formal analysis, Data curation, and Investigation. S.S.: Formal analysis, Visualization, Validation and Reviewing and Editing. V.V.: Formal analysis, Data curation, Visualization, and Investigation. J.M.: Supervision, Conceptualization, methodology, Formal analysis, Data curation, Visualization, and Validation. All authors have read and agreed to the published version of the manuscript.

**Funding:** This research received no external funding.

**Data Availability Statement:** Not applicable.

**Acknowledgments:** The author RV is thankful to the CSIR –RA (09/201(0424)/19-EMR-1) for the partial support; JM thanks RUSA, UPE, and the DST—PURSE programme MK University, for providing the XRD, SEM, and IV facilities.

**Conflicts of Interest:** The authors declare no conflict of interest.

## References


1. Andreani, L.C.; Bozzola, A.; Kowalczewski, P.; Liscidini, M.; Redorici, L. Silicon solar cells: Toward the efficiency limits. *Adv. Phys. X* **2019**, *4*, 1548305. [CrossRef]
2. Almomani, M.S.; Ahmed, N.M.; Rashid, M.; Ibnaouf, K.H.; Aldaghri, O.A.; Madkhali, N.; Cabrera, H. Performance improvement of graded bandgap solar cell via optimization of energy levels alignment in Si quantum dot, TiO2 nanoparticles, and porous Si. *Photonics* **2022**, *9*, 843. [CrossRef]
3. Battaglia, C.; Cuevas, A.; de Wolf, S. High-efficiency crystalline silicon solar cells: Status and perspectives. *Energy Environ. Sci.* **2016**, *9*, 1552–1576. [CrossRef]
4. Rajesh, G.; Muthukumarasamy, N.; Velauthapillai, D.; Mohanta, K.; Ragavendran, V.; Batabyal, S.K. Photoinduced electrical bistability of sputter deposited CdZnTe thin films. *Mater. Res. Express* **2018**, *5*, 026412. [CrossRef]
5. Selvapriya, R.; Abhijith, T.; Ragavendran, V.; Sasirekha, V.; Reddy, V.S.; Pearce, J.M.; Mayandi, J. Impact of coupled plasmonic effect with multishaped silver nanoparticles on efficiency of dye sensitized solar cells. *J. Alloy. Compd.* **2022**, *894*, 162339. [CrossRef]
6. Selvapriya, R.; Abhijith, T.; Ragavendran, V.; Sasirekha, V.; Reddy, V.S.; Mayandi, J. Screen printed multifunctional $TiO_2$ photoanode with plasmonic Ag nanoparticles for performance enhancement of dye sensitized solar cell. *Mater. Lett.* **2020**, *276*, 128194. [CrossRef]
7. Sasikala, R.; Kandasamy, M.; Suresh, S.; Ragavendran, V.; Sasirekha, V.; Pearce, J.M.; Murugesan, S.; Mayandi, J. Enhanced dye-sensitized solar cell performance using strontium titanate perovskite integrated photoanodes modified with plasmonic silver nanoparticles. *J. Alloy. Compd.* **2021**, *889*, 161693. [CrossRef]
8. He, L.; Jiang, C.; Wang, H.; Lai, D.; Rusli. High efficiency planar Si/organic heterojunction hybrid solar cells. *Appl. Phys. Lett.* **2012**, *100*, 073503. [CrossRef]
9. Zhao, Y.; Xie, D.; Xu, J.; Feng, T.; Zhang, X.; Tianling, R.; Zhu, M.; Zhu, H. PEDOT: PSS/planar-Si hybrid solar cells with 12.70% effciency. In Proceedings of the PIERS, Guangzhou, China, 25–28 April 2014.
10. Wang, H.; Wang, J.; Prakoso, A.B.; Hong, L.; Tan, Y.H.; Tan, C.S.; Rusli, R. High-efficiency planar thin-film Si/PEDOT: PSS hybrid solar cell. *IEEE J. Photovolt.* **2016**, *6*, 217–222. [CrossRef]
11. Nagamatsu, K.A.; Avasthi, S.; Jhaveri, J.; Sturm, J.C. Fellow, a 12% efficient silicon/PEDOT: PSS heterojunction solar cell fabricated at <100 °C. *IEEE J. Photovolt.* **2014**, *4*, 260–264.
12. Avasthi, S.; Lee, S.; Loo, Y.; Sturm, J.C. Role of majority and minority carrier barriers silicon/organic hybrid heterojunction solar cells. *Adv. Mater.* **2011**, *23*, 5762–5766. [CrossRef] [PubMed]
13. He, L.; Jiang, C.; Wang, H.; Lai, D.; Rusli. Si nanowires organic semiconductor hybrid heterojunction solar cells toward 10% efficiency. *ACS Appl. Mater. Interfaces* **2012**, *4*, 1704–1708. [CrossRef] [PubMed]
14. Sailor, M.J.; Ginsburg, E.J.; Gorman, C.B.; Kumar, A.; Grubbs, R.H.; Lewis, N.S. Thin films of n-Si/Poly-(CH3)3Si-Cyclooctatetraene: Conducting-polymer solar cells and layered structures. *Science* **1990**, *249*, 1146–1149. [CrossRef] [PubMed]
15. Wang, W.; Schiff, E.A. Polyaniline on crystalline silicon heterojunction solar cells. *Appl. Phys. Lett.* **2007**, *91*, 133504. [CrossRef]
16. El-Nahass, M.M.; Zeyada, H.M.; Abd-El-Rahman, K.F.; Darwish, A.A.A. Fabrication and characterization of 4-tricyanovinyl-N,N-diethylaniline/p-silicon hybrid organic–inorganic solar cells. *Sol. Energy Mater. Sol. Cells* **2007**, *91*, 1120–1126. [CrossRef]
17. McGillivray, D.; Thomas, J.P.; Abd-Ellah, M.; Heinig, N.F.; Leung, K.T. Performance enhancement by secondary doping in PEDOT: PSS/Planar-Si hybrid solar cells. *ACS Appl. Mater. Interfaces* **2016**, *8*, 34303–34308. [CrossRef]
18. Thomas, J.P.; Leung, K.T. Defect-minimized PEDOT:PSS/Planar-Si solar cell with very high efficiency. *Adv. Funct. Mater.* **2014**, *24*, 4978–4985. [CrossRef]

19. Friedel, B.; Keivanidis, P.E.; Brenner, T.J.K.; Abrusci, A.; McNeill, C.R.; Friend, R.H.; Greenham, N.C. Effects of layer thickness and annealing of PEDOT:PSS layers in organic photodetectors. *Macromolecules* **2009**, *42*, 6741–6747. [CrossRef]
20. Shiu, S.-C.; Chao, J.-J.; Hung, S.-C.; Yeh, C.-L.; Lin, C.-F. Morphology dependence of silicon nanowire/Poly (3,4 ethylene dioxythiophene):Poly (styrene sulfonate) heterojunction solar cells. *Chem. Mater.* **2010**, *22*, 3108–3113. [CrossRef]
21. Pietsch, M.; Jäckle, S.; Christiansen, S. Interface investigation of planar hybrid nSi/PEDOT: PSS solar cells with open circuit voltages up to 645 mV and efficiencies of 12.6 %. *Appl. Phys. A* **2014**, *115*, 1109–1113. [CrossRef]
22. He, L.; Jiang, C.; Wang, H.; Lai, D.; Tan, Y.H.; Tan, C.S.; Rusli. Effects of nanowire texturing on the performance of Si/organic hybrid solar cells fabricated with a 2.2 μm thin-film Si absorber. *Appl. Phys. Lett.* **2012**, *100*, 103104. [CrossRef]
23. Wang, H.; Wang, J.; Hong, L.; Tan, Y.H.; Tan, C.S. Thin film silicon nanowire/PEDOT: PSS hybrid solar cells with surface treatment. *Nanoscale Res. Lett.* **2016**, *11*, 311. [CrossRef]
24. Zhang, C.; Zhang, Y.; Guo, H.; Zhang, Z.; Zhang, C. Hole-transporting layer treatment of planar hybrid n-Si/PEDOT:PSS solar cells with power conversion efficiency up to 14.5%. *Nanostructured Sol. Cells* **2017**, *2017*, 3192197. [CrossRef]
25. Zaumseil, P. High-resolution characterization of the forbidden Si 200 and Si 222 reflections. *J. Appl. Cryst.* **2015**, *48*, 528–532. [CrossRef] [PubMed]
26. Singh, P.; Nakra, R.; Sivaiah, B.; Sardana, S.K.; Prathap, P.; Rauthan, C.M.S.; Srivastava, S.K. Effect of ethylene glycol doping on performance of PEDOT:PSS/μT-n-Si heterojunction solar cell. *AIP Conf. Proc.* **2018**, *1961*, 020003. [CrossRef]
27. LiHe, n.; Jiang, C.; Wang, H.; Lei, H.; Lai, D.; Rusli. 11.3% efficient planar Si-PEDOT: PSS hybrid solar cell with a thin interfacial oxide. In Proceedings of the 2012 38th IEEE Photovoltaic Specialists Conference, Austin, TX, USA, 3–8 June 2012. [CrossRef]
28. Schmidt, J.; Titova, V.; Zielke, D. Organic-silicon heterojunction solar cells: Open-circuit voltage potential and stability. *Appl. Phys. Lett.* **2013**, *103*, 183901. [CrossRef]
29. Zhang, Y.; Zu, F.; Lee, S.-T.; Liao, L.; Zhao, N.; Sun, B. Heterojunction with organic thin layers on silicon for record efficiency hybrid solar cells. *Adv. Funct. Mater.* **2014**, *4*, 1300923. [CrossRef]
30. Zhu, Y.; Song, T.; Zhang, F.; Lee, S.T.; Suna, B. Efficient organic-inorganic hybrid Schottky solar cell: The role of built-in potential. *Appl. Phys. Lett.* **2013**, *102*, 113504. [CrossRef]
31. Jiang, X.; Zhang, P.; Zhang, J.; Wang, J.; Li, G.; Fang, X.; Yang, L.; Chen, X. High performance of PEDOT:PSS/n-Si solar cells based on textured surface with AgNWs electrodes. *Nanoscale Res. Lett.* **2018**, *13*, 53. [CrossRef]
32. Zhang, C.; Zhang, Y.; Guo, H.; Jiang, Q.; Dong, P.; Zhang, C. Efficient planar hybrid n-Si/PEDOT:PSS solar cells with power conversion efficiency up to 13.31% achieved by controlling the SiOx interlayer. *Energies* **2018**, *11*, 1397. [CrossRef]
33. Yoon, S.-S.; Khang, D.-Y. High efficiency (>17%) Si-Organic hybrid solar cells by simultaneous structural, electrical, and interfacial engineering via low-temperature processes. *Adv. Energy Mater.* **2017**, *8*, 170265. [CrossRef]
34. van Trinh, P.; Anh, N.N.; Cham, N.T.; Tu, l.; van Hao, N.; Thang, B.H.; van Chuc, N.; Thanh, C.T.; Minh, P.N.; Fukata, N. Enhanced power conversion efficiency of an n-Si/PEDOT:PSS hybrid solar cell using nanostructured silicon and gold nanoparticles. *RSC Adv.* **2022**, *12*, 10514–10521. [CrossRef] [PubMed]

*electronics*

*Article*

# Photoluminescence Properties of InAs Quantum Dots Overgrown by a Low-Temperature GaAs Layer under Different Arsenic Pressures

**Sergey Balakirev [1], Natalia Chernenko [1], Natalia Kryzhanovskaya [2], Nikita Shandyba [1], Danil Kirichenko [1], Anna Dragunova [2], Sergey Komarov [2], Alexey Zhukov [2] and Maxim Solodovnik [1,*]**

[1] Institute of Nanotechnologies, Electronics and Equipment Engineering, Southern Federal University, 347922 Taganrog, Russia
[2] International Laboratory of Quantum Optoelectronics, HSE University, 190008 St. Petersburg, Russia
* Correspondence: solodovnikms@sfedu.ru

**Abstract:** We studied the influence of the arsenic pressure during low-temperature GaAs overgrowth of InAs quantum dots on their optical properties. In the photoluminescence spectrum of quantum dots overgrown at a high arsenic pressure, we observed a single broad line corresponding to unimodal size distribution of quantum dots. Meanwhile, two distinct peaks (~1080 and ~1150 nm) at larger wavelengths are found in the spectra of samples with quantum dots overgrown at a low arsenic pressure. We attributed this phenomenon to the high-pressure suppression of atom diffusion between InAs islands at the overgrowth stage, which makes it possible to preserve the initial unimodal size distribution of quantum dots. The same overgrowth of quantum dots at the low arsenic pressure induces intensive mass transfer, which leads to the formation of arrays of quantum dots with larger sizes. Integrated photoluminescence intensity at 300 K is found to be lower for quantum dots overgrown at the higher arsenic pressure. However, a difference in the photoluminescence intensity for the high- and low-pressure overgrowths is not so significant for a temperature of 77 K. This indicates that excess arsenic incorporates into the capping layer at high arsenic pressures and creates numerous nonradiative recombination centers, diminishing the photoluminescence intensity.

**Keywords:** molecular beam epitaxy; A3B5; semiconductor nanostructures; quantum dots; photoluminescence

**Citation:** Balakirev, S.; Chernenko, N.; Kryzhanovskaya, N.; Shandyba, N.; Kirichenko, D.; Dragunova, A.; Komarov, S.; Zhukov, A.; Solodovnik, M. Photoluminescence Properties of InAs Quantum Dots Overgrown by a Low-Temperature GaAs Layer under Different Arsenic Pressures. *Electronics* **2022**, *11*, 4062. https://doi.org/10.3390/electronics11234062

Academic Editor: Lucas Lamata

Received: 22 November 2022
Accepted: 6 December 2022
Published: 6 December 2022

**Publisher's Note:** MDPI stays neutral with regard to jurisdictional claims in published maps and institutional affiliations.



## 1. Introduction

Semiconductor quantum dots (QDs) are considered to significantly improve the efficiency of optoelectronic and nanophotonic devices because they are spatially limited in all three dimensions, which leads to their strong quantum confinement [1–4]. InAs QD lasers exhibit a low threshold current density [5,6], high internal quantum efficiency [7,8] and high-temperature stability of threshold current and slope efficiency [9,10]. However, output power and modulation bandwidth of QD lasers are still smaller than those characteristics of quantum well lasers [11]. Furthermore, the performance of promising QD-based quantum information devices strongly depends on the geometry and composition of QDs [12].

Photoluminescence (PL) spectra are normally taken from overgrown QDs whose geometrical parameters differ from those of uncovered QDs observed on the surface. Therefore, it is very essential to study the processes of QD capping. Previous studies by transmission electron microscopy (TEM) and cross-sectional scanning tunneling microscopy (X-STM) methods demonstrated that intensive mass transfer and intermixing occurs during overgrowth of InAs QDs [11,13–16]. It was shown that the deposition of only one monolayer (ML) of GaAs above the array of QDs leads to a significant change in their size and density and to a clear bimodal size distribution [17]. Another study demonstrated an abrupt decrease in the average height of InAs QDs from 10 to 1.5 nm after their overgrowth by a

2-nanometer-thick GaAs layer [14]. The active mass transfer observed in these experiments is caused by a necessity to minimize the system energy by strain relaxation.

A lot of techniques were used to preserve the configuration of initial QDs. For example, overgrowth of InAs QDs by a GaAs layer at a temperature of 300 °C made it possible to maintain a pyramidal shape of QDs with sharp heterointerfaces [13]. However, such low-temperature capping may result in poor structural and optical quality of QDs. Various overgrowth temperature sequences were also used to estimate their influence on the QD characteristics. The results indicate that the overgrowth temperature must be at the same level as or below the QD growth temperature to avoid intermixing and flushing of the tops of QDs [11,16,18]. It was also shown that capping at low growth rates leads to more intensive intermixing and QD height reduction than at high growth rates [15]. This indicates that the material of QDs and capping layers tends to intermix and transform into a single continuous layer when not prevented by specific techniques. For the same reason, any growth interruption leads to unceasing dissolving of QDs in the matrix material [13]. However, much less attention has been paid to the dependence of the QD PL spectra on the arsenic beam equivalent pressure during overgrowth, which is also expected to have an impact on the processes of QD dissolving.

In this paper, we studied the influence of the arsenic pressure ($P_{As}$) during overgrowth of InAs QDs by a low-temperature GaAs (LT-GaAs) layer with different thicknesses ($H$) on the PL spectra of these structures. We revealed that a high arsenic pressure makes it possible to achieve a single broad line in the PL spectrum corresponding to unimodal size distribution of QDs, whereas the spectrum has two separate lines (~1080 and ~1150 nm) at a low arsenic pressure for both capping layer thicknesses (5 and 10 nm). An additional longwave line (~1150 nm) is attributed to the contribution of largened QDs. Their formation can be inhibited by increasing the arsenic flux, which suppresses mass transfer and reduces the contribution of large QDs in the PL spectrum.

## 2. Materials and Methods

The samples were grown in a SemiTEq STE 35 solid-source molecular beam epitaxy system on GaAs(001) epi-ready substrates. Standard effusion cells were used as group-III component sources. For group-V arsenic, a valved cracker cell with $As_4$ mode was used. After a standard procedure of the oxide removal, a 250-nanometer-thick GaAs buffer layer was grown at 580 °C. A single InAs QD layer was sandwiched between a 100-nanometer-thick GaAs inner layer and a 50-nanometer-thick $Al_{0.33}Ga_{0.67}As$ outer layer grown at 580 °C and 600 °C, respectively. The QDs were formed by the Stranski–Krastanov growth mode after deposition of 2.5 MLs of InAs at 500 °C and 0.05 ML/s. The formation of InAs three-dimensional islands was verified by the streaky–spotty transition of the reflection high-energy electron diffraction pattern. After the QD formation, a growth interruption of 30 s under the arsenic flux was applied. Then, four different low-temperature (500 °C) GaAs capping procedures were carried out for samples 1–4: (1) 5 nm at $P_{As} = 3 \times 10^{-5}$ Pa; (2) 10 nm at $P_{As} = 3 \times 10^{-5}$ Pa; (3) 5 nm at $P_{As} = 1 \times 10^{-5}$ Pa; (4) 10 nm at $P_{As} = 1 \times 10^{-5}$ Pa. Thereafter, the substrate temperature was elevated to 580 °C to grow a high-temperature GaAs layer with a thickness of 95 nm for samples 1 and 3 and 90 nm for samples 2 and 4. Above the $Al_{0.33}Ga_{0.67}As$ cladding layer, a 10-nanometer-thick high-temperature GaAs layer was grown, and the structure was annealed at 610 °C to reduce defects. For reference, an additional sample was grown with an array of uncovered QDs formed straightly on a GaAs buffer layer (sample 0).

A schematic QD PL structure grown in this study is illustrated in Figure 1.

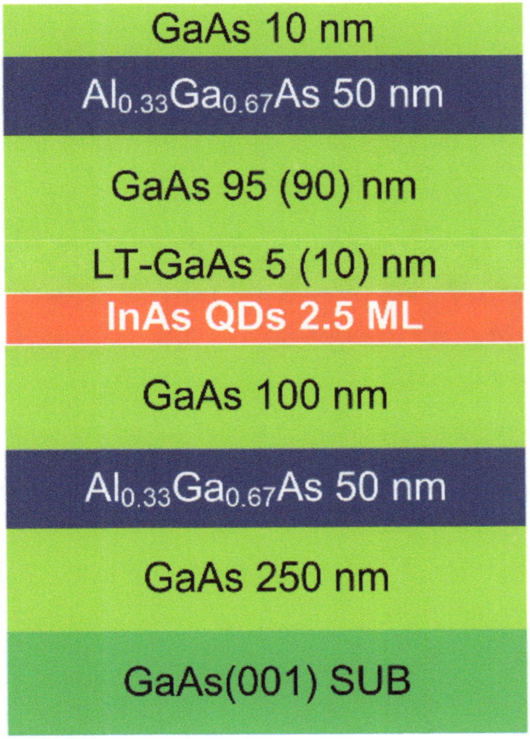


**Figure 1.** Schematic representation of a typical PL structure in which the LT-GaAs capping layer differs in the thickness and the arsenic pressure provided during the growth.

The reference sample was studied in FEI Nova Nanolab scanning electron microscope (SEM) and NT-MDT Ntegra atomic force microscope (AFM) for the characterization of shape, size and surface density of uncovered QDs.

For PL studies, the samples were placed in a flow Janis ST-500 cryostat to control the sample temperature ($T_{PL}$) in the range of 77–300 K. The PL was excited by a YLF:$Nd^{+3}$ laser operating in the cw mode ($\lambda$ = 527 nm). The laser power was varied in the range from 0.084 to 40 mW. PL signal was detected using an SOL instrument's MS 5204i monochromator and a single-channel InGaAs detector using synchronous detection (SRS 830 Stanford Research Systems).

### 3. Results and Discussion

The QD growth conditions were chosen so that the surface density of QDs was sufficiently large to provide high PL intensity, but not too large to stimulate QD coalescence. It is commonly known that InAs dots arranged quite close to each other merge due to Ostwald ripening [7,11]. A value of the surface density of uncovered QDs that we observe after 30 s interruption at the growth temperature before quenching is $4.1 \times 10^{10}$ cm$^{-2}$ (Figure 2). Meanwhile, the density of the coalesced QDs is found to be $3.8 \times 10^{8}$ cm$^{-2}$. An average diameter of uncovered QDs is 23 nm, and their average height is 5.5 nm. QDs with these parameters are typically supposed to have separate energy levels contributing to distinct peaks in the PL spectrum after the capping procedure [14,19].

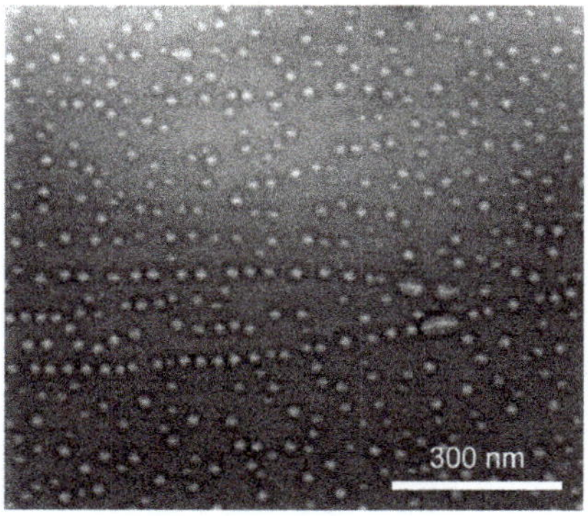


**Figure 2.** SEM image of an array of uncovered InAs QDs formed on the GaAs surface (sample 0).

PL spectra taken from the overgrown structures demonstrate their strong dependence on the arsenic pressure used during overgrowth of QDs by the LT-GaAs layer (Figure 3a). For each room-temperature spectrum, the lines corresponding to volume GaAs and InAs wetting layer are at ~870 and 935 nm, respectively, which is in good agreement with previous studies [20]. However, longer-wavelength lines corresponding to the energy levels of QDs differ in their position and relative intensity (Table 1).

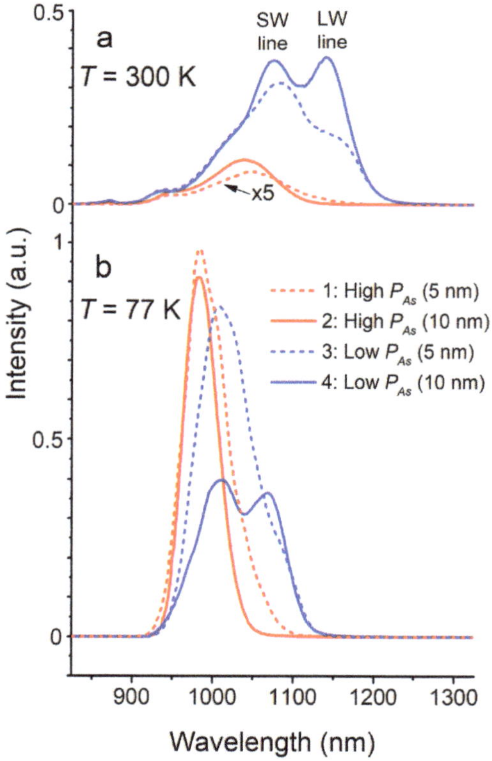


**Figure 3.** 300 K (**a**) and 77 K (**b**) PL spectra of structures with InAs QDs overgrown by the GaAs layer under various growth conditions according to the sample numbers in the legend: red lines: $P_{As} = 3 \times 10^{-5}$ Pa; blue lines: $P_{As} = 1 \times 10^{-5}$ Pa; solid lines: 10 nm; dashed lines: 5 nm.

**Table 1.** Summary of the PL parameters of InAs QDs covered by the LT-GaAs layer under various growth conditions.

| $T_{PL}$ (K) | $P_{As}$ (Pa) | $H$ = 5 nm | | | $H$ = 10 nm | | |
|---|---|---|---|---|---|---|---|
| | | SW Line (nm) | LW Line (nm) | Integrated Intensity (a.u.) | SW Line (nm) | LW Line (nm) | Integrated Intensity (a.u.) |
| 300 | $3 \times 10^{-5}$ | 1049 | - | 0.03 | 1038 | - | 0.16 |
| | $1 \times 10^{-5}$ | 1083 | ~1160 | 0.61 | 1077 | 1140 | 0.76 |
| 77 | $3 \times 10^{-5}$ | 983 | - | 0.88 | 983 | - | 0.64 |
| | $1 \times 10^{-5}$ | 1009 | ~1080 | 1.00 | 1011 | 1070 | 0.62 |

When QDs are overgrown at $P_{As} = 3 \times 10^{-5}$ Pa, an unambiguous single broad line is observed in the room-temperature PL spectrum at 1049 and 1038 nm (shortwave (SW) line) for the LT-GaAs thicknesses of 5 and 10 nm, respectively (Figure 3a), whereas at $P_{As} = 1 \times 10^{-5}$ Pa, two longer wave lines are pronounced. The first line (SW) is approximately the same for both LT-GaAs thicknesses and shifted to ~1080 nm. The longwave (LW) line is at a wavelength of ~1160 and ~1140 nm for 5 and 10 nm LT-GaAs, respectively. Although the appearance of an additional LW line is often attributed to an array of coalesced QDs [11], we consider that its contribution is not sufficient to give a pronounced PL peak, taking into account the two-orders-of-magnitude-lower surface density of coalesced QDs compared with basic ones. Moreover, the coalescence of QDs before capping is equally inherent in each of the PL structures under consideration, but the additional LW lines appear only after the overgrowth at $P_{As} = 1 \times 10^{-5}$ Pa. Therefore, this behavior is not due to the initial size distribution of InAs islands, but to mass transfer processes during the overgrowth.

The red shift of the SW line, as well as the appearance of an additional LW line in the low-pressure LT-GaAs PL spectra, is due to the redistribution of the QD material and formation of new groups of QD sizes. At the high arsenic pressure, the diffusion of group-III atoms is suppressed, and therefore, InAs islands are predominantly conserved. Although the shape, size and composition of InAs QDs can change due to the indium segregation and strain-relaxing mass transfer, their uniformity in size is assumed to be retained. At the same time, an InAs QD covered by an intermediate InGaAs layer with changing In content is formed as a result of QD and LT-GaAs intermixing. As shown by TEM and X-STM studies [11,13], this form of QDs is typical for capping at the QD growth temperature. When the temperature is reduced to ultracold 300 °C, the QD shape remains almost perfectly pyramidal [13]. Meanwhile, an increase in the temperature causes more intensive intermixing, which leads to blurring of the original pyramidal shape of QDs.

Similar behavior is observed when the arsenic pressure during the overgrowth is decreased. The diffusion length of group-III atoms increases, leading to more intensive mass transfer and coalescence of QDs located close to each other. As a result, two types of QDs appear in the volume and contribute to LW lines: 1140 nm for 10 nm LT-GaAs and ~1160 nm for 5 nm LT-GaAs (Figure 3a). These lines are attributed to the appearance of QDs of quite a large size which contain an increased number of energy levels in the potential well, providing lower energy electron transitions with emission at longer wavelengths.

Integrated room-temperature PL intensity of QDs obtained at the low arsenic pressure during the overgrowth essentially exceeds the intensity of QDs which were overgrown at the high arsenic pressure (Table 1). However, this difference is not so significant at 77 K PL (Figure 3b). This indicates that excess arsenic accumulates in the capping layer grown at the high arsenic pressure and acts as additional nonradiative recombination centers for charge carriers migrating faster over the semiconductor volume at room temperature than at 77 K. It also should be noted that the integrated room-temperature PL intensity of QDs overgrown by 5 nm LT-GaAs is weaker than for 10 nm LT-GaAs, especially for the high arsenic pressure. It can be associated with higher defectiveness of thinner capping layers, which has been reported elsewhere. After the low-temperature overgrowth of QDs by a layer of 2–7 nm thickness, shallow depressions are formed above the tops of QDs and preserved on the surface at a stage of substrate heating up to a high temperature used for the overgrowth by a thick GaAs layer [7,11,14,17]. After the high-temperature overgrowth of the layer with these depressions, multiple dislocations are produced in the area above QDs [7,11]. In our study, the defectiveness of the structure increases when the 5 nm LT-GaAs layer is grown at the high arsenic pressure because, in this case, the excess arsenic incorporated into the thin capping layer (5 nm) desorbs, leading to the expansion of the depressions above QDs. Although the concentration of excess arsenic atoms decreases, an increased density of dislocations prevents efficient radiative recombination of carriers in QDs to a greater extent. However, the overgrowth of QDs by the thicker LT-GaAs layer (10 nm) allows inhibition of the depression formation and subsequent generation of dislocations [11].

The intensity maximum of the LW line in the room-temperature spectra for the low arsenic pressure is observed to be higher than the intensity maximum of the SW line in the case of the 10 nm capping layer, but it is lower in the case of the 5-nanometer-thick layer (Figure 3a). The integrated PL intensity from QDs overgrown by the 10 nm layer is also higher than from QDs overgrown by the 5 nm layer (Table 1). This behavior can be explained by the difference in the surface density of QDs of various sizes formed during the diffusion of In atoms within the overgrown layer mixed with the QD material. In the case of the capping layer of a smaller thickness (5 nm), as noted earlier, a greater number of dislocations are formed in the layer after the high-temperature overgrowth than in the case of the overgrowth by the thicker LT-GaAs layer (10 nm). As a result, the integrated PL intensity at room temperature decreases, as does the intensity of the LW line, which corresponds to larger QDs having a greater number of dislocations in their neighborhood. Therefore, charge carriers in larger QDs are more prone to nonradiative transitions than in smaller QDs. This assumption is confirmed by the fact that the integrated PL intensity from QDs overgrown by the 5 nm layer increases with a decrease in the PL temperature to 77 K (Figure 3b).

However, for the 10-nanometer-thick capping layer, the integrated PL intensity remains at the same level, with a slight decrease when moving from 300 to 77 K (Table 1). As mentioned earlier, this indicates lower defectiveness of the material compared with the structure with the 5 nm layer. However, it is important to clarify the reason for the reduction in the integrated PL intensity from QDs at 77 K in comparison with the 10 nm layer. This regularity may be due to the continuation of intensive diffusion of In atoms at the stage of heating the 5 nm layer, which leads to an increase in the surface density of smaller QDs responsible for the SW line. Although the density of such QDs overgrown by the 5 nm layer is higher, the integrated PL intensity of QDs at room temperature becomes lower than in the case of the 10 nm layer due to the higher imperfection of the material. When QDs are overgrown by the thicker LT-GaAs layer (10 nm), the QD density remains almost unchanged during the heating, which is confirmed by the retention of two distinct lines in the PL spectrum (SW and LW lines).

In order to estimate the nature of the separate SW and LW lines, the excitation intensity was varied in the range from 0.084 to 40 mW for the two PL temperatures of 300 K and 77 K for the samples with the low-pressure overgrowth by LT-GaAs layers of different thicknesses (Figure 4).

For the room-temperature PL spectra of QDs overgrown at the low arsenic pressure, an increase in the ratio of SW to LW lines' maximum intensity is observed, with increasing excitation power for both LT-GaAs layer thicknesses: 5 nm (Figure 4a) and 10 nm (Figure 4b). This is attributed to the fact that low-energy electron transitions in QDs of larger sizes are more favorable than shortwave transitions in smaller QDs. When the first ones reach saturation at a certain excitation intensity, the probability of high-energy transitions in smaller QDs becomes higher, contributing to the SW line of the PL spectrum.

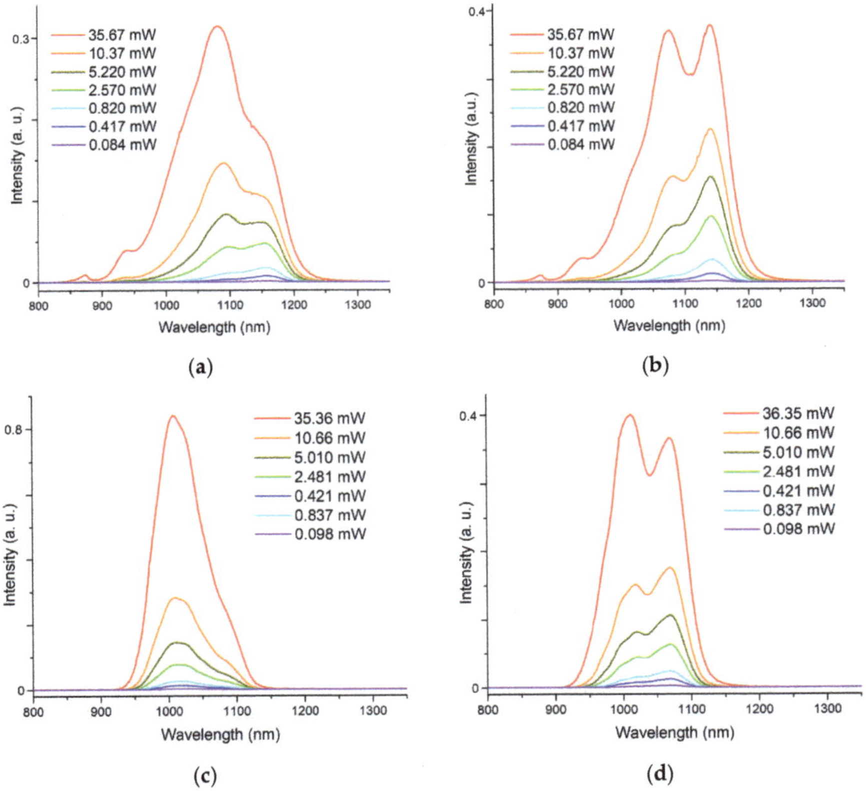


**Figure 4.** Excitation intensity-dependent PL spectra of QDs overgrown at the low arsenic pressure: (**a**) $H = 5$ nm (300 K); (**b**) $H = 10$ nm (300 K); (**c**) $H = 5$ nm (77 K); (**d**) $H = 10$ nm (77 K).

A similar situation is observed in the cryogenic PL spectra, where the ratio of the SW to LW lines' maximum intensity rises with increasing excitation power for the samples with both LT-GaAs thicknesses, i.e., 5 nm (Figure 4c) and 10 nm (Figure 4d). However, two LW lines become prominent at wavelengths of ~1020 and ~1080 nm. This fact confirms an assumption about the saturation of low-energy electron transitions, but it also demonstrates that appearance of a new group of QDs with an intermediate size (emitting at ~1020 nm) is possible in the case of the 5 nm LT-GaAs layer. Thus, the diffusion of In atoms within the thin capping layer is more rapid when the substrate heating occurs than within the thicker capping layer. Observation of both (SW and LW) lines in the cryogenic PL spectra at any excitation power confirms that the SW line is not due to the transitions from an excited level but due to the transitions in a certain type of small-sized QDs.

## 4. Conclusions

The influence of the arsenic pressure during low-temperature GaAs overgrowth of InAs QDs on their optical properties was revealed in this study. We demonstrated that overgrowth at a high arsenic pressure provides a single broad line in the QD PL spectrum, whereas two distinct lines are observed in the PL spectrum of QDs overgrown by the LT-GaAs layer at a low arsenic pressure. This behavior is explained by the strain-compensating mass transfer induced by the QD overgrowth. While the diffusion of In atoms is suppressed at the high arsenic pressure, formation of different size groups of QDs occurs during the low-pressure overgrowth. However, the integrated room-temperature PL intensity of QDs is significantly lower for QDs overgrown at the high arsenic pressure because excess arsenic incorporates into the capping layer in this case and leads to the enhancement of nonradiative recombination of charge carriers.

We also reveal a decrease in the integrated intensity of a thinner LT-GaAs capping layer (5 nm) as compared with a thicker layer (10 nm). We attribute it to one more source of nonradiative recombination centers, which has dislocations produced in the area of QD tops. These dislocations are caused by the formation of depressions during the low-temperature overgrowth and expand during the substrate heating of the thin layer (5 nm). Meanwhile, this pit formation can be inhibited using the low-temperature overgrowth by the thicker layer (10 nm). The higher defectiveness of the 5 nm LT-GaAs layer is also found for the overgrowth at the high arsenic pressure, which is associated with the desorption of the excess arsenic, inducing the depression formation.

The presence of the SW line at each excitation power in the cryogenic PL spectra for the samples with the low-pressure overgrowth confirmed the assumption that it is not caused by the electron transitions from an excited energy level, but it is due to the excitation of smaller-sized QDs. The SW line's maximum intensity increases when the low-energy transitions from larger QDs reach saturation, giving rise to the shorter-wavelength transitions from smaller QDs. The analysis of the ratio of relative maximum intensities of the SW and LW lines also indicated a difference in the surface density of QDs obtained after the overgrowth by LT-GaAs layers of different thicknesses. Because the LW line's intensity is relatively lower than the SW line's intensity for the 5 nm LT-GaAs, we can conclude that the concentration of dislocations near larger QDs is higher than that near smaller QDs.

**Author Contributions:** Conceptualization, S.B. and M.S.; methodology, S.B. and M.S.; validation, S.B., N.K., A.Z. and M.S.; formal analysis, S.B., N.K. and M.S.; investigation, S.B., N.C., N.K., N.S., D.K., A.D., S.K. and M.S.; resources, S.B., N.K., A.Z. and M.S.; writing—original draft preparation, S.B. and M.S.; writing—review and editing, S.B., N.K., A.Z. and M.S.; visualization, S.B., N.C., N.K., N.S., D.K., A.D., S.K. and M.S.; supervision, M.S.; project administration, N.K., A.Z. and M.S.; funding acquisition, N.K., A.Z. and M.S. All authors have read and agreed to the published version of the manuscript.

**Funding:** The research was supported by the Strategic Academic Leadership Program of the Southern Federal University ("Priority 2030") and by the project "Mirror Laboratories", HSE University, RF.

**Data Availability Statement:** Not applicable.

**Conflicts of Interest:** The authors declare no conflict of interest. The funders had no role in the design of the study; in the collection, analyses, or interpretation of data; in the writing of the manuscript; or in the decision to publish the results.

## References


1. Cusack, M.; Briddon, P.; Jaros, M. Electronic Structure of InAs/GaAs Self-Assembled Quantum Dots. *Phys. Rev. B Condens. Matter Mater. Phys.* **1996**, *54*, R2300–R2303. [CrossRef] [PubMed]
2. Johansen, J.; Stobbe, S.; Nikolaev, I.S.; Lund-Hansen, T.; Kristensen, P.T.; Hvam, J.M.; Vos, W.L.; Lodahl, P. Size Dependence of the Wavefunction of Self-Assembled InAs Quantum Dots from Time-Resolved Optical Measurements. *Phys. Rev. B Condens. Matter Mater. Phys.* **2008**, *77*, 073303. [CrossRef]
3. Li, S.Y.; He, L. Recent Progresses of Quantum Confinement in Graphene Quantum Dots. *Front. Phys.* **2022**, *17*, 33201. [CrossRef]
4. Pryor, C. Geometry and Material Parameter Dependence of InAs/GaAs Quantum Dot Electronic Structure. *Phys. Rev. B* **1999**, *60*, 2869–2874. [CrossRef]
5. Moiseev, E.I.; Kryzhanovskaya, N.V.; Zubov, F.I.; Mikhailovskii, M.S.; Abramov, A.N.; Maximov, M.V.; Kulagina, M.M.; Guseva, Y.A.; Livshits, D.A.; Zhukov, A.E. Record Low Threshold Current Density in Quantum Dot Microdisk Laser. *Semiconductors* **2019**, *53*, 1888–1890. [CrossRef]
6. Stintz, A.; Liu, G.T.; Li, H.; Lester, L.F.; Malloy, K.J. Low-Threshold Current Density 1.3-Mm InAs Quantum-Dot Lasers with the Dots-in-a-Well (DWELL) Structure. *IEEE Photonics Technol. Lett.* **2000**, *12*, 591–593. [CrossRef]
7. Liu, W.-S.; Chang, H.; Liu, Y.-S.; Chyi, J.-I. Pinholelike Defects in Multistack 1.3μm InAs Quantum Dot Laser. *J. Appl. Phys.* **2006**, *99*, 114514. [CrossRef]
8. Ledentsov, N.N.; Kovsh, A.R.; Zhukov, A.E.; Maleev, N.A.; Mikhrin, S.S.; Vasil'ev, A.P.; Semenova, E.S.; Maximov, M.V.; Shernyakov, Y.M.; Kryzhanovskaya, N.V.; et al. High Performance Quantum Dot Lasers on GaAs Substrates Operating in 1.5 μm Range. *Electron. Lett.* **2003**, *39*, 1126. [CrossRef]

9. Nishi, K.; Takemasa, K.; Sugawara, M.; Arakawa, Y. Development of Quantum Dot Lasers for Data-Com and Silicon Photonics Applications. *IEEE J. Sel. Top. Quantum Electron.* **2017**, *23*, 1–7. [CrossRef]
10. Zhukov, A.E.; Kryzhanovskaya, N.V.; Moiseev, E.I.; Nadtochiy, A.M.; Dragunova, A.S.; Maximov, M.V.; Zubov, F.I.; Kadinskaya, S.A.; Berdnikov, Y.; Kulagina, M.M.; et al. Impact of Self-Heating and Elevated Temperature on Performance of Quantum Dot Microdisk Lasers. *IEEE J. Quantum Electron.* **2020**, *56*, 1–8. [CrossRef]
11. Okumura, S.; Fujisawa, K.; Naruke, T.; Nishi, K.; Onishi, Y.; Takemasa, K.; Sugawara, M.; Sugiyama, M. Impact of Low-Temperature Cover Layer Growth of InAs/GaAs Quantum Dots on Their Optical Properties. *Jpn. J. Appl. Phys.* **2022**, *61*, 085503. [CrossRef]
12. Arakawa, Y.; Holmes, M.J. Progress in Quantum-Dot Single Photon Sources for Quantum Information Technologies: A Broad Spectrum Overview. *Appl. Phys. Rev.* **2020**, *7*, 021309. [CrossRef]
13. Gong, Q.; Offermans, P.; Nötzel, R.; Koenraad, P.M.; Wolter, J.H. Capping Process of InAs/GaAs Quantum Dots Studied by Cross-Sectional Scanning Tunneling Microscopy. *Appl. Phys. Lett.* **2004**, *85*, 5697–5699. [CrossRef]
14. García, J.M.; Medeiros-Ribeiro, G.; Schmidt, K.; Ngo, T.; Feng, J.L.; Lorke, A.; Kotthaus, J.; Petroff, P.M. Intermixing and Shape Changes during the Formation of InAs Self-Assembled Quantum Dots. *Appl. Phys. Lett.* **1997**, *71*, 2014–2016. [CrossRef]
15. Utrilla, A.D.; Grossi, D.F.; Reyes, D.F.; Gonzalo, A.; Braza, V.; Ben, T.; González, D.; Guzman, A.; Hierro, A.; Koenraad, P.M.; et al. Size and Shape Tunability of Self-Assembled InAs/GaAs Nanostructures through the Capping Rate. *Appl. Surf. Sci.* **2018**, *444*, 260–266. [CrossRef]
16. Lian, G.D.; Yuan, J.; Brown, L.M.; Kim, G.H.; Ritchie, D.A. Modification of InAs Quantum Dot Structure by the Growth of the Capping Layer. *Appl. Phys. Lett.* **1998**, *73*, 49–51. [CrossRef]
17. Ferdos, F.; Wang, S.; Wei, Y.; Larsson, A.; Sadeghi, M.; Zhao, Q. Influence of a Thin GaAs Cap Layer on Structural and Optical Properties of InAs Quantum Dots. *Appl. Phys. Lett.* **2002**, *81*, 1195–1197. [CrossRef]
18. Songmuang, R.; Kiravittaya, S.; Schmidt, O.G. Shape Evolution of InAs Quantum Dots during Overgrowth. *J. Cryst. Growth* **2003**, *249*, 416–421. [CrossRef]
19. Nishi, K.; Saito, H.; Sugou, S.; Lee, J.-S. A Narrow Photoluminescence Linewidth of 21 MeV at 1.35 Mm from Strain-Reduced InAs Quantum Dots Covered by In0.2Ga0.8As Grown on GaAs Substrates. *Appl. Phys. Lett.* **1999**, *74*, 1111–1113. [CrossRef]
20. Kosogov, A.O.; Werner, P.; Gösele, U.; Ledentsov, N.N.; Bimberg, D.; Ustinov, V.M.; Egorov, A.Y.; Zhukov, A.E.; Kop'ev, P.S.; Bert, N.A.; et al. Structural and Optical Properties of InAs–GaAs Quantum Dots Subjected to High Temperature Annealing. *Appl. Phys. Lett.* **1996**, *69*, 3072–3074. [CrossRef]

*electronics*

MDPI

*Article*

# Development of Quantum Protocol Modification CSLOE–2022, Increasing the Cryptographic Strength of Classical Quantum Protocol BB84

**Larissa V. Cherckesova [1], Olga A. Safaryan [1,*], Alexey N. Beskopylny [2,*] and Elena Revyakina [1]**

[1] Department of Cyber Security of Information Systems, Don State Technical University, 344003 Rostov-on-on, Russia
[2] Department of Transport Systems, Faculty of Roads and Transport Systems, Don State Technical University, 344003 Rostov-on-Don, Russia
* Correspondence: safari_2006@mail.ru (O.A.S.); besk-an@yandex.ru (A.N.B.); Tel.: +7-(863)-238-15-18 (O.A.S.); +7-(863)-273-84-54 (A.N.B.)

**Abstract:** Quantum cryptography protocols make it possible not only to ensure the protection of data transmitted in a communication channel from unauthorized access by intruders, but also to detect the existence of any attempted interception. This scientific direction is currently relevant, since it is related to the problem of security and data protection in current information and communication networks. The article is devoted to quantum cryptography; it describes the development of quantum protocols as quantum key distribution systems. Grounded on the laws of quantum mechanics, the elaboration of modifications of secure data transfer protocols is shown. The authors considered the best-known protocol to be BB84 of quantum key distribution; a more modern modification of this protocol is BB84 Info-Z. Comparative analysis of these has also been carried out. It has been established that the BB84-Info-Z quantum protocol works more efficiently than BB84 since its lower error threshold allows the interceptor to obtain much less information about the secret key. The authors put forward a new idea to improve the BB84 protocol (which has been quite outdated for almost 40 years), due to the increase in modern requirements for quantum cryptography protocols. The modification is called CSLOE-2022. It enables significant intensification of cryptographic strength and the entanglement degree of the interceptor (cryptanalyst), which greatly complicates the very possibility of intercepting information. The ultimate goal of the CSLOE-2022 modification is to complicate the eavesdropping process so much that it can be considered completely useless for an attacker in terms of wasting time and resources. The modification allows exceeding the known speed limit of key generation without repeaters since it uses two sources, the phases of which, in addition to the hundreds of kilometers of fiber between them, are very difficult to stabilize. Comparison of the protocols by working distance showed that for BB84, this distance does not exceed 70 km; for BB84-Info-Z it is similar, at no more than 70 km, and the modification of CSLOE-2022 proposed by the authors theoretically allows increasing the working distance of the quantum protocol to 511 km (7.3 times).

**Keywords:** quantum protocol; quantum cryptography; quantum key distribution; error threshold; modification of the quantum protocol BB84

**Citation:** Cherckesova, L.V.; Safaryan, O.A.; Beskopylny, A.N.; Revyakina, E. Development of Quantum Protocol Modification CSLOE–2022, Increasing the Cryptographic Strength of Classical Quantum Protocol BB84. *Electronics* **2022**, *11*, 3954. https://doi.org/10.3390/electronics11233954

Academic Editor: Lucas Lamata

Received: 22 October 2022
Accepted: 25 November 2022
Published: 29 November 2022



## 1. Introduction

The article is devoted to quantum cryptography. It particularly concerns quantum cryptographic protocols, which scientists have been investigating for almost forty years. Quantum cryptography is considered technology capable of adding to the new, unique look of telecommunication networks of the future.

However, no one can predict with full confidence what the formed quantum infrastructure will look like, and to what outcome it may lead. Today, quantum cryptography allows us not only to provide increased protection against unauthorized access to transmitted information, but also to reveal the very existence of such attempt with higher probability. In modern information-based society, this aspect is extremely important, since humanity is striving to achieve, in fact, complete digitalization.

In parallel with this process, the global modernization of technological equipment, computer software and hardware is rapidly taking place; load on telecommunications networks is increasing. Volumes of secret data, confidential information, and personal data of users are increasing. At the same time, the number of computer security incidents is inevitably increasing. Attackers are modernizing the types of threats and ways to implement them, causing serious problems with information security and data protection.

Maintaining the reliability and operability of technical equipment in communication networks is becoming an increasingly difficult task. Recent cyberattacks in various countries of the world have partially paralyzed the public activity of citizens and caused serious problems in government apparatus. Many countries face similar incidents of computer security every day, and in some cases, users become practically helpless. National governments are extremely concerned about this situation, which threatens their critical infrastructure.

The moment is not far off when the process of modernization of the available means of information protection will come to a standstill, and then the transition to safer quantum technologies will become the ubiquitous inevitability.

Therefore, many countries of the world are optimistic for quantum cryptography technologies. In this regard, the process of transition to quantum technologies seems inevitable, and the attitude to scientific research in this area has become much more serious.

The article proposed by the authors is aimed at analyzing the foundations of quantum cryptography protocols, comparing their cryptographic strength, and the possibilities for their modification. Two well–known quantum protocols are taken as examples: the outdated, but still competitive BB84, and its modification BB84 Info-Z, which has its own distinctive features. Understanding the grounds and principles of quantum protocol functioning is becoming a necessity for many information security specialists. The priority directions of the development of quantum key distribution systems are demonstrated, and the construction of safe and secure data transmission protocols grounded on quantum mechanics laws is shown.

This article puts forward new assumptions about the possibility of universal improvement of quantum protocols on the example of modification of the quantum protocol BB84. The application of the idea put forward by the authors is based on the theory of man-made reconstruction of photons and their further use, which allows significantly reducing the danger of potential threats, and in some cases, to avoid situations related to vulnerabilities of quantum protocols. This method is combined; together with the application of false photon states and the creation of many traps, in order to increase the computing resources spent on cyberattacks by hackers, in the end, it nullifies this attack, making it useless. An attacker, or "eavesdropper", will find a much smaller opportunity to guess the desired data transfer qubit, as well as the photon polarization state. The simulated copies, in addition to their main task, will become good bait for hackers, capable of driving him into the trap, thereby confirming his presence in the communication channel. In this case, the process of disconnecting the communication channel will not be mandatory.

## 2. Protocol BB84

This quantum protocol is named after the first letters of the surnames of its creators and the year of its publication [1]. The BB84 protocol is designed to transmit secret information encoded in binary. Figure 1 shows how, in the BB84 protocol, one bit can be encoded in the polarization state of photon. Binary 0 is defined as the polarization of $0^\circ$ in the rectilinear coordinate basis, or $45^\circ$ on a diagonal basis. Binary 1 can also be equal to $90^\circ$ on a rectilinear coordinate basis, or $135^\circ$ on a diagonal basis. Thereby, one bit can be represented by the polarization of a photon in one of two coordinate bases.

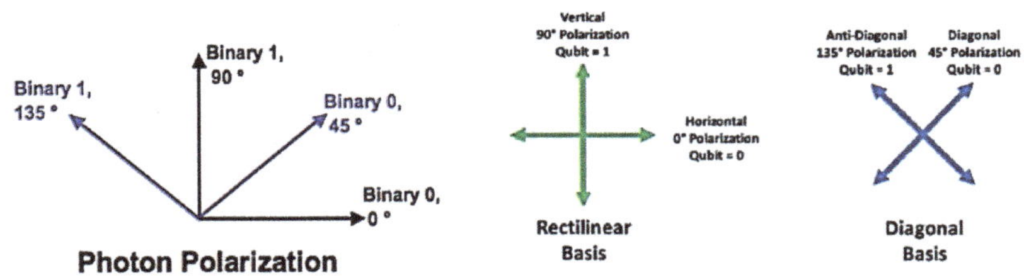


**Figure 1.** Encoding of the bit in protocol BB84.

Currently, the participants in the transmission of the data process, Alice and Bob, agree to approve some big number n, the threshold of error pa and the linear correction code of error C, with the parity check matrix PC of the order r × n.

The linear key generation function (to enhance confidentiality), represented by $P_K$ matrix of order $m \times n$, is also agreed upon. Both matrices can be known in advance or can be determined during the execution of the protocol, and then can be sent via the classical channel. In turn, the matrix $(r + m) \times n$, the rows of which are the rows $P_C$ and $P_K$, if taken together, must have the rank $r + m$. Alice randomly selects the sequences of bits: strings $i$ from $2n$ – bit, $b \in F_2{}^{2n}$, where $F_2$ defines the field of 2 elements {0; 1}, that is, field of integer numbers modulo 2. Then, the state $\left|i^b\right\rangle = \left|i_1^{b_1}\right\rangle \ldots \left|i_{2n}^{b_{2n}}\right\rangle$ is encoded. For each bit, the coordinate basis is randomly selected, rectilinear or diagonal, with which the bit will be encoded. When transmitting the photon from Alice, Bob will inform her about the photon receiving, but will not measure it.

For each photon that Bob receives, he will measure the polarization of the photon on the randomly selected coordinate basis, applying it to his own state. If Bob chose the same coordinate basis of his state for particular photon as Alice did, when he performed the transformation $H^b = H^{b_1} \otimes \ldots \otimes H^{b_{2n}}$, he switches to the state $|i\rangle = |i_1 \ldots i_{2n}\rangle$. Bob should measure the same polarization in the line $i^B$, and thus output the bit that Alice sent correctly, in the case of absence of noise and signs of eavesdropping on the communication line from the side of an attacker.

To detect eavesdropping, Alice will randomly select $n$ bits that will be used to detect the presence of attackers. By selecting a $2n$–bit string containing $n$ units exactly, Alice must ensure that equality $|s| = n$ was fulfilled. Alice sends to Bob $s$ bits publicly, such, that $s_j = 0$, which are applicated for testing, and the rest of the bits are employed for generating of the final key. Let us denote the corresponding substrings, appropriate for testing process values $i$ and $b$, while the substrings appropriate for the generation of a key will be defined as $i_s$ and $b_s$. For every $j \in [1 \ldots 2n]$, such, that the value $s_j = 0$, the participants of the data transfer process Alice and Bob publish the value of the bit with $j$th number.

If Bob chose the incorrect basis of coordinates, then his outcome and, consequently, the bit that he received will be random. It is worth noting that if the mismatch of the bits determined in the published values of the $j$th bit by comparison exceeds the $np_a$, then they interrupt the protocol execution. The preliminarily fixed parameter $p_a$ of the protocol, in fact, is the relationship of permitted bit flips, intended for the testing process.

Alice and Bob save the values of the residue $n$ bits in the strict confidence. The string of Alice is designated as $x = i_s$ and is called the *information string*. The appropriate string of bits on the side Bob is determined as $x^B$.

At the second stage, Bob must inform Alice, via any unprotected channel of communication, what coordinate basis he has used to measure every photon. Then, Alice informs Bob if he chose the proper coordinate basis for each photon. At this stage, Alice and Bob throw out the bits appropriating the photons that Bob has measured with another coordinate basis. On the condition that no errors occurred, or no one managed the photons, Alice and Bob now should possess the same string of bits, which is named the *sifted key*.

The example demonstrated below (Figure 2) illustrates the bits that Alice selected; the coordinate bases in which she encoded her bits; and the coordinate bases that Bob applicated for his dimension. In addition, Figure 2 shows the resulting *sifted key*, after the stage when Bob and Alice rejected their bits, as indicated above. However, Alice and Bob, before they finish, have to approve a subset of random bits for the comparison of the provision of the consistency of their actions. For verification, Alice sends Bob an $r$–bit string of error correction. Bob will use $\xi$ to compare or rectify his string $x^B$ as necessary. String $\zeta = xP_C^T$ is named the *syndrome* of *string x* (relative to $P_C$). If bits correspond, then they are rejected, and other bits create the common secret $m$–bit key.

| Alice's bit | 0 | 1 | 1 | 0 | 1 | 0 | 0 | 1 |
|---|---|---|---|---|---|---|---|---|
| Alice's basis | + | + | X | + | X | X | X | + |
| Alice's polarization | ↑ | → | ↖ | ↑ | ↖ | ↗ | ↗ | → |
| Bob's basis | + | X | X | X | + | X | + | + |
| Bob's measurement | ↑ | ↗ | ↖ | ↗ | → | ↗ | → | → |
| Public discussion | | | | | | | | |
| Shared Secret key | 0 | | 1 | | | 0 | | 1 |

**Figure 2.** Sifted key.

If there is no noise and there are no other dimension errors, a mismatch in any of the compared bits would specify the availability of an *interceptor* presenting in the quantum channel. The reason for this is that if attacker Eve was trying to define the key value, then she would not have any other choice but to measure the photons dispatched by Alice before transmitting these photons to Bob.

Therefore, Eve should attach the separate quantum probe (at quantum sensing [2–6]), which, as she assumed, is in a pure state, and apply the unitary transformation $U_j$ to the composite system. This is true because the theorem on prohibition of cloning (anti–cloning theorem) [7] guarantees that it is impracticable to reproduce a particle of an unknown state.

In view of the fact that Eve will not know which coordinate bases Alice have applicated to encode the bit while she discussed her measurements with Bob, she will be forced to guess. If she measures the states of photons in the wrong coordinate bases, then, as the uncertainty principle of Heisenberg guarantees, the information that was encoded in another coordinate basis will be lost.

Therefore, when a photon reaches Bob, his dimension will be random, and he will receive incorrect information in 50% of cases. Taking into account that the interceptor Eve will choose the basis of measurement incorrectly in 50% cases, on average, then 25% the bits measured by Bob will discern from the bits of Alice. If interceptor Eve captures all possible bits, then, after comparisons of $n$–bit by Alice and Bob, they would decrease the possibility that attacker Eve can remain unnoticed by a value of $\frac{3}{4}n$.

Therefore, the probability that the interceptor–attacker Eve has investigated the secret is insignificant if relatively long bit sequences are compared and identified.

### 3. Protocol BB84–Info–Z

Modification of BB84 and the considered quantum protocol BB84–Info–Z, is analogous to BB84 [8], except that it applies the following differences:

- Generalized numbers of bits $n$, $n_z$ and $n_x$ (n–the numbers of informational bits, where $Z$ and $X$ are the test bits, accordingly);
- Section $P = (s, z, b)$ to divide n-bit string $i$ into three non–overlapping sets ($I$, $T_Z$ $T_X$);
- Two special thresholds, which are separate ($p_{A,z}$ and $p_{A,x}$) in place of one threshold ($p_A$).

Before starting the quantum protocol functioning, Alice and Bob must select some general parameters or public parameters:

- numbers denoted as $n$, $n_Z$ and $n_X$ (specified as the relationship $N = n + n_Z + n_X$);
- thresholds of errors denoted as $P_{A,Z}$ and $P_{A,X}$, $r \times n$ (which correspond to the linear code of error correction C);
- matrices of privacy enhancement $m \times n$ (representing the linear key of generation function).

It is important that all rows $R + M$ of the previously considered $P_C$ and $P_K$ matrices are assembled as linearly independent.

Alice randomly selects the section $\mathcal{P} = (s, z, b)$ of the $N$–bit strings, randomly choosing $N$–bit strings denoted as, $z$, $b \in F_2^N$, which satisfy the conditions:

$$\mid s \mid = n; \mid z \mid = n_z; \mid B \mid = n_X; \text{and} \mid s + z + b \mid = N.$$

Thereby, the section $\mathcal{P}$ splits the set of indexes denoted as $\{1, 2, \ldots, N\}$ into three non-overlapping and disjointed sets:

- I (information bits, where $s_j = 1$) size $n$;
- $T_Z$ (test bits Z, where $z_j = 1$) with size $n_z$;
- $T_X$ (test bits X, where $b_j = 1$) with size $n_x$.

Alice selects $N$–bit strings randomly, where $i \in F_2^N$, and performs the dispatch of $N$ qubits $\left|i_2^{b_1}\right\rangle, \left|i_2^{b_2}\right\rangle, \ldots, \left|i_N^{b_N}\right\rangle$, one by one, through the quantum information channel. At the same time, Alice uses coordinate basis Z to send the information of test Z–bits, as well as X coordinate basis to send X–bits. Initially, Bob stores every qubit he has received in the quantum memory, without measuring this qubits.

Next, Alice dispatches the string of bits $b = b_1 \ldots b_N$ via the classical channel to Bob. Bob gauges and measures every qubit that he has received and saved. When measuring $i$th qubit, Bob measures it in the Z–coordinate basis if $b_i = 0$, and measures it in the X–coordinate basis if $b_i = 1$. This string of bits that Bob measured is designated as $i^B$. If noise and eavesdropping are absent, then the bit string is equal to $i^B = i$.

After that, Alice sends Bob the string of bits, designated as $s$. The information bits (which will be applied for the final key generating) are $n$ bits $s_j = 1$, while test Z and X bits (which will be applied for testing) are $n_Z + n_X$ with $s_j = 0$. Substrings are denoted by $i$ and $b$, and correspond to information bits $i_S$ and $b_S$, accordingly.

Next, Alice and Bob publish together the values of bits, which they obtain for all test bits Z and X, and then the bit values are compared. If, for Alice and Bob, more than $n_Z \cdot p_{a,Z}$ test Z–bits are, or more than $n_X \cdot p_{a,X}$ test X–bits between them are different, then they interrupt this protocol, where $p_{a,Z}$ and $p_{a,X}$ are preliminarily coordinated thresholds of errors. Alice and Bob keep values of residual $n$ bits (information bits where $s_j = 1$) a strict secret. The bit chain of Alice is designated as $x = i_s$, and the bit string of Bob is denoted as $x^B$.

Then, the syndrome $x$ should be dispatched from Alice to Bob (regarding code C correction of errors and its check on parity of the $P_C$ matrix), which includes $r$ bits and is determined as $\xi = xP_C^T$. Using the value $\xi$, Bob rectifies the errors in his string of bits $x^B$ (it is similarly $x$). The final key is $m$–bit sequence and is determined as $k = xP_K^T$. Alice and Bob, together, calculate it. It is obvious that the protocols are very similar.

Let us consider their security against collective attacks [9,10], which are one of the most powerful theoretical cyberattacks.

## 4. Description of Cyberattack of Eve and the Properties of It

To every $j$ qubit $\left|i_j^{b_j}\right\rangle_{T_j}$ shipped by Alice ($1 \leq j \leq N$), Eve attaches the separate quantum probe (at quantum sensing), which, as it assumed, is in pure state, and applies the unitary transformation $U_j$ to the composite system. Then, she stores her quantum probes in the quantum memory for the subsequent measurements, and dispatches to Bob his part of the system [11].

Therefore, for every qubit there is a certain trial Hilbert space and a certain unitary transformation $U_j$; they are determined by Eve in advance and, thus, are corrected and fixed for all feasible variants and options of $i$, $b$ and $s$.

*4.1. Cyberattack of Eve on the Separate Qubit*

Because the cyberattack is bitwise, it is possible to focus the analysis on some selected fixed qubit, temporarily discarding subindex $j$ and expressing the general impact of the actions of Eve on the concrete qubit relative to the coordinate basis $\left|0^b\right\rangle$, $\left|1^b\right\rangle$

$$U\left|0^E\right\rangle\left|0^b\right\rangle = \left|E_{00}^b\right\rangle\left|0^b\right\rangle + \left|E_{01}^b\right\rangle\left|1^b\right\rangle = \left|\phi_0^b\right\rangle. \tag{1}$$

$$U\left|0^E\right\rangle\left|1^b\right\rangle = \left|E_{10}^b\right\rangle\left|0^b\right\rangle + \left|E_{11}^b\right\rangle\left|1^b\right\rangle = \left|\phi_1^b\right\rangle. \tag{2}$$

where $\left|E_{00}^b\right\rangle, \left|E_{01}^b\right\rangle, \left|E_{10}^b\right\rangle$ and $\left|E_{11}^b\right\rangle$ represent the vectors, or non–normalized states, in the trial Hilbert space of Eve, respective to this concrete qubit. Because the transformation $U$ is unitary, then $\left|\phi_0^b\right\rangle$ and $\left|\phi_1^b\right\rangle$ have norm 1 and are orthogonal. This means that

$$\left\langle E_{00}^b \middle| E_{00}^b\right\rangle + \left\langle E_{01}^b \middle| E_{01}^b\right\rangle = 1. \tag{3}$$

$$\left\langle E_{10}^b \middle| E_{10}^b\right\rangle + \left\langle E_{11}^b \middle| E_{11}^b\right\rangle = 1. \tag{4}$$

$$\left\langle E_{00}^b \middle| E_{10}^b\right\rangle + \left\langle E_{01}^b \middle| E_{11}^b\right\rangle = 0 \quad \left\langle E_{10}^b \middle| E_{00}^b\right\rangle + \left\langle E_{11}^b \middle| E_{01}^b\right\rangle = 0. \tag{5}$$

*4.2. Spreading the Cyberattack to the Several Qubits–Collective Cyberattack*

For every qubit $\in [1 \ldots 2n]$ Eve uses the transformation $Uj$ in the space $\mathcal{H}_j^E \otimes \mathcal{H}_2$, where $\mathcal{H}_j^E$ is Eve's trial space, $\mathcal{H}_2$– is the space of qubits. The coordinate basis bj, expressed relative to Eve's point of view, is obtained by tracking Bob from the $\left|\phi_0^{b_j}\right\rangle_j$ and $\left|\phi_1^{b_j}\right\rangle_j$, resulting in the following density matrices:

$$\left(\rho_0^{b_j}\right)_j = \left|E_{00}^{b_j}\right\rangle_j\left\langle E_{00}^{b_j}\right| + \left|E_{01}^{b_j}\right\rangle_j\left\langle E_{01}^{b_j}\right| \tag{6}$$

$$\left(\rho_1^{b_j}\right)_j = \left|E_{10}^{b_j}\right\rangle_j\left\langle E_{10}^{b_j}\right| + \left|E_{11}^{b_j}\right\rangle_j\left\langle E_{11}^{b_j}\right|. \tag{7}$$

If Alice dispatches the string $i$ applying the coordinate basis $b$, then the global state of Eve is the tensor product of these states $\left(\rho_{i_j}^{b_j}\right)_j$. After revealing the test bits [12], Eve requires only those values $\left(\rho_{i_j}^{b_j}\right)_j$, or which $s_j = 1$. The set $\{j \mid s_j = 1\}$ has $n$ elements and is global. The global corresponding values $s$, $b$, $x$ can now be saved as:

$$\rho_x^{b_s} = \left(\rho_{i_{j_1}}^{b_{j_1}}\right)_{j_1} \otimes \ldots \otimes \left(\rho_{i_{j_n}}^{b_{j_n}}\right)_{j_n} = \otimes_{l=1}^{n}\left(\rho_{i_{j_l}}^{b_l}\right)_{j_l} \tag{8}$$

*4.3. Probability of Errors*

Supposing that the qubit is under cyberattack by transformation $U$, as determined in (1) and (2), then an error emerges if Alice dispatches 0 and Bob measures 1, or if Alice dispatches 1 and Bob measures 0. Let $k$ be the value Bob has measured, $i$ the value dispatched by Alice for the concrete qubit, and $b$ the coordinate basis applied by Alice for encoding $i$.

Then, the possibility of Bob measuring the error is expressed as:

$$p_\epsilon^b \triangleq \frac{1}{2}\left[\left\langle E_{01}^b | E_{01}^b \right\rangle + \left\langle E_{10}^b | E_{10}^b \right\rangle\right]. \quad (9)$$

*4.4. Probability of Errors Occurrence in the Conjugate Basis*

Now we study the expression $p_e^{\overline{b}}$, where $\overline{b} = 1 - b$ (when $\overline{0} = 1, \overline{1} = 0$), which responds to the basis, conjugating to the basis that was given and specified as $b$.

Cyberattack $U$ is described usually by formulas (1) and (2), in the coordinate basis $b$.

To compute the probability of error when Alice encodes $i_j$ as $\left|i_j^{\overline{b}}\right\rangle$ instead $\left|i_j^b\right\rangle$, it is necessary to represent $U$ in the coordinate basis $\overline{b}$. From the Equation (10), it is known that in this circumstance, the probability of errors is determined by the expression:

$$p_\epsilon^{\overline{b}} = \frac{1}{2}\left[\left\langle E_{01}^{\overline{b}} \middle| E_{01}^{\overline{b}} \right\rangle + \left\langle E_{10}^{\overline{b}} \middle| E_{10}^{\overline{b}} \right\rangle\right] \quad (10)$$

Using the fact that

$$|0\rangle^{\overline{b}} = \frac{1}{\sqrt{2}}\left[\left|0^b\right\rangle + \left|1^b\right\rangle\right],\ |1\rangle^{\overline{b}} = \frac{1}{\sqrt{2}}\left[\left|0^b\right\rangle - \left|1^b\right\rangle\right]$$

and using linearity $U$, expressions are derived from (1) and (2):

$$U\left|0^E\right\rangle\left|0^{\overline{b}}\right\rangle = \frac{1}{\sqrt{2}}\left(\left|E_{00}^b\right\rangle + \left|E_{10}^b\right\rangle\right)\left|0^b\right\rangle + \frac{1}{\sqrt{2}}\left(\left|E_{01}^b\right\rangle + \left|E_{11}^b\right\rangle\right)\left|1^b\right\rangle \quad (11)$$

$$U\left|0^E\right\rangle\left|1^{\overline{b}}\right\rangle = \frac{1}{\sqrt{2}}\left(\left|E_{00}^b\right\rangle - \left|E_{10}^b\right\rangle\right)\left|0^b\right\rangle + \frac{1}{\sqrt{2}}\left(\left|E_{01}^b\right\rangle - \left|E_{11}^b\right\rangle\right)\left|1^b\right\rangle \quad (12)$$

Replacing $\left|0^b\right\rangle$ and $\left|1^b\right\rangle$ in the right parts of expressions on their magnitudes from the positions $\left|0^{\overline{b}}\right\rangle$ and $\left|1^{\overline{b}}\right\rangle$ I, i.e., $\left|0^b\right\rangle = \frac{1}{\sqrt{2}}\left[\left|0^{\overline{b}}\right\rangle + \left|1^{\overline{b}}\right\rangle\right]$ and $\left|1^b\right\rangle = \frac{1}{\sqrt{2}}\left[\left|0^{\overline{b}}\right\rangle + \left|1^{\overline{b}}\right\rangle\right]$ I, we receive

$$\begin{aligned} U\left|0^E\right\rangle\left|0^{\overline{b}}\right\rangle &= \frac{1}{2}\left[\left|E_{00}^b\right\rangle + \left|E_{10}^b\right\rangle + \left|E_{01}^b\right\rangle + \left|E_{11}^b\right\rangle\right]\left|0^{\overline{b}}\right\rangle \\ &+ \frac{1}{2}\left[\left(\left|E_{00}^b\right\rangle - \left|E_{11}^b\right\rangle\right) + \left(\left|E_{10}^b\right\rangle - \left|E_{01}^b\right\rangle\right)\right]\left|1^{\overline{b}}\right\rangle \end{aligned} \quad (13)$$

$$\begin{aligned} U\left|0^E\right\rangle\left|1^{\overline{b}}\right\rangle &= \frac{1}{2}\left[\left|E_{00}^b\right\rangle - \left|E_{11}^b\right\rangle - \left|E_{10}^b\right\rangle - \left|E_{01}^b\right\rangle\right]\left|0^{\overline{b}}\right\rangle \\ &+ \frac{1}{2}\left[\left(\left|E_{00}^b\right\rangle - \left|E_{11}^b\right\rangle\right) + \left(\left|E_{10}^b\right\rangle - \left|E_{01}^b\right\rangle\right)\right]\left|1^{\overline{b}}\right\rangle \end{aligned} \quad (14)$$

where components for $\left|E_{01}^{\overline{b}}\right\rangle$ and $\left|E_{10}^{\overline{b}}\right\rangle$ are enclosed in parentheses, we can easily see that:

$$p_\epsilon^{\overline{b}} = \frac{1}{2}\left[\left\langle E_{01}^{\overline{b}} \mid E_{01}^{\overline{b}} \right\rangle + \left\langle E_{10}^{\overline{b}} \mid E_{10}^{\overline{b}} \right\rangle\right] = \frac{1}{4}\left[\left(\langle E_{00}^b| - \langle E_{11}^b|\right)\left(\left|E_{00}^b\right\rangle - \left|E_{11}^b\right\rangle\right) + \left(\left\langle E_{10}^b\right| - \left\langle E_{01}^b\right|\right)\left(\left|E_{10}^b\right\rangle - \left|E_{01}^b\right\rangle\right)\right]$$

Let us distribute this result applying Equalities $\langle\phi|\psi\rangle = \left\langle\overline{\psi|\phi}\right\rangle$ and $\mathcal{Z} + \overline{\mathcal{Z}} = 2Re(\mathcal{Z})$ for $\mathcal{Z} \in C$ (where the complex conjugation is denoted above the line [13]). Using the Equalities (3) and (4), we obtain the following expressions:

$$\begin{aligned} p_e^{\overline{b}} &= \frac{1}{4}\left[2 - \left\langle E_{00}^b \mid E_{11}^b \right\rangle - \left\langle E_{11}^b \mid E_{00}^b \right\rangle - \left\langle E_{01}^b \mid E_{10}^b \right\rangle - \left\langle E_{10}^b \mid E_{01}^b \right\rangle\right] \\ p_e^{\overline{b}} &= \frac{1}{2}\left[1 - \mathrm{Re}\left(\left\langle E_{00}^b \mid E_{11}^b \right\rangle + \left\langle E_{01}^b \mid E_{10}^b \right\rangle\right)\right] \end{aligned} \quad (15)$$

This expression will be applied to relate the perturbation caused by Eve while Alice is encoding $\mathrm{i_j}$ bits in the coordinate basis $\overline{\mathrm{b_j}}$, so $\mathrm{s_j} = 1$, with the information which Eve obtains, while Alice will encode it in this coordinate basis.

According to the principle of «Information against perturbation» [14], the more information Eve receives while the encoding is carried out in the coordinate basis b, the more interference it causes while the bits are encoding and checking in conjugate coordinate basis. Therefore, it is possible to limit Eve's knowledge about the key by limiting the allowable rate (frequency) of errors in this quantum protocol.

## 5. Security Confirmation of Classical Protocol BB84 against the Collective Cyberattacks

*Proof of Security*

Let us choose such code $\frac{d_{r,m}}{2n} > p_a + \epsilon$, or some small value $\epsilon$; then the expression $2m\sqrt{P\left[\left(\frac{|C_I|}{n} \geq \frac{d_{r,m}}{2n}\right) \wedge \left(\frac{|C_T|}{n} \leq p_a\right)\right]}$ will be less than value $P\left[\left(\frac{|C_I|}{n} > p_a + \epsilon\right) \wedge \left(\frac{|C_T|}{n} \leq p_a\right)\right]$, which by itself is exponentially small in n.

We can apply Heffding's selection from [8] (the theorem) for every specific row $c_1 \ldots c_{2n}$, appropriate to all qubit measurements in some valid coordinate basis $b$. Let $\overline{X} = \frac{|C_I|}{n}$ be the average value of selection respective to incorrect bits of information;

$\mu = \frac{|C_I|+|C_T|}{2n}$–is the mathematical expectation $\overline{X}$, which is equal to the expression $2\mu - \overline{X} \leq p_a$, or is identical to inequality $\overline{X} - \mu \geq \mu - p_a$.

To the rows $\left(\frac{|C_I|}{n} > p_a + \epsilon\right)$ and $\left(\frac{|C_T|}{n} \leq p_a\right)$ we will rewrite the conditions as:

$$(\overline{X} - \mu > \epsilon + p_a - \mu) \wedge (\overline{X} - \mu \geq \mu - p_a) \tag{16}$$

whence it follows that using Heffding's theorem [15], the relation is obtained:

$$P\left[\left(\frac{|C_I|}{n} > p_a + \epsilon\right) \wedge \left(\frac{|C_T|}{n} \leq p_a\right)\right] \leq P\left[\overline{X} - \mu > \frac{\epsilon}{2}\right] \leq e^{-\frac{1}{2}n\epsilon^2} \tag{17}$$

It is necessary to make sure that the rate of errors in the bits of information is less than the maximum speed at which the error correction code can process.

This condition is necessary for the key to be reliable.

## 6. Security Proof for the Protocol BB84–Info–Z against the Collective Cyberattacks

### 6.1. General Collective Cyberattack of Eve

Let us assume that, before executing the quantum key distribution (QKD) protocol, Eve chooses to carry out a collective cyberattack [8]. Let the $j$th qubit be given, sent by Alice to Bob. Eve attaches the quantum probe state (at quantum sensing) and applies some unitary operator $U_j$ to the composite system. Then, Eve holds in her quantum memory subsystem $E_j$, which is the state of her quantum probe; next, she dispatches to Bob the subsystem $T_j$. This subsystem is the qubit dispatched from Alice to Bob (it may be modified by Eve's cyberattack $U_j$).

The biggest common collective cyberattack $U_j$ of Eve is directed on the $j$th qubit, presented by orthonormal coordinate basis. The cyberattack described as:

$$U_j\left|0^E\right\rangle E_j\left|0^{b_j}\right\rangle T_j = \left|E_{00}^{b_j}\right\rangle E_j\left|0^{b_j}\right\rangle T_j + \left|E_{01}^{b_j}\right\rangle E_j\left|1^{b_j}\right\rangle T_j \tag{18}$$

$$U_j\left|0^E\right\rangle E_j\left|0^{b_j}\right\rangle T_j = \left|E_{10}^{b_j}\right\rangle E_j\left|0^{b_j}\right\rangle T_j + \left|E_{11}^{b_j}\right\rangle E_j\left|1^{b_j}\right\rangle T_j \tag{19}$$

where $\left|E_{00}^{b_j}\right\rangle E_j$, $\left|E_{01}^{b_j}\right\rangle E_j$, $\left|E_{10}^{b_j}\right\rangle E_j$, and $\left|E_{11}^{b_j}\right\rangle E_j$ are non–normalized states in Eve's quantum probe of system $E_j$, that was fastened and fixed to $j$th qubit.

So, it can be noticed that the quantum probe state can change the initial state of the composite system product, $|0^E\rangle E_j \left|i_j^{b_j}\right\rangle T_j$, to an entangled state. This implies that Eve's cyberattack can cause the entangling of her quantum probe with Bob's probe (at quantum sensing).

Firstly, this can clarify the situation, and bring some kind of information about the state of Bob; secondly, it is the reason for the disturbance, and can be discovered by him. The information received the day before and the disturbance that Eve caused, by their nature, are interrelated with their relationship—this is the main reason why the QKD protocols are secure and safe.

### 6.2. Proof of Security

As was mentioned earlier [8], the random variable $\widetilde{C}_i$ matches the errors in the string of bits in information bits if they were encoded in the coordinate basis X. Bits of TEST–X are also encoded in basis X. The $C_{TX}$ random variable matches the string of bits of the errors, on these bits. Hence, it is possible to consider the choice of $n$–bits indexes of information (Info) and $n_x$–bits TEST–X as a random selection (after $n$, $n_z$ and $n_x$ numbers; and bit indexes TEST–Z, which was selected already) and to apply the theorem of Heffding [9].

Hence, for every string of bits $c_1 \ldots c_{n+nx}$, which consists of errors in the bits $n + n_x$ Info and TEST–X, if Info bits were coded in the coordinate basis X, then we can use Heffding's theorem: let us take the sample with size $n$ without changing from combination $c_1, \ldots, c_{n+nx}$. In the above discussion [8], the following theorem is actually proven:

**Theorem.** *Let the values $\delta > 0$ and $R > 0$; for infinite number of values n, the vectors family is given, which is linearly independent $\{v_1^n, \ldots, v_{r_n+m_n}^n\}$, such, that $\delta < \frac{d_{r_n,m_n}}{n}$, $\frac{m_n}{n} \leq R$. Hence, for anyone $p_{a,Z}, p_{a,X} > 0$, such, that $p_{a,x} + \epsilon_{sec} \leq \frac{\delta}{2}$, and for any $n$, $n_z$, $n_x > 0$ and two final keys k, k′ that are the $m_n$–bit keys, the distance between the states of Eve, appropriate to k and k′, meets the following requirements and boundaries:*

$$\langle \Delta_{Eve}^{(p_{a,z},p_{a,x})}(k,k') \rangle \leq 2Rne^{-(\frac{n_x}{n+n_x})^2 n\epsilon_{sec}^2}. \tag{20}$$

### 6.3. Reliability

By itself, the security is insufficient; it is also necessary that the key be reliable. More specifically, it must be identical for Alice and Bob [16]. It assumes it a need to ensure that quantity of errors in the Info bits was less than the maximum error number, which can be rectified by the error correction code. To do this, it is necessary that the code for error correction can really correct the errors. Consequently, the final key reliability, having exponentially low failure probability, is provided by the inequality below:

$$P\left[\left(\frac{|C_I|}{n} > p_{a,z} + \epsilon_{rel}\right) \wedge \left(\frac{|C_{T_z}|}{n_z} \leq p_{a,z}\right)\right] \leq e^{-2(\frac{n_z}{n+n_z})^2 n\epsilon_{rel}^2} \tag{21}$$

The choice of the Info bits indexes, and the bits of TEST–Z, is random separation of $n + n_Z$ bits into two subsets, with sizes $n$ and $n_Z$ (provided that the bits indexes of TEST–X was already selected). Thus, it matches Heffding's sample.

## 7. Protocol CSLOE–2022 (BB84–CSLOE–2022)

It is recommended to become acquainted with the new modification proposed by the authors and named CSLOE–2022, for the old but still effective quantum protocol using the quantum distribution of the key–BB84. For the protocol BB84, it is possible to modify significantly the cryptographic strength and degree of entanglement of the listener, which in their perspective will complicate the possibility of intercepting information from such an interceptor (or cryptanalyst) targeting confidential messages.

It is known that after coordination of the coordinate bases in the classic protocol BB84, an interceptor can receive accurate information about the transmitted state.

The final purpose of modification is to complicate eavesdropping process to the point of uselessness, in terms of spending time and resources, as well as to confirm the guesses about the real possibility of using such a method. The idea consists in the following: it is known that the process of replicating the quantum state, recorded as $\psi \rightarrow \psi \otimes \psi$, (cloning) can be performed perfectly, with probability of 1, then and only then if the coordinate basis to which $\psi$ refers is understood and known [17]. Otherwise, the ideal cloning is not possible, since the copies are not perfect. These are contents and consequences of the theorem prohibiting quantum information cloning. This circumstance will be useful.

If it is impossible to reproduce an exact clone of a photon, in order to obtain information from it, it is necessary to measure the characteristics of the original. The only way to measure the characteristics of the photon is to use a detector of single photons. However, as soon the photon reaches the detector, it transmits energy and disappears. That is, the measurement destroys the photon itself [9]. It is worth considering that each photon is unique [18]. However, it is possible to create some kind of photon (kind of similarity). It is known, thanks to quantum teleportation, that it is possible to obtain an exact copy of a photon [17], which, in turn, can be used also to construct such a similarity. For simplicity, we call it a «pseudo-photon». The interceptor will perceive such a pseudo-photon either as a real photon with its own specific set of characteristics, or as some distortion in the channel.

To detect and recognize such a clone in the communication channel, it is necessary to make considerable efforts. In the theory, at the first stage, as in the classic protocol BB84, Alice will communicate with Bob via the quantum communication channel. In turn, Alice will transmit to Bob the modified sequence consisting of cloned pseudo–photons and forming a dictionary (glossary) for each bit with corresponding polarization.

As mentioned earlier, the pseudo-photon will be a kind of photon created artificially [4]. In each concrete case, it is possible to form new sequences that generate the dynamic dictionary, thereby reducing the repetition during encryption.

Knowing that each photon is unique [18], one pseudo-photon cannot be used for each bit, but the whole group with a certain range of values that will be corrected, rectified, adjusted, shifted or expanded can [19].

Each bit, or their sequence, even if they are repeated, will have random pseudo-photons from a certain range of values that are attributed to the concrete bit or bit sequence. When sending the dictionary for decoding, it is possible to send it in parts, to maintain secrecy, and to avoid declassification.

If listening is detected at this first step, the dictionary can be expanded and the intercepted part discarded, as was described earlier [20], or data transmitting can be stopped and a new dictionary created. After successful transmission of the dictionary, it is possible to start sending encoded messages via the communication channel in which real photons, as well as their created copies, will alternate and have absolutely random positions in the sequence. It is worth noting that the protocol can be further complicated. For example, four quantum states can be used to encode bits in two coordinate bases, which corresponds to the quantum protocol BB84 (4 + 2) [21]. Further, such a protocol works according to the classical scenario, but using the dictionary–glossary.

Every time Bob receives a qubit, he reports it to Alice, but does not measure it. Subsequently, for each photon and pseudo-photon which Bob obtains, he will measure the value of polarization on the randomly selected coordinate basis, applying it to his state. If Bob has chosen the same coordinate basis of state for the particular photon, then when he performs transformation $H^b$, as Alice does, then he goes into the same state. Then, Bob must measure the same polarization in the string $i^B$. Hence, he can output properly the bit that Alice was going to dispatch in that case, if there is no noise orsigns of eavesdropping in the communication channel.

On the second step, Bob must tell Alice, using any unsecured communication channel, which coordinate basis he applied to quantify and to measure every photon. Then, Alice informs Bob which photons were real by sending the encrypted ranges, and informs him whether he has chosen the right coordinate basis for every original photon.

On this step, Alice and Bob reject the bits corresponding to pseudo-photons and the bits that Bob have quantified and measured in other coordinate bases. If there were no errors and nobody manipulated the photons, Bob and Alice must acquire the same bits string that is named as the sifted key (Table 1). The example below (Figure 3) demonstrates the bits which Alice selected [22]; the coordinate bases in which she encoded it; and bases that Bob used to measure them. In addition, the sifted key, obtained in the result, is demonstrated after Bob and Alice rejected their bits.

**Table 1.** Sifted key.

| | | | | | | | | |
|---|---|---|---|---|---|---|---|---|
| Bit of Alice | 0 | 1 | 1 | 0 | 1 | 0 | 0 | 1 |
| Basis of Alice | × | + | + | × | + | × | + | × |
| Polarization of Alice | / | ↑ | → | \ | → | \ | ↑ | / |
| Basis of Bob | × | × | × | + | + | × | + | + |
| Measurement of Bob | / | \ | / | → | → | \ | ↑ | → |
| Public discussion | × | + | + | * | + | × | + | * |
| Shared secret key | 0 | | | | 1 | 0 | 0 | |

↑, →, \, /—is the state of polarization of photon 1, 0 at angles $0^\circ$, $45^\circ$, $90^\circ$, $135^\circ$.

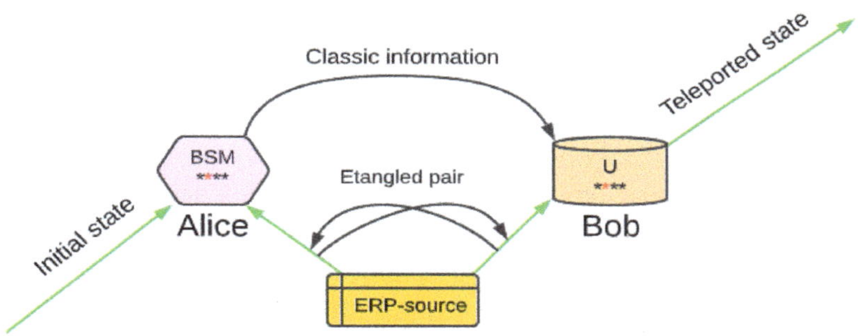


**Figure 3.** Results of the protocol CSLOE–2022 work.

The principle of imperfect replication is not new, and it is used often in telecommunications practice. In fact, information transmitted in optical fibers is encoded in the light state; thus, this process is quantum coding [23]. The information is strengthened several times on the way from source to receiver; hence, its quality must deteriorate. However, the telecommunication signal includes a large amount of photons being prepared in an identical quantum state. Strengthening in telecommunications boils down to creating several new copies of $\psi$ from $\psi \otimes N$.

That is, the theorem of prohibition of cloning is applied to the amplification of the telecommunication signals, due to the fact that spontaneous radiation is usually available in the amplifiers. Nevertheless, the copy is practically ideal, as the stimulated radiation is the dominant effect. In addition, the sensitivity of modern equipment is quite high, and at this stage is such that quantum limitation must be achieved in the nearest future. Thus, information encoding following the theorem of the prohibition of cloning can also be useful for quantum key distribution protocols [24].

The impossibility of exact and accurate copying of quantum information cannot negate the whole conception of the quantum information. On the contrary, it serves as a demonstration and illustration of power. It is impossible to copy the state of a certain quantum system for intelligent information coding completely using a set of states which are non–orthogonal.

Therefore, if a similar system reaches the receiver undisturbed, that proves that no opponent has copied it. This means that, due to the theorem of the prohibition of cloning, quantum information supports the tools to complete such tasks, which are impossible to solve applying only familiar information. For example, the detection of any eavesdropping device on the communication channel is possible only with the application of the ideas of quantum cryptography.

### 7.1. Cloning Methods

Let us consider the possible methods of non-ideal cloning of discrete quantum systems, which are performed in the protocol BB84.

There are several variants of such machines:

(1) Optimal symmetric universal quantum copying machine (UQCM) proposed by Vladimir Bužek (Buzhek, in various sources is spelled differently)–Mark Hillery (BH) in 1996;
(2) Symmetric universal quantum copying machine (UQCM) proposed by Nickolas Gisin–Serge Massar and their scientific group in 1997;
(3) Asymmetric universal quantum copying machine (copier)–UQCM.

The symmetric UQCM qubits cloning for $1 \to 2$, developed by Buzhek and Hillery (BH) accepts the cloning qubit as input, and uses the separate qubit as an auxiliary qubit [25]. The action of such a universal quantum–copying machine in the computational base of initial original qubit is described by the expression:

$$\begin{aligned} |0\rangle|\mathrm{R}\rangle|\mathrm{M}\rangle &\to \sqrt{\frac{2}{3}}|0\rangle|0\rangle|1\rangle - \sqrt{\frac{1}{3}}|\psi^{+}\rangle|0\rangle \\ (-|1\rangle)|E\rangle|M\rangle &\to \sqrt{\frac{2}{3}}|1\rangle|1\rangle|0\rangle - \sqrt{\frac{1}{6}}|\psi^{+}\rangle|1\rangle \end{aligned} \tag{22}$$

where $\psi^{+}\rangle = \frac{1}{\sqrt{2}}[|1\rangle|0\rangle + |0\rangle|1\rangle]$. By its linearity, these ratios cause the next action on the most common and general input state: $|\psi\rangle = \alpha|0\rangle + \beta|1\rangle$:

$$|\psi\rangle|R\rangle|M\rangle \to \sqrt{\frac{2}{3}}|\psi\rangle|\psi\rangle\left|\psi^{\perp}\right\rangle - \sqrt{\frac{1}{6}}\left[|\psi\rangle\left|\psi^{\perp}\right\rangle + \left|\psi^{\perp}\right\rangle|\psi\rangle\right]|\psi\rangle \tag{23}$$

where $\left|\psi^{\perp}\right\rangle = \alpha * |1\rangle - \beta * |0\rangle$.

As can be seen from the Equation (23), $\alpha$ and $\beta$ can swap their places. In addition, for all input states, this transformation has the similar shape $|\psi\rangle$.

Therefore, the quantum copying machine (QCM) is universal and symmetric. For the original and the copy, its partial states are expressed as:

$$\rho_A = \rho_B = \frac{5}{6}|\psi\rangle\langle\psi| + \frac{1}{6}\left|\psi^{\perp}\right\rangle\left\langle\psi^{\perp}\right| = \frac{1}{2}\left(1 + \frac{2}{3}\hat{m} + \vec{\sigma}\right) \tag{24}$$

The symmetric universal quantum copying machine for NM qubits was developed by Nickolas Gisin and Serge Massar in 1997. It generalizes the Buzhek–Hillery UQCM, and its correctness is determined by the following expression:

$$F_{N\to M} = \frac{MN + M + N}{M(N+2)} \; (d = 2) \tag{25}$$

which reproduces $F_{1\to 2} = \frac{5}{6}$ for $N = 1$ and $M = 2$.

Gisin and Massar provided numerical proofs of the optimality of their universal quantum copying machine. Later, Dagmar Bruss, Artur Eckert and Chiara Machiavelli gave analytical proof for this optimality in 1998. They suggested that the output state appertains to the $M$ qubits symmetric subspace. Further, Reinhard Werner generalized this outcome for the quantum systems of whatever dimension [26].

The universal and asymmetric cloning relate to such a state of affairs where the resulting clones may have varying accuracy in reproduction.

We should concentrate on universal cloning $1 \to 1 + 1$. Several researchers studied cases that are more general in 2005. Let us look at some of these ideas with their experimental implementation. In the detailed research of cloning $1 \to 1 + 1$, Chi–Sheng Niu and Robert Griffiths, in 1998, obtained, in particular, the optimal asymmetric universal quantum copying machine $1 \to 1 + 1$. Nickolas Cerf in 1998 and 2000 independently obtained the same result. He used the algebraic method, along with Vladimir Bužek and Mark Hillery in 1998 [27], who developed the method of quantum schemes, which was improved compared to the previous construction used for symmetric cloning.

Optimality is illustrated by confirming and proving the accuracy of reproduction of two clones, $F_A$ and $F_B$, that confirms the inequality of non–cloning:

$$\sqrt{(1-F_A)(1-F_B)} \geq \frac{1}{2} - (1-F_A) - (1-F_B) \tag{26}$$

The authors managed to extend the development of their schemes far beyond the individual case of qubits. Based on the above, to improve the cryptographic strength of protocol BB84, the universal quantum copying machine (UQCM) of Bužek–Hillery is best suited, as it is the simplest quantum copier to implement all the concepts discussed above.

### 7.2. Comparison of Protocols

Since the protocols BB84, BB84–Info–Z and the CSLOE–2022 modification proposed by the authors have a lot in common, several parameters can be distinguished for their comparison. Let us start by comparing the threshold of errors.

This parameter is necessary to determine whether there was eavesdropping. In the practical implementation of the protocol of quantum keys distribution (QKD), the disadvantages of individual separate components will always show and manifest themselves, and some qubits will be unsuitable for the forming of a secret key.

In addition, listening to the quantum channel makes changes to the transmitted qubits, which also prevents them from being used when forming the secret key.

In the case of the classic protocol BB84, the threshold value of the error rate is 11% [28]. For an ideal model, the number of bits received as the result, the final secret key (R) for the bit of the sifted key, is expressed by the relationship:

$$R = 1 - 2\,H(QBER), \tag{27}$$

where the value $H$ is the binary entropy of Shannon and QBER is number of errors measured by Bob. The dependence of $R$ on QBER is demonstrated in the Figure 4:

This may not always be the case. The quality of execution of the equipment implementing the protocol may allow reducing the threshold.

In the protocol BB84–Info–Z, in addition to the information qubits responsible for key generation, the test qubits X and Z are used [7].

They are necessary for the eavesdropping check. The error threshold for this modification is slightly smaller, at 7.56% [9] (Figure 5).

For the modification proposed by the authors of CSLOE–2022, the threshold of errors is theoretically similar to the original one, since during key formation, imperfect copies of photons used to achieve the entanglement of a listener (eavesdropping interceptor) are discarded, and then the originals are checked, as in the classical protocol. On the other hand, when trying to listen, it is unlikely that the photons will be distorted, since their quantity is significantly lower than the number of copies.

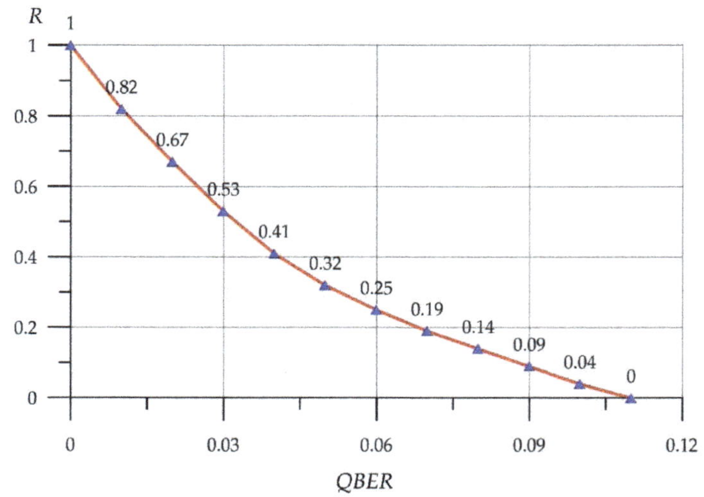


**Figure 4.** Error threshold for protocol BB84.

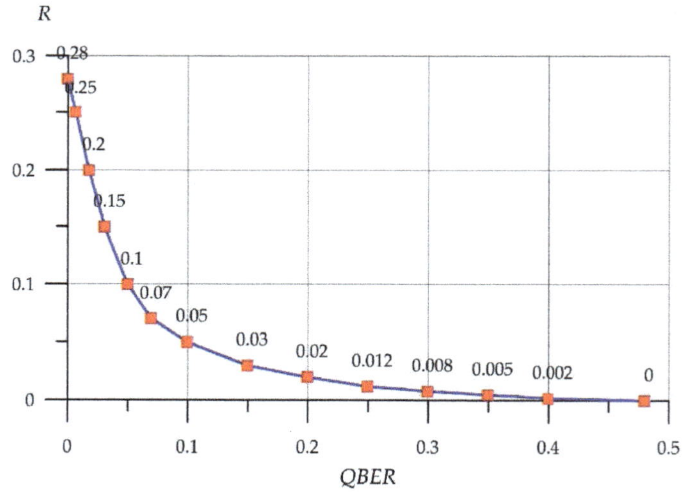


**Figure 5.** Safe zone of asymptotic rates of errors for BB84–Info–Z protocol.

From publication [17] it is known that the code of Steane (CSS), otherwise known as Calderbank–Shor–Steane, is the tool of quantum error correction, introduced by Andrew Steane in 1996. Steane's code uses the classical binary Hamming's code to correct the errors of the qubit flip ($X$–errors) and the double Hamming code to correct the errors of the phase flip ($Z$–errors), in practice allowing:

$$1 - 2 * H(\delta)]n \tag{28}$$

qubits, where $\delta$–number of measured errors, and $n$–length of qubits sequence.

If expression (28) would be written as: $f(x) = 1 - 2 * H(x)$, then the chart (graph) will intersect the axis $X$ at the point 0.11, which gave us the threshold of errors of 11% in the classical quantum protocol BB84 [29].

For the protocol CSLOE–2022, the dependence of the number of bits of the final resulting secret key $R$ from the quantity of recorded and fixed errors QBER is preserved, as is shown in Figure 6, since all photons, both the original and their copies in the form of pseudo-photons, will be distorted during transmission. Thus, the comparison of protocols by the threshold of errors can be observed in Table 2:

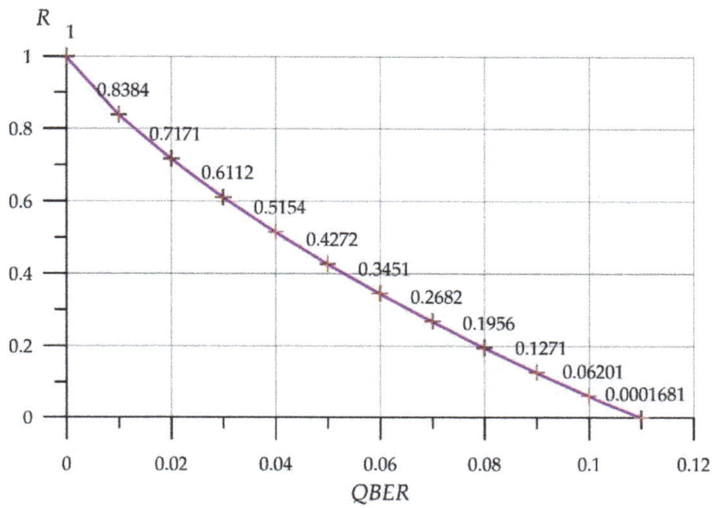


**Figure 6.** The threshold of errors for the modification CSLOE–2022 proposed by the authors.

**Table 2.** Comparative characteristics of protocols by threshold of errors.

| Protocol | Error Threshold |
|---|---|
| BB84 | 11% |
| BB84–Info–Z | 7.56% |
| CSLOE–2022 | 11% |

The protocol BB84–Info–Z differs for the better from all the others, since a much smaller threshold of errors will allow the eavesdropping intruder to receive significantly less information about the secret key.

Another important parameter is the working distance. The protocols of quantum keys distribution (QKD) operate with single photons, which can be distorted during transmission. Therefore, the working distance is relatively small. For example, for protocol BB84, it is about 70 km [30]. When implementing the protocols DPS and COW, it became possible to reach distances of 250 and 307 km, respectively, but their safety has not yet been proven.

Based on the findings of the research, it was suggested that in the case of using pseudo-photons, it is impossible to declare the numbers of photon losses correctly. If the losses occur from the number of pseudo-photons, and they are used as the trap and entanglement, they may have completely different behavior and completely different interactions, which can lead to both a decrease in losses, or an increase. To declare this confidently without conducting a series of experiments is incorrect and unpredictable. In the case of the protocols BB84 and BB84–Info-Z, the loss of photons is particularly significant. However, the feature of the CSLOE–2022 protocol is its modularity and application of pseudo-photons, which fundamentally distinguishes it from its predecessors.

Therefore, it is possible to make another hypothesis to rely on it, and refer to the fact that the method is based on the cloning of pseudo-photons. At large distances, the transmission of information with photons is carried out with large losses of ($\eta \simeq 0$), and this limitation can be exceeded only by using quantum repeaters. In turn, they will have to be used without limitation for the transmission of recreated pseudo-photons. In addition, it is necessary to be able to combine them with ordinary photons which can be transmitted without repeaters at all.

It is known that quantum communications guarantee the reliable transmission of quantum information and efficient distribution of entanglement, together with the generation of completely secure keys. However, it should be borne in mind that, at long distances, photon transmissions incur significant losses. Quantum repeaters can surpass this limitation. The article [31] discusses the theoretical aspects of the possible limit for ensuring the transmission of information without repeaters.

Relying on this article is possible if the objective is to increase the distance between repeaters. However, the studies themselves given in [31] are difficult to use for the clear practical application without conducting of series of experiments, and especially in combination of multi-qubit states, which is what some scientists are trying to practice now.

However, in the case considered by the authors, pseudo-photons are used (that is, recreated from real photons with clear differences), which, in turn, may display completely different, unpredictable behavior. This can lead to both a decrease in the losses or an increase in them. In the case of real photons, for the channel with losses, the article [31] proves that $Q_2 = K = -\log_2(1-\eta)$, where $\eta$ is the coefficient of transmission. In particular, the transmission capacity of the secret key of the channel with losses is the maximum speed achievable with any optical implementation of the quantum key distribution (QKD). At large distances, that is, with large losses, $\eta \simeq 0$, finding the optimal scaling of losses and the speed of $K \simeq 1.44\ \eta$ secret bits per channel use, the fundamental limitation is obtained, which, at this point in time, can be overcome, practically, only with repeaters.

A team of Chinese physicists succeeded in transmitting a secret quantum key at a distance of 511 km in real-world conditions [20]. They were thus able to implement the quantum line of data transmission outside laboratory conditions.

Physicists continue to search for different ways to increase the distance to hundreds and thousands of kilometers, developing repeaters for existing protocols, as well as new ones. For example, the protocol of the twin fields TF (Twin Field). Unlike the standard protocol, for example, BB84, in which Alice directly sends photons to Bob, the protocol TF [32] includes an additional Charlie node, which is located between Alice and Bob. In that case, Alice and Bob carry out the transfer and transmit their information on the weak coherent pulse to Charlie, who compares them and announces whether the received bits coincide or not.

However, Charlie has no information about the bits he has received; he can only compare them and declare whether they have coincided at the moment or not, so Charlie is considered to be an untrusted node.

This approach allows exceeding the known limit of the key generation rate without repeaters. It uses two photon sources, the phases of which, in addition to hundreds of kilometers of optical fiber between them, are not easy to stabilize.

Thus, we can summarize the comparison of protocols BB84, BB84–Info–Z and CSLOE–2022 by working distance and record the results in Table 3:

**Table 3.** Working distance of protocols.

| Protocol | Working Distance |
|---|---|
| BB84 | 70 km |
| BB84–Info–Z | 70 km |
| CSLOE–2022 | 511 km |

## 8. Conclusions

The article shows that the quantum protocol BB84–Info–Z is protected completely from collective attacks—one of the most powerful cyberattacks.

It is found that the results of the quantum protocol BB84 have much in common with the BB84–Info–Z quantum protocol, with two significant exceptions:

1. The rate of errors should be checked separately so that it remains below the threshold values $p_{a,\,z}$ and $p_{a,\,x}$ for bits TEST–Z and TEST–X; accordingly, when in the quantum protocol BB84, the threshold value of error rate $p_a$ is applied to all bits of TEST jointly [7].
2. The indexes and information indicators of the interceptor Eve (in security terms) and probability of the error-correcting code failure (in reliability terms) differ from the indexes and indicators in the case of classical quantum protocol BB84 [8].

It can be concluded that if the quantum protocol BB84 is modified so that the bits Info would be only at the coordinate basis Z, this will not weaken its reliability and security (against the collective cyberattacks, at least). It will not even change the threshold of the asymptotic rate of errors [8].

Protocol BB84–Info-Z can be applied safely to distribute the secret key; its security has ideal implementation, and it is protected against collective cyberattacks.

It is shown that modification of classical quantum protocol BB84 (CSLOE–2022) proposed by the authors could be used for quantum key distribution, since the applied principle of imperfect copying does not violate the laws of physics, but allows increasing the cryptographic strength of the protocol. So far, this modification is just an idea, and it needs to be proven by a series of practical experiments using specific equipment.

However, it is known that the principle of imperfect copying has already been applied experimentally in communication channels, as is mentioned in [33–47], which makes it possible to transmit information over much longer distances.

Nevertheless, for quantum key distribution, there are other requirements for the quality of performance of individual components.

Therefore, these new ideas are being put forward, and engineers are given specific tasks in order to prove the correctness of the new hypotheses.

Further, the authors plan to research the optical quantum memory of photons and pseudo-photons for the possibility of recording information into them. Based on the data obtained, it is necessary to identify theoretical limitations as well as to find the ways to level and neutralize them, both for photons and for pseudo-photons recreated on their basis.

**Author Contributions:** Conceptualization, L.V.C., O.A.S., A.N.B. and E.R.; methodology, L.V.C. and O.A.S.; software, O.A.S. and E.R.; validation, L.V.C., O.A.S. and E.R.; formal analysis, L.V.C., O.A.S. and E.R.; investigation, L.V.C. and O.A.S.; resources, A.N.B.; data curation, L.V.C. and O.A.S.; writing—original draft preparation, O.A.S. and A.N.B.; writing—review and editing, O.A.S. and A.N.B.; visualization, O.A.S.; supervision, L.V.C. and A.N.B.; project administration, L.V.C. and E.R.; funding acquisition, A.N.B. All authors have read and agreed to the published version of the manuscript.

**Funding:** This research received no external funding.

**Data Availability Statement:** The study did not report any data.

**Conflicts of Interest:** The authors declare no conflict of interest.

## References

1. Bennett, C.H.; Brassard, G. Quantum Cryptography: Public Key Distribution and Coin Tossing. In Proceedings of the IEEE International Conference on Computers, Systems and Signal Processing, Bangalore, India, 10–12 December 1984.
2. Aiello, C.D.; Hirose, M.; Cappellaro, P. Composite-pulse magnetometry with a solid-state quantum sensor. *Nat. Commun.* **2013**, *4*, 1419. [CrossRef] [PubMed]
3. Kröll, S.; Gallo, K.; Hennrich, M. *Research in Quantum Sensing*; Wallenberg Centre for Quantum Technology, Science Advances; Chalmers University of Technology: Gothenburg, Sweden, 2022. [CrossRef]

4. Riexinger, F.; Kutas, M.; Haase, B.; Bortz, M. *General Simulation Method for Quantum-Sensing Systems*; Institute for Industrial Mathematics ITWM, Department of Physics and Research Center OPTIMAS: Kaiserslautern, Germany, 2022. Available online: https://arxiv.org/pdf/2112.07243v1.pdf (accessed on 28 November 2022).
5. Schowengerdt, R. *Remote Sensing. Models and Methods for Image Processing*; Elsevier: Amsterdam, The Netherlands, 2007.
6. Molotkov, S.N. On the Resistance of Quantum Cryptography Systems with Phase-Time Coding to Active Probing Attacks. *J. Exp. Theor. Phys.* **2008**, *158*, 1011–1031.
7. Park, J.L. The concept of transition in quantum mechanics. *Found. Phys.* **1970**, *1*, 23–33. [CrossRef]
8. Boyer, M.; Liss, R.; Mor, T. Composable Security against Collective Attacks of Modified BB84 QKD Protocol with Information Only in One Basis. *Theor. Comput. Sci.* **2020**, *801*, 96–109. [CrossRef]
9. Biham, E.; Mor, T. Security of Quantum Cryptography against Collective Attacks. *Phys. Rev. Lett.* **1997**, *78*, 2256–2259. [CrossRef]
10. Boyer, M.; Gelles, R.; Mor, T. Security of the Bennett–Brassard Quantum Key Distribution Protocol against Collective Attacks. *Algorithms* **2009**, *2*, 790–807. [CrossRef]
11. Vercruysse, D.; Sapra, N.; Yang, K.; Vukovi, J. Inverse–Designed Photonic Crystal Devices for Optical Beam Steering. *arXiv* **2021**, arXiv:2102.00681. [CrossRef]
12. Buckley, S.; Radulaski, M.; Zhang, J.; Petykiewicz, J.; Biermann, K.; Vučković, J. Nonlinear Frequency Conversion Using High Quality Modes in GaAs Nanobeam Cavities. *Opt. Lett.* **2014**, *39*, 5673–5676. [CrossRef]
13. Cerf, N.; Ipe, A.; Rottenberg, X. Cloning of Continuous Quantum Variables. Ecole Polytechnique, CP 165. *Phys. Rev. Lett.* **2000**, *85*, 1754–1757. [CrossRef]
14. Fuchs, C.; Peres, A. Quantum–State Disturbance versus Information Gain: Uncertainty Relations for Quantum Information. *Phys. Rev.* **1996**, *53*, 2038–2045. [CrossRef]
15. Skori'c, B.; Wolfs, Z. Diagrammatic Security Proof for 8–State Encoding. *arXiv* **2021**, arXiv:2103.01936v1.
16. Morimae, T. *Quantum Randomized Encoding, Verifification of Quantum Computing, No–Cloning, and Blind Quantum Computing*; Yukawa Institute for Theoretical Physics, Kyoto University: Kyoto, Japan, 2020.
17. Schimpf, C.; Reindl, M.; Huber, D.; Lehner, B.; Silva, S.; Manna, S.; Vyvlecka, M.; Walther, P. Quantum Cryptography with Highly Entangled Photons from Semiconductor Quantum Dots. *arXiv* **2020**, arXiv:2007.12726v1. [CrossRef] [PubMed]
18. Tan, X. *Introduction to Quantum Cryptography*; IntechOpen: London, UK, 2013. [CrossRef]
19. Shor, P.; Preskill, J. *Simple Proof of Security of the BB84 Quantum Key Distribution Protocol*; AT&T Labs Research: Florham Park, NJ, USA, 2000.
20. Huttner, B.; Imoto, N.; Gisin, N.; Mor, T. Quantum cryptography with coherent states. *Phys. Rev. A* **1995**, *51*, 1863–1869. [CrossRef] [PubMed]
21. Djordjevic, I. *Quantum Information Processing, Quantum Computing, and Quantum Error Correction*, 2nd ed.; Academic Press: Cambridge, MA, USA, 2021. [CrossRef]
22. Bouwmeester, D.; Pan, J.-W.; Mattle, K.; Eibl, M.; Weinfurter, H.; Zeilinger, A. Experimental quantum teleportation. *Nature* **1997**, *390*, 575–579. [CrossRef]
23. Houça, R.; Belouad, A.; El Choubabi, B.; Kamal, A.; El Bouziani, M. Quantum teleportation via a two-qubit Heisenberg XXX chain with x-component of Dzyaloshinskii–Moriya interaction. *J. Magn. Magn. Mater.* **2022**, *563*, 169816. [CrossRef]
24. Yang, L.; Liu, Y.C.; Li, Y.S. Quantum Teleportation of Particles in an Environment. *Chin. Phys. B* **2020**, *29*, 060301. [CrossRef]
25. Chen, J.; Zhang, C.; Liu, Y.; Li, Y.; Liu, H.; Jiang, H.; Chen, T.; Zhang, Q.; Pan, J. Twin–Field Quantum Key Distribution over 511 km Optical Fiber Linking two Distant Metropolitans areas. *Nat. Photon.* **2021**, *15*, 570–575. [CrossRef]
26. Nang Paing, S.; Setiawan, J.W.; Tariq, S.; Talha Rahim, M.; Lee, K.; Shin, H. Counterfactual Anonymous Quantum Teleportation in the Presence of Adversarial Attacks and Channel Noise. *Sensors* **2022**, *22*, 7587. [CrossRef] [PubMed]
27. Gisin, N. *Quantum Randomness. Non–Locality, Teleportation and Other Quantum Wonders*; Alpina non–fiction: Moscow, Russia, 2018; 208p.
28. Safaryan, O.A.; Lemeshko, K.S.; Beskopylny, A.N.; Cherckesova, L.V.; Korochentsev, D.A. Mathematical Analysis of Parametric Characteristics of the Consensus Algorithms Operation with the Choice of the Most Priority One for Implementation in the Financial Sphere. *Electronics* **2021**, *10*, 2659. [CrossRef]
29. Wang, Y.; Hu, M.-L. Quantum Teleportation and Dense Coding in Multiple Bosonic Reservoirs. *Entropy* **2022**, *24*, 1114. [CrossRef] [PubMed]
30. Wen, X.; Chen, Y.; Zhang, W.; Jiang, Z.L.; Fang, J. Blockchain Consensus Mechanism Based on Quantum Teleportation. *Mathematics* **2022**, *10*, 2385. [CrossRef]
31. Pirandola, S.; Laurenza, R.; Ottaviani, C.; Banchi, L. Fundamental limits of repeaterless quantum communications. *Nat. Commun.* **2017**, *8*, 15043. [CrossRef] [PubMed]
32. Lucamarini, M.; Yuan, Z.; Dynes, J.; Shields, A. Overcoming the Rate–Distance Limit of Quantum Key Distribution without Quantum Repeaters. *Nature* **2018**, *557*, 400–403. [CrossRef] [PubMed]
33. Yan, F.; Wang, D. Probabilistic and controlled teleportation of unknown quantum states. *Phys. Lett. A* **2003**, *316*, 297–303. [CrossRef]
34. Khawasik, M.; El-Sayed, W.G.; Rashad, M.Z.; Younes, A. A Secured Half-Duplex Bidirectional Quantum Key Distribution Protocol against Collective Attacks. *Symmetry* **2022**, *14*, 2481. [CrossRef]

35. Cardoso-Isidoro, C.; Delgado, F. Shared Quantum Key Distribution Based on Asymmetric Double Quantum Teleportation. *Symmetry* **2022**, *14*, 713. [CrossRef]
36. Blunt, N.S.; Camps, J.; Crawford, O.; Izs'ak, R.; Leontica, S.; Mirani, A.; Moylett, A.E.; Scivier, S.A.; S¨underhauf, C.; Schopf, P.; et al. A Perspective on the Current State–of–the–art of Quantum Computing for Drug. *arXiv* **2022**, arXiv:2206.00551.
37. Chamberland, C.; Noh, K.; Arrangoiz–Arriola, P.; Campbell, E.T.; Hann, C.T.; Iverson, J.; Putterman, H.; Bohdanowicz, T.C.; Flammia, S.T.; Keller, A.; et al. Building a Fault–Tolerant Quantum Computer Using Concatenated Cat Codes, PRX Quantum 3. *arXiv* **2022**, arXiv:2012.04108.
38. Chamberland, C.; Campbell, E.T. Universal Quantum Computing with Twist–Free and Temporally Encoded Lattice Surgery. *PRX Quantum* **2022**, *3*, 010331. [CrossRef]
39. Kivlichan, I.D.; Gidney, C.; Berry, D.W.; Wiebe, N.; McClean, J.; Sun, W.; Jiang, Z.; Rubin, N.; Fowler, A.; Aspuru–Guzik, A.; et al. Improved Fault-Tolerant Quantum Simulation of Condensed–Phase Correlated Electrons via Trotterization, Quantum 4. *arXiv* **2020**, arXiv:1902.10673. [CrossRef]
40. Lu, D.; Li, Z.; Yu, J.; Han, Z. A Verifiable Arbitrated Quantum Signature Scheme Based on Controlled Quantum Teleportation. *Entropy* **2022**, *24*, 111. [CrossRef] [PubMed]
41. Hermans, S.L.; Pompili, M.; Beukers, H.K.C.; Baier, S.; Borregaard, J.; Hanson, R. Qubit Teleportation between Non–nbeighbouring Nodes in a Quantum Network. *Nature* **2022**, *605*, 663–668. [CrossRef]
42. Sun, Q.-C.; Mao, Y.-L.; Chen, S.-J.; Zhang, W.; Jiang, Y.-F.; Zhang, Y.-B.; Miki, S.; Yamashita, T.; Terai, H.; Jiang, X.; et al. Quantum teleportation with independent sources and prior entanglement distribution over a network. *Nat. Photon.* **2016**, *10*, 671–675. [CrossRef]
43. Xu, J.; Chen, X.; Xiao, H.; Wang, P.; Ma, M. A Performance–Consumption Balanced Scheme of Multi-Hop Quantum Networks for Teleportation. *Appl. Sci.* **2021**, *11*, 10869. [CrossRef]
44. Wu, H.; Liu, X.; Zhang, H.; Ruan, X.; Guo, Y. Performance Analysis of Continuous Variable Quantum Teleportation with Noiseless Linear Amplifier in Seawater Channel. *Symmetry* **2022**, *14*, 997. [CrossRef]
45. Benatti, F.; Floreanini, R.; Marzolino, U. Entanglement and Non-Locality in Quantum Protocols with Identical Particles. *Entropy* **2021**, *23*, 479. [CrossRef]
46. Raj, R.; Banerjee, S.; Panigrahi, P.K. Remote State Design for Efficient Quantum Metrology with Separable and Non-Teleporting States. *Quantum Rep.* **2021**, *3*, 228–241. [CrossRef]
47. Liss, R.; Mor, T. Quantum Communication—Celebrating the Silver Jubilee of Teleportation. *Entropy* **2020**, *22*, 628. [CrossRef] [PubMed]

*electronics*

MDPI

*Article*

# Investigation of Charge Transport Properties in VTP: PC71BM Organic Schottky Diode

**Nur Adilah Roslan [1,2,*], Azzuliani Supangat [2] and Suresh Sagadevan [3,*]**

[1] Department of Physics, Faculty of Science, Universiti Putra Malaysia, Serdang 43400, Selangor, Malaysia
[2] Low Dimensional Materials Research Centre, Department of Physics, Faculty of Science, University of Malaya, Kuala Lumpur 50603, Malaysia
[3] Nanotechnology & Catalysis Research Centre, University of Malaya, Kuala Lumpur 50603, Malaysia
* Correspondence: nur.adilah@upm.edu.my (N.A.R.); drsureshsagadevan@um.edu.my (S.S.)

**Abstract:** In this work, the charge transport properties of organic vanadyl 3,10,17,24-tetra-tert-butyl-1,8,15,22-tetrakis(dimethylamino)-29H,31H phthalocyanine (VTP) were investigated. The I-V profile demonstrated by single VTP shows a rectifying behavior, and Schottky diode parameters such as the ideality factor, barrier height, shunt, and series resistance were calculated. Further, the charge transport behavior of single-layer VTP and its blend with phenyl C71 butyric acid methyl ester ($PC_{71}BM$) was evaluated using the I-V conventional method and diode analysis. In addition, the optimized diode properties of VTP: $PC_{71}BM$ were chosen to evaluate its photovoltaic effect. The current density-voltage (J-V) characteristics were evaluated in both dark and light conditions to determine the key parameters of the photovoltaic effect. The results indicate the optimized VTP: the $PC_{71}BM$ composite blend yielded a relatively low photovoltaic efficiency. However, due to the presence of extended ligands, it gives a very good sensitivity when applied in the organic photodetector device, as reported in our previous work.

**Keywords:** organic electronic devices; VTP; J-V characteristics; photoactive films

**Citation:** Roslan, N.A.; Supangat, A.; Sagadevan, S. Investigation of Charge Transport Properties in VTP: PC71BM Organic Schottky Diode. *Electronics* **2022**, *11*, 3777. https://doi.org/10.3390/electronics11223777

Academic Editor: Lucas Lamata

Received: 7 October 2022
Accepted: 14 November 2022
Published: 17 November 2022



## 1. Introduction

Organic electronic devices have gradually made their way into the commercial market to replace conventional inorganic electronic devices. The desire to improve the performance of organic semiconductor materials has prompted many researchers to look for new materials. Metal phthalocyanines (MPcs), in particular, have sparked considerable interest due to their enticing benefits, which include cost efficiency, eco-friendly organic material, and thermal and chemical stability. Organic materials have been widely evaluated in recent decades to overcome several shortcomings of their inorganic counterparts in optoelectronic applications. Some unique properties of organic materials, such as an enhanced sensitivity, cost-effectiveness, environmental friendliness, and the suitability of simple fabrication methods such as spin coating, drop casting, dip coating, and spray coating directly from liquid solutions, are advantageous for the fabrication of organic photovoltaics (OPVs) [1,2]. Over the years, bulk heterojunction (BHJ) OPV and organic photodiodes (OPDs) have been extensively researched, in which an acceptor and a donor are blended in the photoactive medium to develop a donor/acceptor (D/A) interface at the molecular level [3–5]. MPcs containing heterocyclic macrocyclic organic compounds have been thoroughly evaluated for many applications [6,7]. Several MPcs, including VOPcPhO [8–13], were used to evaluate OPVs and OPDs [8–11].

Organic devices such as organic field effect transistors (OFET) and organic photovoltaics (OPVs) can lead to the next level of advancement by using a fundamental device such as a metal/organic/metal-based Schottky diode [14]. One electrode in a Schottky diode is Ohmic (injecting), and the other is blocking [14]. The slope and intercept of the J-V characteristic in the linear and saturation regions can be used to calculate the ideality factor and carrier mobility, respectively [14]. A Schottky diode is a metal-semiconductor that is in contact with rectification properties, similar to a p-n junction diode [15]. A Schottky diode is a majority carrier device, as it has a fast switching speed [15]. In theory, both p-type and n-type semiconductors can be used to fabricate Schottky diodes from a wide range of semiconducting materials, including organic semiconductors [15]. To make a Schottky contact with semiconductors, metals such as Pd, Pt, Au, Ti, Al, W, Cr, molybdenum, and a few metal silicides, are used [15]. Various Schottky diode configurations have previously been studied, fabricated, and tested [16,17].

In this study, a vanadium metal complex MPc, known as VTP, is proposed. With the exception of its different extended ligands, the aromatic structure of VTP is similar to that of VOPcPhO. Essentially, the ligands in VTP have the potential to change the molecule's photoabsorption sensitivity [18,19]. VTP has yet to be used as the active layer in the fabrication of OPVs. Based on the combination of their absorption properties, this study incorporated a blend of two organic compounds; a network structure is bound to improve charge separation and photogenerated charge carrier transport [20]. VTP is introduced as a donor material with distinct absorption properties, especially in the lower visible region [21]. The VTP and $PC_{71}BM$ have relatively exceptional hole-transport and electron-transport abilities, respectively [22], making them highly sought donor and acceptor components. The absorption of both materials plays an important role in capturing specific light wavelengths within the visible range. Light absorbance in the lower wavelength region of the visible spectrum is expected to increase due to the synergic absorption profile of the VTP and $PC_{71}BM$ matrix. Earlier, we reported a composite blend of VTP:$PC_{71}BM$ as an organic photodetector (OPD). The electrical characteristics of the OPD demonstrated improved sensing performances, and this could be attributed to the judicious selection of donor and acceptor components, which represented VTP and $PC_{71}BM$, respectively [19]. Therefore, in this study, the VTP: $PC_{71}BM$ composites were utilized to study the charge transport properties and to investigate the photovoltaic effect of the composite blend.

## 2. Experimental Section

VTP and $PC_{71}BM$ were purchased from Sigma Aldrich (St. Louis, MO, USA) and used without modification. Each material was separately dissolved in chloroform to produce a 10 mg/mL solution, which was then continuously stirred for hours. VTP and $PC_{71}BM$ were mixed in three different volume ratios (1.0:1.0, 1.0:1.2, and 1.0:1.4), then stirred in a nitrogen-filled glove box for an hour. The experimental approach was briefly discussed in our earlier publication [19].

## 3. Results and Discussion

Figure 1 depicts (a) the molecular structure of VTP and $PC_{71}BM$, (b) the schematic diagram of OPVs and OPDs, (c) a cross-section Field Effect Scanning Electron Microscopy (FESEM) image of the prepared samples, (d) the energy level diagram reported previously and (e) schematic diagram of energy levels for Schottky diode [19,23]. According to Figure 1c, the thicknesses of the PEDOT: PSS, active layer (VTP: $PC_{71}BM$), and aluminum (Al) were 40, 65, and 75 nm, respectively.

The measurement of current-voltage (I-V) characteristics in a dark setting is extremely useful for evaluating junction properties. This analysis produced significant parameters such as the ideality factor, rectification ratio (RR), reverse saturation current, barrier height, shunt and series resistance. The illumination setting was omitted due to minor fluctuations in the light intensity that could cause significant noise to the device, making production difficult. In the case of dark I-V measurements, charge carriers are injected into the circuit using current rather than light-generated carriers. Figure 2 depicts the semi-logarithmic plot of I-V curves of the ITO/PEDOT: PSS/VTP/Al device, which was analyzed in dark conditions at room temperature. According to Figure 2, the device displayed a rectifying behavior that resembled a non-linear and asymmetric diode at room temperature (300 K). The RR is the forward current to reverse current ($I_F/I_R$) ratio at a given applied voltage.

Its value is determined by the bias, which reflects the increased charge injection into the active layer. The RR value of the tested device was 1.6 at ± 0.5 V, which is attributable to the formation of the space charge layer at the interface [24]. The recorded RR value is comparable with the MPc group, as reported in past studies [20,25,26]. Predominantly, in relation to the Schottky barrier diode, the thermionic emission theory predicts that the *I-V* characteristics at the forward bias are given as follows:

$$I = I_o \exp\left(\frac{qV}{nkT}\right)\left[1 - \exp\left(\frac{-qV}{nkT}\right)\right] \tag{1}$$

where $I_o$ is given as follows:

$$I_o = AA^*T^2 \exp\left(\frac{-q\Phi_b}{k_B T}\right) \tag{2}$$

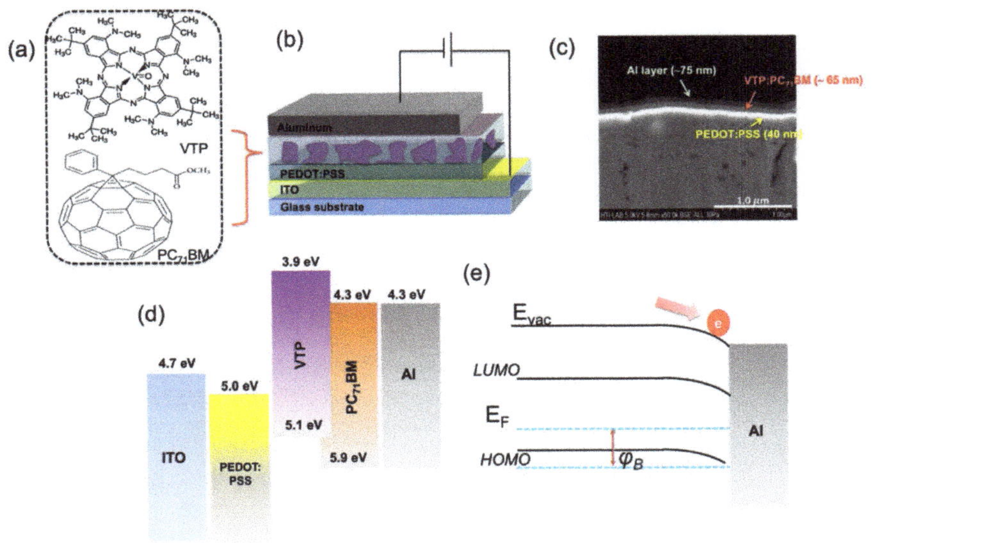


**Figure 1.** (**a**) The molecular structure of VTP (Roslan et al., 2018) [19] and $PC_{71}BM$ (**b**) device structure [19], (**c**) cross-section image of the prepared sample [19], (**d**) Energy level diagram [19], and (**e**) schematic diagram of energy levels of Schottky diode.

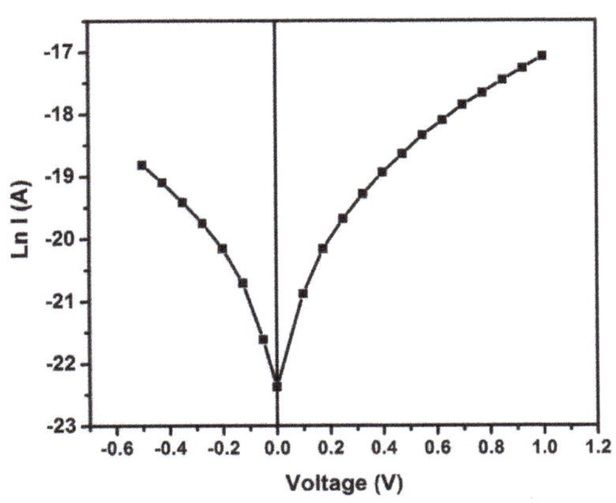


**Figure 2.** Current-voltage characteristics of VTP-based solar cell in semi-logarithmic scale.

$I_o$ denotes the saturation current, $V$ refers to the forward bias voltage, $\Phi_b$ reflects the zero-bias barrier height, $k_B$ exemplifies the Boltzmann constant, $T$ represents the temperature in Kelvin, $A$ depicts the active area, and $n$ is the ideality factor of the diode device. $A^*$ signifies the effective Richardson constant, which is attainable from the Richardson–Dushman correlation ($A = 4\pi emk^2/h^3$), where $m$ is the mass of the electron, $e$ is the elementary charge, and $h$ is Plank's constant. Scott and Malliaras [27] asserted that most organic semiconductor materials are ideally found at $10^{-2}$ A/cm$^2$ K$^2$ [27]. The reverse saturation current is obtained from the y-intercept of the semi-log *I-V* curve, which resulted in 0.2 nA upon weighing in its natural log.

$$\Phi_b = \frac{k_B T}{q} \ln\left(\frac{AA^*T^2}{I_o}\right) \tag{3}$$

The effective barrier height of the junction $\Phi_b$ was obtained from Equation (3), which refers to the contact that is present between the metal and semiconductor interface. This was calculated using Equation (2), with the value of the reverse saturation current at 0.673 eV. The conformity of the diode device to pure thermionic emission was determined by incorporating the ideality factor, $n$, by using the same semi-log *I-V* characteristics. For an ideal diode, the $n$ value reflects unity, despite it deviating from its ideal value at times, wherein the observed value is greater than unity. The value of the ideality factor in this study was obtained from the slope of the linear region in the forward bias of semi-log *I-V* curves using the following equation:

$$n = \frac{q}{kT} \frac{dV}{d(\ln I)} \tag{4}$$

The ideality factor recorded in this study is 2.93. The $n$ value exceeded '2' due to the following reasons: the non-homogeneous barrier, the non-homogeneous thickness of the organic film, temperature [28], and the prevalent current in the single junction photovoltaic device because of recombination [29]. Prior studies reported that the ideality factor of Pc derivatives such as CuPc, MgPc, and AlPc is massive, primarily due to recombination issues [30].

The shunt and series resistance ($R_{sh}$ and $R_s$, respectively) are essential factors that dictate the performance of a device. These resistance values are obtained from the plot of the junction resistance, $R_J$, versus the voltage, $V$; $R_J = \frac{dV}{dJ}$. Massive shunt resistance is required to decrease the leakage of current via cells (e.g., pinholes), apart from suppressing the recombination of charge carriers at the device interface [31–33]. This scenario, however, contradicts the low series resistance that reflects internal resistance, which allows and enhances high current to flow through the device, whereby a low $R_s$ denotes a low resistivity of the organic material [34,35]. Figure 3 illustrates that the values of the shunt resistance, $R_{sh}$, and series resistance, $R_s$, can be retrieved from the graph of the junction resistance ($R$) versus voltage ($V$). The resistance decreased exponentially with an increment in the forward bias. The $R_{sh}$ corresponded to the resistance within the vicinity of zero bias, thus resulting in 745 $\Omega$/cm$^2$. The $R$ curve was saturated upon a continuous supply of forward bias. Therefore, the applied external electric field compensated for the rectifying potential barrier between the cathode and anode, whereby the current that flowed through the device was restricted by $R_s$. Hence, the $R_s$ value was deduced by extrapolating the saturated part of the $R$ curve towards the interception with the resistance axis, in which the result was recorded at 32 $\Omega$/cm$^2$ [36].

A similar analysis was executed for the blended film to determine its $R_{sh}$ and $R_s$, as exhibited in Figure 4. Meanwhile, Table 1 represents the estimated values of both the shunt and series resistance values of all the diode devices. As a result, the value of $R_s$ decreased from 19.40 to 10.47 and 7.42 $\Omega$/cm$^2$ by increasing the concentrations of $PC_{71}BM$, respectively. This reduction reflects the increasing surface roughness morphology of the highest doping $PC_{71}BM$, mainly because the $R_s$ value is closely related to the active layer morphology, intrinsic resistance, and thickness of the active layer [32]. The processing additives improved the device performance by dissolving the aggregates of $PC_{71}BM$ in the VTP: $PC_{71}BM$ matrix. This facilitates the integration of the $PC_{71}BM$ molecule into the VTP, thus resulting in a greater donor–acceptor interface. The enhanced surface interface between D/A in the active layer contributes to an improved migration of excitons between the interface and to a favorable charge separation. It is worth highlighting here that $R_{sh}$ is correlated with pin holes and traps that are present in the thin film morphology, hence causing charge carrier recombination and current leakage. According to Table 1, the $R_{sh}$ value increased from 1636.12 to 1724.40 and, finally, to 3838.74 $\Omega/\text{cm}^2$, with an increment of $PC_{71}BM$. The increment in $R_{sh}$ value reflects that the incorporation of $PC_{71}BM$ in the active layer generated additional percolating pathways that helped the charge carriers to diffuse at a longer distance.

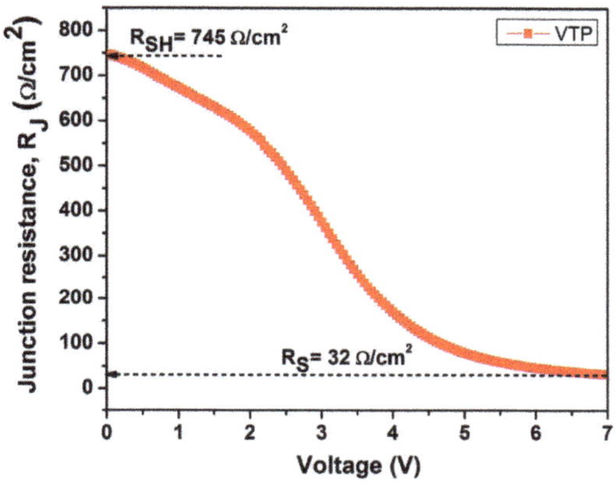


**Figure 3.** Junction resistance versus biased voltage for single junction VTP device.

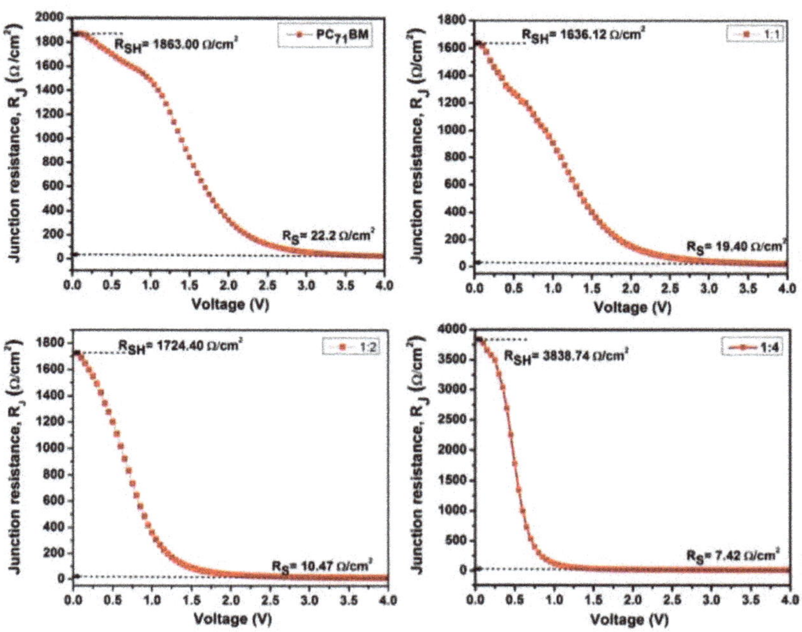


**Figure 4.** Junction resistance versus biased voltage for $PC_{71}BM$ and blended thin films of diode devices.

**Table 1.** Extracted $R_s$ and $R_{sh}$ values from junction resistance versus biased voltage.

| Resistance ($\Omega/cm^2$) | VTP | $PC_{71}BM$ | 1:1 | 1:2 | 1:4 |
|---|---|---|---|---|---|
| $R_{series}$, $R_s$ | 32.00 | 22.20 | 19.40 | 10.47 | 7.42 |
| $R_{shunt}$, $R_{sh}$ | 745.00 | 1863.00 | 1636.12 | 1724.40 | 3838.74 |

Studies pertaining to charge transport in organic thin film devices can be classified into two types: charge carrier injection at the interface and charge carrier transport in the bulk system. The characterization of the forward bias *I-V* curves in dark conditions yielded significant characteristics in light of the transport mechanism, which was responsible for the conduction process that took place at the electrode interface. The effective carrier mobility for all devices was determined by adopting the space-charge-limited conduction (SCLC) approach with a positive voltage of up to 10 *V* in dark conditions. Figure 5 displays the current density versus voltage curves excluding illumination, which was re-plotted in a double logarithmic scale to identify the mechanism that dominated the transport charge in the VTP diode. It is common for the double log forward bias *J-V* to reveal the power law behavior of $J \sim V^m$, where 'm' denotes the slope of each region. The $m$ value varies with the injection level and is linked with the distribution of trapping centers. The region $m = 1$ reflects the ohmic region, while $m = 2$ is the SCLC region and $m > 2$ represents the trapped-charge limited-current region [37].

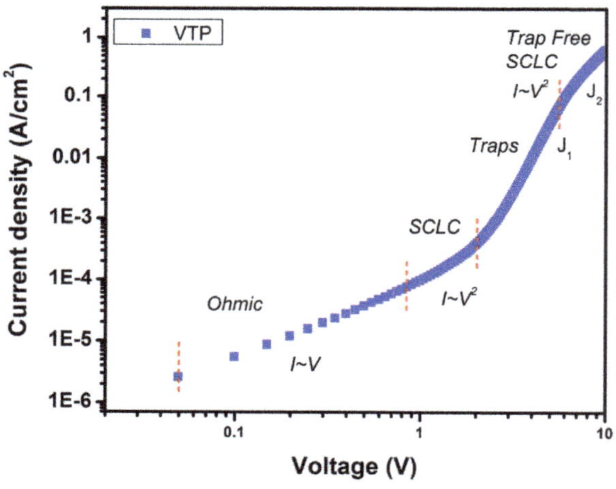


**Figure 5.** The double log scale of *J-V* characteristics of the VTP-based diode device measured in dark conditions.

Figure 5 illustrates the characteristics of the VTP-based diode device in order to elucidate the conduction mechanism. The double logarithmic graph is distributed in four regions (marked with dotted red lines), with various regions having varied slopes: ohmic, SCLC, traps, and trap-free SCLC regions (TF-SCLC). At the low voltage region, the slope was ~1.2, signifying the ohmic conduction mechanism. The current increased slowly with the applied voltage, as the injected effective-charge carrier density was lower than the background thermal carrier density [38]. In region 2 (intermediate voltage range), the conduction mechanism of the device was dominated by SCLC, wherein the slope was ~2.08. In this scenario, the voltage was more significant than that of region 1. The density of the injected free charge in this region was higher than the thermally generated free-charge carrier density; hence, there was an increment in the current. Typically, the formation of SCLC is attributed to the low mobility of charge carriers in an organic thin film, thus localizing the injected carriers by the trap states to limit the carrier conduction of the device [39]. The presence of traps was due to structural defects that derived from the non-uniformity and sub-atomic structure of the VTP layer. Under such conditions, the mobility of the charge carriers for the VTP diode devices was determined using the following equation:

$$J = \frac{9}{8}\varepsilon\varepsilon_O\mu\left(\frac{V^2}{d^3}\right) \tag{5}$$

where $J$ reflects the current density, $V$ is the voltage, $d$ denotes the active cell thickness, $\varepsilon$ represents the permittivity of free space, $\varepsilon_O$ signifies the relative dielectric constant ($\varepsilon_O = 8.854 \times 10^{-12}\ Fm^{-1}$), and $\mu$ stands for the charge carriers' mobility. Table 2 shows the estimated charge carrier mobility. Next, the exponential increase of current in region 3 at higher voltages gave the slope ~5.6. This condition was dominated by the trap-filled limit (TFL) mechanism, whereby all deep traps were filled by the injected electrons, hence making the existing trap sites be fully occupied. As for region 4, where a high voltage was applied, the slope of the plot decreased sharply (~3.5) as the diode device approached the TFL at a high injection level. At the high injection level, the conduction mechanism was similar to that in TF-SCLC. Here, the injected carriers occupied most of the traps, and the accumulation of the space charge near the interfacing electrode culminated in the generation of a field that hindered further rejection. The VTP diode between the interfaces' contact functioned as if there was no trap in it, and the current varied with the square of the voltage [40–42]. The estimated current density in this region is expressed via Child's law [34,43]:

$$J = \frac{9}{8}\varepsilon\varepsilon_O\mu\theta\left(\frac{V^2}{d^3}\right) \tag{6}$$

where $\theta$ denotes the trapping factor defined as $(J_1/J_2)$, in which $J_1$ and $J_2$ are the initial and final values of the current density in the TF-SCLC region, as portrayed in Figures 6 and 7, respectively. Figure 6 portrays similar analyses applied to the remaining thin films, while Table 2 summarizes the estimated values of the charge mobility for all the diode devices.

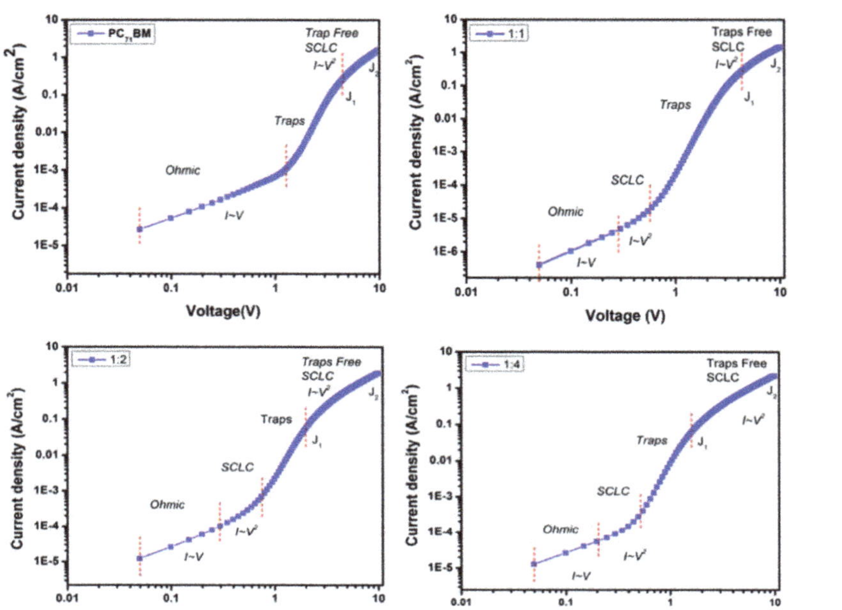


**Figure 6.** The double log of $J$-$V$ characteristics of $PC_{71}BM$ and blended film measured in dark conditions.

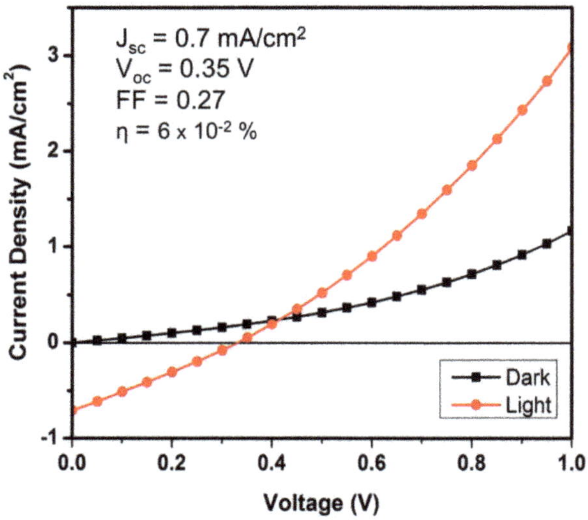


**Figure 7.** The *J-V* characteristics of ITO/PEDOT: PSS/VTP: $PC_{71}BM$/Al photovoltaic device under 100 $mWcm^{-2}$ illumination.

**Table 2.** Extracted mobility ($cm^2/Vs$) values for VTP, $PC_{71}BM$, and their varied ratios from diode measurements.

| REGIME / OPVs | SCLC | TF-SCLC |
|---|---|---|
| | Mobility ($cm^2/V{\cdot}s$) | Mobility ($cm^2/V{\cdot}s$) |
| VTP | $4.24 \times 10^{-7}$ | $5.00 \times 10^{-5}$ |
| $PC_{71}BM$ | – | $1.20 \times 10^{-4}$ |
| 1:1 (VTP: $PC_{71}BM$) | $2.02 \times 10^{-7}$ | $7.06 \times 10^{-5}$ |
| 1:2 | $1.07 \times 10^{-6}$ | $7.55 \times 10^{-5}$ |
| 1:4 | $2.38 \times 10^{-6}$ | $7.63 \times 10^{-5}$ |

Figure 7 shows the *I-V* performance in the dark and under light settings for the ITO/PEDOT: PSS/VTP: $PC_{71}BM$/Al photovoltaic device. The photovoltaic characteristics, such as $J_{SC}$, $V_{OC}$, *FF*, and PCE ($\eta$), are portrayed in Figure 7. The efficiency displayed by the device is attributable to the increased charge generation, which relies on the broader absorption range. Another factor that contributed to the efficiency in performance was the dissociation of excitons in the VTP: $PC_{71}BM$ device due to the BHJ system. This condition limits the single-layer and bilayer solar cells, primarily because the typical travel path of carriers is below 20 nm before recombination [44]. This particular blend system yielded an efficiency of up to ~0.06%. This is perhaps due to a low absorption intensity within the visible-range spectrum, which is insufficient to enhance charge separation, thus contributing to the low conversion from light energy to electrical energy. This corresponds to the low-charge carrier mobility displayed by the device.

The potential use of VTP:$PC_{71}BM$ has been utilized in the OPD device, as reported in our previous work [19]. The presence of extended ligands in the macrocycle of VTP helped improve the sensitivity of the photodetector so as to yield a good device.

## 4. Conclusions

In summary, the present study expands upon previous research into the use of VTP: $PC_{71}BM$ as a potential application in optical devices. The *I-V* profile demonstrated by single VTP shows a rectifying behavior, and Schottky diode parameters such as the ideality factor, barrier height, shunt, and series resistance were calculated. Further, the optimized composite blend of VTP: $PC_{71}BM$ was utilized in relation to its photovoltaic effect. This demonstrates that even though the same composite blend shows great potential for organic photodetectors, as reported previously, it does not however yield a good efficiency in organic photovoltaics. This is probably due to the relatively low mobility of the composite blend, which contributes to a low absorption intensity within the visible-range spectrum. The findings from the present studies are expected to pave the way for more comprehensive investigations of composite blends in organic devices.

**Author Contributions:** N.A.R. Conceptualization, methodology, Formal analysis, Data curation, Visualization, and Original draft preparation. A.S. Supervision, Formal analysis, Data curation, Visualization, and Validation S.S.: Formal analysis, Visualization, Validation and Reviewing and Editing. All authors have read and agreed to the published version of the manuscript.

**Funding:** This research is funded by GP-IPM/2021/9707300 under Universiti Putra Malaysia.

**Conflicts of Interest:** The authors declare no conflict of interest.

## References

1. Ahmad, Z.; Sayyad, M.H.; Yaseen, M.; Aw, K.C.; M-Tahir, M.; Ali, M. Potential of 5,10,15,20-Tetrakis (3′,5′-di-tertbutylphenyl) porphyrinatocopper (II) for a multifunctional sensor. *Sens. Actuators B Chem.* **2011**, *155*, 81–85. [CrossRef]
2. Fakir, M.S.; Ahmad, Z.; Sulaiman, K. Modification of optical band gap and surface morphology of NiTsPc thin films. *Chin. Phys. Lett.* **2012**, *29*, 126802. [CrossRef]
3. Ahmad, Z.; Suhail, M.H.; Muhammad, I.I.; Al-Rawi, W.K.; Sulaiman, K.; Zafar, Q.; Hamzah, A.S.; Shaameri, Z. MEH-PPV/Alq3-based bulk heterojunction photodetector. *Chin. Phys. B* **2013**, *22*, 100701. [CrossRef]
4. Arbab, E.A.; Taleatu, B.A.; Mola, G.T. Ternary molecules blend organic bulk heterojunction solar cell. *Mater. Sci. Semicond. Process.* **2015**, *40*, 158–161. [CrossRef]
5. Koster, L.J.; Smits, E.C.P.; Mihailetchi, V.D.; Blom, P.W. Device model for the operation of polymer/fullerene bulk heterojunction solar cells. *Phys. Rev. B* **2005**, *72*, 085205. [CrossRef]
6. Hamui, L.; Sánchez-Vergara, M.E. Innovative Implementation of an Alternative Tetrathiafulvene Derivative for Flexible Indium Phthalocyanine Chloride-Based Solar Cells. *Micromachines* **2021**, *12*, 633. [CrossRef] [PubMed]
7. Roslan, N.A.; Abdullah, S.M.; Halizan, M.Z.M.; Bawazeer, T.M.; Alsenany, N.; Alsoufi, M.S.; Majid, W.H.A.; Supangat, A. VTP as an Active Layer in a Vertical Organic Field Effect Transistor. *J. Electron. Mater.* **2018**, *47*, 2184–2191. [CrossRef]
8. Ahmad, Z.; Abdullah, S.M.; Sulaiman, K. Bulk heterojunction photodiode: To detect the whole visible spectrum. *Measurement* **2013**, *46*, 2073–2076. [CrossRef]
9. Farooq, A.; Karimov, K.S.; Ahmed, N.; Ali, T.; Khalid Alamgir, M.; Usman, M. Copper phthalocyanine and metal free phthalocyanine bulk heterojunction photodetector. *Phys. B Condens. Matter* **2015**, *457*, 17–21. [CrossRef]
10. Hamdan, K.S.; Abdullah, S.M.; Sulaiman, K.; Zakaria, R. Effects of silver nanoparticles towards the efficiency of organic solar cells. *Appl. Phys. A* **2013**, *115*, 63–68. [CrossRef]
11. Zafar, Q.; Fatima, N.; Karimov, K.S.; Ahmed, M.M.; Sulaiman, K. Realizing broad-bandwidth visible wavelength photodiode based on solutionprocessed ZnPc/PC71BM dyad. *Opt. Mater.* **2017**, *64*, 131–136. [CrossRef]
12. Abdullah, S.M.; Ahmad, Z.; Aziz, F.; Sulaiman, K. Investigation of VOPcPhO as an acceptor material for bulk heterojunction solar cells. *Org. Electron.* **2012**, *13*, 2532–2537. [CrossRef]
13. Zafar, Q.; Ahmad, Z.; Sulaiman, K.; Hamzah, A.S.; Rahman, Z.A. A MEHPPV/VOPcPhO composite based diode as a photodetector. *Sens. Actuators A Phys.* **2014**, *206*, 138–143. [CrossRef]
14. Rani, V.; Yadav, S.; Ghosh, S. Organic Schottky Diode: Characterization of Traps. *AIP Conf. Proc.* **2015**, *1665*, 120023. [CrossRef]
15. Srivastava, A.; Chakrabarti, P. An organic Schottky diode (OSD) based on a-silicon/polycarbazole contact. *Synth. Met.* **2015**, *207*, 96–101. [CrossRef]
16. Mott, N.F. Note on the contact between a metal and an insulator or semi-conductor. *Proc. Camb. Philos. Soc.* **1938**, *34*, 568–572. [CrossRef]
17. Rideout, V.L. A review of the theory, technology and applications of metal-semiconductor rectifiers. *Thin Solid Film.* **1978**, *48*, 261–399. [CrossRef]
18. Guo, L.; Ma, G.; Liu, Y.; Mi, J.; Qian, S.; Qiu, L. Optical and non-linear optical properties of vanadium oxide phthalocyanine films. *Appl. Phys. B* **2002**, *74*, 253–257. [CrossRef]
19. Roslan, N.A.; Abdullah, S.M.; Abd Majid, W.H.; Supangat, A. Investigation of VTP:PC71BM organic composite as highly responsive organic photodetector. *Sens. Actuators A Phys.* **2018**, *279*, 361–366. [CrossRef]
20. Zafar, Q.; Aziz, F.; Sulaiman, K. Eco-benign visible wavelength photodetector based on phthalocyanine-low bandgap copolymer composite blend. *RSC Adv.* **2016**, *6*, 13101–13109. [CrossRef]

21. Aziz, F.; Ahmad, Z.; Najeeb, M.A.; Malik, H.A.; Abdullah, S.M.; Touati, F.; Sulaiman, K. Colloidal distribution of the PCPDTBT and VOPcPhO in the organic amalgam thin films and their optical properties. *Appl. Phys. A* **2017**, *123*, 773. [CrossRef]
22. Ebenhoch, B.; Thomson SA, J.; Genevičius, K.; Juška, G.; Samuel, I.D.W. Charge carrier mobility of the organic photovoltaic materials PTB7 and PC71BM and its influence on device performance. *Org. Electron.* **2015**, *22*, 62–68. [CrossRef]
23. Wang, H.; Wang, X.; Fan, P.; Yang, X.; Yu, J. Enhanced power conversion efficiency of P3HT:PC71BM bulk heterojunction polymer solar cells by doping a high-mobility small organic molecule. *Int. J. Photoenergy* **2015**, *2015*. [CrossRef]
24. Ahmad, Z.; Sayyad, M.H.; Wahab, F.; Sulaiman, K.; Shahid, M.; Chaudry, J.A.; Munawar, M.A.; Aziz, F. Enhancement of electronic and charge transport properties of NiPc by potassium-tetrasulpho group. *Phys. B Condens. Matter* **2013**, *413*, 21–23. [CrossRef]
25. Benhaliliba, M.; Ocak, Y.S.; Benouis, C.E. Effect of metal on characteristics of MPc organic diodes. *J. Nano Electron. Phys.* **2014**, *6*, 04009-1–04009-3.
26. Aziz, F.; Ahmad, Z.; Abdullah, S.M.; Sulaiman, K.; Sayyad, M.H. Photovoltaic effect in single-junction organic solar cell fabricated using vanadyl phthalocyanine soluble derivative. *Pigment. Resin Technol.* **2015**, *44*, 26–32. [CrossRef]
27. Scott, J.C.; Malliaras, G.G. Charge injection and recombination at the metal–organic interface. *Chem. Phys. Lett.* **1999**, *299*, 115–119. [CrossRef]
28. Al-Ta'ii, H.M.J.; Amin, Y.M.; Periasamy, V. Humidity influenced capacitance and resistance of an Al/DNA/Al Schottky diode irradiated by alpha particles. *Sci. Rep.* **2016**, *6*, 25519. [CrossRef]
29. Yakuphanoglu, F. Photovoltaic properties of hybrid organic/inorganic semiconductor photodiode. *Synth. Met.* **2007**, *157*, 859–862. [CrossRef]
30. Rajaputra, S.; Vallurupalli, S.; Singh, V.P. Copper phthalocyanine based Schottky diode solar cells. *J. Mater. Sci. Mater. Electron.* **2007**, *18*, 1147–1150. [CrossRef]
31. Ito, M.; Palanisamy, K.; Kumar, A.; Murugesan, V.S.; Shin, P.-K.; Tsuda, N.; Yamada, J.; Ochiai, S. Characterization of the organic thin film solar cells with active layers of PTB7/PC71BM prepared by using solvent mixtures with different additives. *Int. J. Photoenergy* **2014**, *2014*. [CrossRef]
32. Murugesan, V.S.; Ono, S.; Tsuda, N.; Yamada, J.; Shin, P.-K.; Ochiai, S. Characterization of organic thin film solar cells of PCDTBT:PC71BM prepared by different mixing ratio and effect of hole transport layer. *Int. J. Photoenergy* **2015**, *2015*. [CrossRef]
33. Servaites, J.D.; Ratner, M.A.; Marks, T.J. Organic solar cells: A new look at traditional models. *Energy Environ. Sci.* **2011**, *4*, 4410–4422. [CrossRef]
34. Khan, S.M.; Sayyad, M.H.; Karimov, K.S. Investigation of temperature dependent electrical properties of p-VOPc/n-si heterojunction under dark conditions. *Ionics* **2011**, *17*, 307–313. [CrossRef]
35. Lim, L.W.; Aziz, F.; Muhammad, F.F.; Supangat, A.; Sulaiman, K. Electrical properties of Al/PTB7-Th/n-Si metal-polymer-semiconductor Schottky barrier diode. *Synth. Met.* **2016**, *221*, 169–175. [CrossRef]
36. Brus, V.V.; Proctor, C.M.; Ran, N.A.; Nguyen, T.Q. Capacitance spectroscopy for quantifying recombination losses in nonfullerene small molecule bulk heterojunction solar cells. *Adv. Energy Mater.* **2016**, *6*, 1502250. [CrossRef]
37. Chiguvare, Z.; Parisi, J.; Dyakonov, V. Current limiting mechanisms in indium-tin-oxide/poly3-hexylthiophene/aluminum thin film devices. *J. Appl. Phys.* **2003**, *94*, 2440–2448. [CrossRef]
38. Missoum, I.; Ocak, Y.; Benhaliliba, M.; Benouis, C.; Chaker, A. Microelectronic properties of organic Schottky diodes based on MgPc for solar cell applications. *Synth. Met.* **2016**, *214*, 76–81. [CrossRef]
39. Shang, D.; Wang, Q.; Chen, L.; Dong, R.; Li, X.; Zhang, W. Effect of carrier trapping on the hysteretic current-voltage characteristics in Ag/La 0.7 Ca 0.3 $MnO_3$/Pt heterostructures. *Phys. Rev. B* **2006**, *73*, 245427. [CrossRef]
40. Güllü, Ö.; Aydoğan, Ş.; Türüt, A. Fabrication and electrical characteristics of Schottky diode based on organic material. *Microelectron. Eng.* **2008**, *85*, 1647–1651. [CrossRef]
41. Janardhanam, V.; Jyothi, I.; Lee, J.-H.; Kim, J.-Y.; Reddy, V.R.; Choi, C.-J. Electrical properties and carrier transport mechanism of Au/n-GaN Schottky contact modified using a copper pthalocyanine (CuPc) interlayer. *Mater. Trans.* **2014**, *55*, 758–762. [CrossRef]
42. Jomaa, T.B.; Beji, L.; Ltaeif, A.; Bouazizi, A. The current–voltage characteristics of heterostructures formed by MEH-PPV spin-coated on n-type GaAs and n-type porous GaAs. *Mater. Sci. Eng. C* **2006**, *26*, 530–533. [CrossRef]
43. Najeeb, M.A.; Abdullah, S.M.; Aziz, F.; Ahmad, Z.; Shakoor, R.; Mohamed, A.; Khalil, U.; Swelm, W.; Al-Ghamdi, A.A.; Sulaiman, K. A comparative study on the performance of hybrid solar cells containing ZnSTe QDs in hole transporting layer and photoactive layer. *J. Nanoparticle Res.* **2016**, *18*, 384–392. [CrossRef]
44. Cai, W.; Gong, X.; Cao, Y. Polymer solar cells: Recent development and possible routes for improvement in the performance. *Solar Energy Mater. Sol. Cells* **2010**, *94*, 114–127. [CrossRef]

electronics

MDPI

*Article*

# K-Shaped Silicon Waveguides for Logic Operations at 1.55 μm

**Amer Kotb [1,2,*] and Kyriakos E. Zoiros [3]**

1 GPL, State Key Laboratory of Applied Optics, Changchun Institute of Optics, Fine Mechanics, and Physics, Chinese Academy of Sciences, Changchun 130033, China
2 Department of Physics, Faculty of Science, University of Fayoum, Fayoum 63514, Egypt
3 Lightwave Communications Research Group, Department of Electrical and Computer Engineering, School of Engineering, Democritus University of Thrace, 67100 Xanthi, Greece
* Correspondence: amer@ciomp.ac.cn

**Abstract:** Silicon has properties that make it the preferable semiconductor material for realizing a wide suite of electronic devices. In this paper, all basic optical logic operations, including XOR, AND, OR, NOT, NOR, XNOR, and NAND, are demonstrated by means of simulation using K-shaped compact silicon waveguides operated at the 1.55 μm telecommunications wavelength. This waveguide comprises three waveguide strips, all made of silicon printed on silica. By adjusting the phase of the incident beams, the pursued logic operations can be realized. To evaluate how well the considered operations are performed, the contrast ratio (CR) is employed as a figure of merit. Compared to other reported waveguides, the suggested K-shaped waveguide achieves higher CRs and a speed of the order of 120 Gb/s.

**Keywords:** logic operations; silicon-on-silica waveguide; contrast ratio

**Citation:** Kotb, A.; Zoiros, K.E. K-Shaped Silicon Waveguides for Logic Operations at 1.55 μm. *Electronics* **2022**, *11*, 3748. https://doi.org/10.3390/electronics11223748

Academic Editor: Lucas Lamata

Received: 24 October 2022
Accepted: 14 November 2022
Published: 15 November 2022



## 1. Introduction

By enabling the execution of signal-processing functionalities without troublesome optoelectronic conversions at the photonic nodes, all-optical gates serve as essential building blocks for the construction of lightwave broadband communications networks [1]. The accomplishment of the many signal processing tasks entirely in the optical domain, such as packet processing [2,3], pseudorandom binary sequence generation [1,4], encryption/decryption [5], error detection and correction [6], arithmetic operations [7,8], construction of optical memory elements [9], digital comparison [10,11], buffering [12], implementation of any other Boolean function [13], and construction of combinational logic circuits [14], is made possible by the XOR, AND, OR, NOT, NOR, NAND, and XNOR logic operations. On the other hand, the development of effective and low-loss platforms at a reasonable cost is claimed by silicon photonics. A type of structure known as silicon-on-silica technology is created by depositing a thin layer of crystalline silicon on an insulating layer, which is silica (silicon dioxide). Due to the significant infrared transparency of silicon and refractive index difference between silicon (i.e., core with $n_{silicon} \approx 3.48$ at 1.55 μm) and silica (i.e., cladding with $n_{silica} \approx 1.444$ at 1.55 μm), silicon-on-silica optical waveguides have unique optical features [15]. Various optical waveguides have been recently used for implementing both all-optical logic gates and all-optical networks [16–23]. Therefore, in this paper, we simulate seven basic logic operations, including XOR, AND, OR, NOT, NOR, NAND, and XNOR, using K-shaped waveguides operated at the telecommunications wavelength of 1.55 μm. This waveguide has four terminals, each of which has an output port and three input ports composed of silicon patterned on silica. It is generally known that silicon has a relatively low optical loss (2 dB/cm) for wavelengths up to 8 μm, but silica's optical loss increases rapidly beyond 3.6 μm [15]. The interferences, both constructive and destructive, which are created by the phase difference between the input beams, are the key for the realization of the considered logic operations. In order to demonstrate how the logic operations are executed, finite-difference time-domain (FDTD) solutions are obtained, using commercially available software, with the convolutional optimally matched layer as an absorbing boundary condition. The logic operations' performance is assessed against the contrast ratio (CR) metric. According to the derived simulation results, the employed waveguide can achieve higher CRs at an extended data rate of 120 Gb/s and, hence, can outperform previously reported designs [16–20].

## 2. K-Shaped Waveguide

In this work, we build a K-shaped waveguide with four terminals made of three silicon slots patterned on a silica substrate. The three input ports are excited by an electromagnetic wave that is polarized in the transverse magnetic mode at 1.55 μm. The wavelength and intensity of the incident beams are identical. The K-shaped silicon-on-silica waveguide is depicted schematically in Figure 1, along with its field intensity distributions.

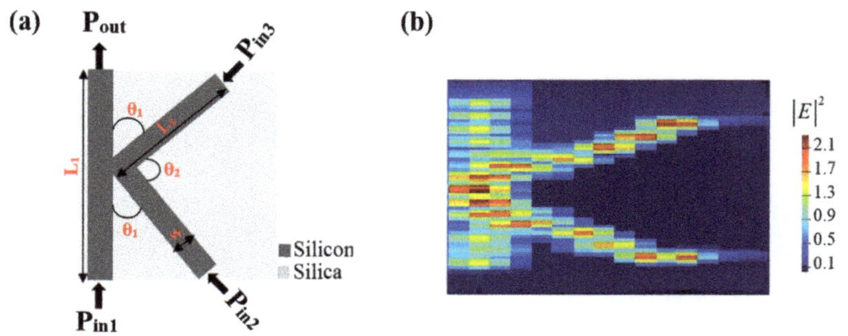


**Figure 1.** (**a**) Schematic depiction and (**b**) field-intensity distributions of K-shaped silicon-on-silica waveguide.

To record the simulation outcomes, the FDTD monitors are employed. Setting the threshold transmission ($T_{th}$) value to 0.12 is necessary at first. The formula for the output transmission (T) is $T = I_{out}/I_{in}$ [16], where $I_{out} = |E_{out}|^2$ is the intensity at $P_{out}$, and $I_{in} = I_1 + I_2 + I_3$ is the sum of the intensities at the three input ports. The input beams must satisfy the requirements for phase-matching in order to maximize T. In essence, this implies ensuring sure that the interacting waves are kept in the proper relative phase throughout the direction of propagation. However, before high CR logic gates can be accomplished, the phase-matching condition necessitates a specific selection of the input wavelength and waveguide characteristics. The phase-matching in silicon waveguides is induced by the contributions of the waveguide birefringence, material dispersion, waveguide dispersion, and cross- and phase-self modulations [24]. It is, therefore, feasible to achieve phase-matching by designing the waveguide such that the birefringence and material dispersion terms cancel one another, according to the phase-matching analysis of silicon waveguides, as reported in [24]. When $T > T_{th}$, $P_{out}$ generates a logical output of '1', while in all other cases, it generates a '0'. The CR is an important metric for logic devices and is defined as $CR(dB) = 10\, ln\left[P^1_{mean}/P^0_{mean}\right]$ [25], where $P^1_{mean}$ and $P^0_{mean}$ represent the mean peak powers of output logic bits '1' and '0', respectively. Compared to other metrics, such as the extinction ratio, the CR offers a better and more accurate evaluation of the performance of the optical logic operations [26]. For the proposed waveguide, Table 1 lists the default parameters' values used in the simulation.

**Table 1.** Simulation parameters.

| Symbol | Definition | Value | Unit |
|---|---|---|---|
| $L_1$ | Length of long slot | 2.5 | μm |
| $L_2$ | Length of short slot | 1.0 | μm |
| W | Width of slot | 0.22 | μm |
| D | Thickness of slot | 0.3 | μm |
| $\theta_1$ | Angle between long and short slots | 50 | degree |
| $\theta_2$ | Angle between short slots | 80 | degree |
| $n_{silicon}$ | Silicon refractive index at 1.55 μm | 3.48 | - |
| $n_{silica}$ | Silica refractive index at 1.55 μm | 1.444 | - |
| $\lambda$ | Operating wavelength | 1.55 | μm |
| $T_{th}$ | Threshold transmission | 0.12 | - |

When all incident beams (i.e., two beams and a reference or clock light) are launched at the three input ports with the same phase of 180°, the normalized spectral transmission (T) and loss as a function of the operating wavelength (λ) are shown in Figure 2. The employed waveguide achieves a high T = 0.852 and a low loss = 0.69 dB/μm at 1.55 μm. Such small propagation losses are a direct result of the scattering at the inner slots' interfaces and absorption within the materials. Figure 2 also shows that this waveguide achieves a high T and a low loss at a wide range of telecommunication wavelengths, from 1.3–1.6 μm.

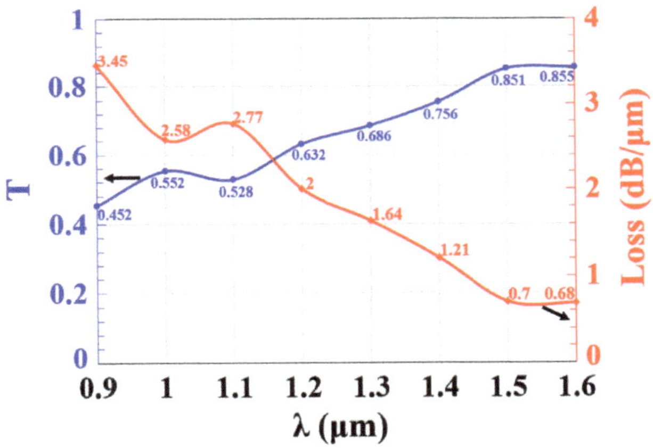


**Figure 2.** Normalized spectral transmission (T) and loss versus operating wavelength (λ), using K-shaped silicon-on-silica waveguide.

Relaxed tolerances are crucial for both manufacturing and operating conditions. Manufacturing tolerances refer to the management of the geometrical dimensions during processing and their ensuing effect on device performance. Operation tolerances describe how the device responds to variations in wavelength, polarization, temperature, input field distribution, and refractive index [27,28]. Most laser sources have a significant practical wavelength tolerance. For example, a 1550 nm fiber laser may have a wavelength tolerance of ±20 nm, resulting in an actual wavelength of 1550 ± 20 nm [29]. Figure 3 shows the dependence of the loss on the wavelength tolerance using the proposed waveguide. This result has been taken based on Equations (4)–(7) from Ref. [30].

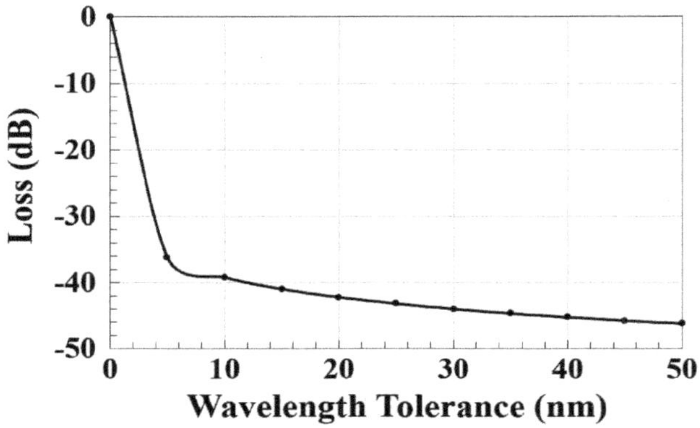


**Figure 3.** Optical loss versus wavelength tolerance, using K-shaped silicon-on-silica waveguide.

In this waveguide, we have used three acute angles, with a sum that is 180°, to perform the K letter. These angles (i.e., $\theta_1$ and $\theta_2$) play an important role in the K-shaped design in order to implement the considered logic gates with high CRs. Thus, the effect of the angle between the long and short slots ($\theta_1$) on a normalized spectral transmission (T) at an operating wavelength of 1550 nm is simulated, as shown in Figure 4. It is clear from Figure 4 that the highest T occurred at $\theta_1 = 50°$ (i.e., $\theta_2 = 80°$), as optimized in this simulation.

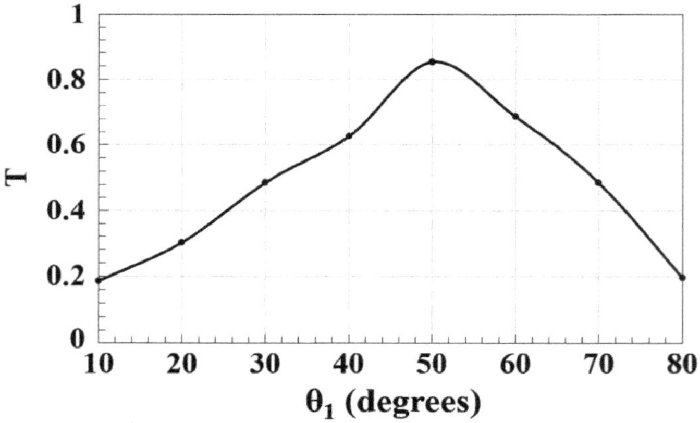


**Figure 4.** Normalized spectral transmission (T) versus angle between long and short slots ($\theta_1$), using K-shaped silicon-on-silica waveguide.

## 3. Logic Operations

### *3.1. XOR*

To carry out the XOR, AND, and OR logic operations, a reference beam (REF) must be injected into $P_{in2}$ of Figure 1, while the other two beams are launched into $P_{in1}$ and $P_{in3}$. The REF (all '1's) is used to introduce a reference phase difference between the input signals, resulting in either constructive or destructive interference. Constructive interference happens when all input beams are injected at the same phase (resulting in an output of '1'); destructive interference happens when they are launched at different phases (resulting in an output of '0'). As a result, for an XOR operation, $P_{out}$ produces a '1' (meaning $T > T_{th}$) because of the constructive interference that occurs between the input beams when the combination of these input beams (01, 10) is injected along with the REF at the same phase (i.e., $\Phi_1 = \Phi_3 = \Phi_{REF} = 180°$). The destructive interference between the incident beams causes a '0' output to be produced at $P_{out}$ (meaning $T < T_{th}$) when the combination (11), with the REF at different phases (i.e., $\Phi_1 = 0°$, $\Phi_3 = 90°$, and $\Phi_{REF} = 180°$), is launched. This results in the XOR logic function. We notice the presence of light at ports having '0' input, which is a natural result because the inner interfaces of the three input ports of the K-shaped waveguide are all opposite, and, therefore, the light is deflected inside them in an outward direction. The XOR field intensity distributions are displayed in Figure 5, using a K-shaped silicon-on-silica waveguide at 1.55 μm.

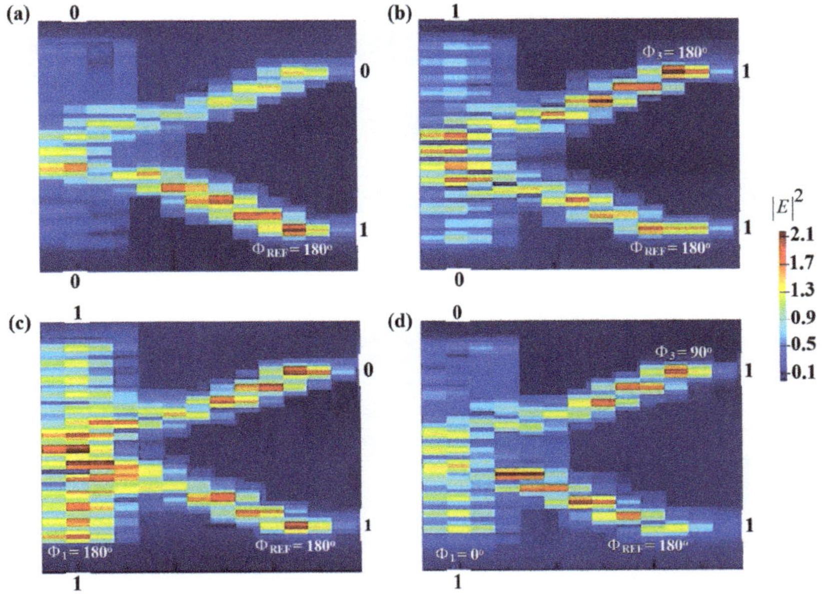


**Figure 5.** XOR field-intensity distributions, using K-shaped silicon-on-silica waveguide at 1.55 μm: (**a**) '00' input, (**b**) '01' input, (**c**) '10' input, and (**d**) '11' input.

Due to the large difference between the mean peak powers of '1' and '0', the suggested waveguide achieves a high CR = 34 dB. The XOR simulation outcomes, employing the K-shaped silicon-on-silica waveguide at 1.55 μm, are shown in Table 2.

**Table 2.** XOR simulation outcomes ($T_{th} = 0.12$).

| $P_{in1}$ | $P_{in3}$ | $P_{in2}$ (REF) | T | $P_{out}$ | CR (dB) |
|---|---|---|---|---|---|
| 0 | 0 | 1 | 0.021 | 0 | |
| 0 | 1 | 1 | 0.464 | 1 | 34 |
| 1 | 0 | 1 | 0.852 | 1 | |
| 1 | 1 | 1 | 0.023 | 0 | |

### 3.2. AND

Similar to the XOR operation, the AND operation involves injecting two beams into $P_{in1}$ and $P_{in3}$ as well as launching the REF (all '1's) from $P_{in2}$. $P_{out}$ creates a '1' output, due to constructive interference, when all incident beams are released into the proposed waveguide at the same phase (i.e., $\Phi_1 = \Phi_3 = \Phi_{REF} = 180°$). In contrast, when these incident beams are injected at a different phase, $P_{out}$ outputs a '0' because of destructive interference. This results in the AND operation. In Figure 6, the AND field intensity distributions are shown, using a K-shaped silicon-on-silica waveguide at 1.55 μm.

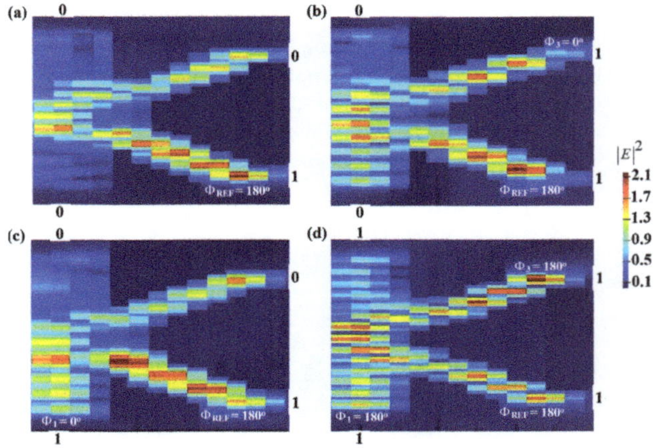


**Figure 6.** AND field intensity distributions, using K-shaped silicon-on-silica waveguide at 1.55 μm: (**a**) '00' input, (**b**) '01' input, (**c**) '10' input, and (**d**) '11' input.

For the logic AND operation, the proposed waveguide achieves CR = 31 dB at 1.55 μm. The further results of the AND simulation are listed in Table 3.

**Table 3.** AND simulation outcomes ($T_{th}$ = 0.12).

| $P_{in1}$ | $P_{in3}$ | $P_{in2}$ (REF) | T | $P_{out}$ | CR (dB) |
|---|---|---|---|---|---|
| 0 | 0 | 1 | 0.021 | 0 | 31 |
| 0 | 1 | 1 | 0.022 | 0 | |
| 1 | 0 | 1 | 0.023 | 0 | |
| 1 | 1 | 1 | 0.521 | 1 | |

## 3.3. OR

When the combination of input beams (01, 10, or 11) is inserted with REF at the same phase of 180°, the result of the $P_{out}$ becomes a '1'. Thus, the OR logic function between the two input beams is realized. Figure 7 depicts the OR field intensity distributions, using the proposed waveguide at 1.55 μm.

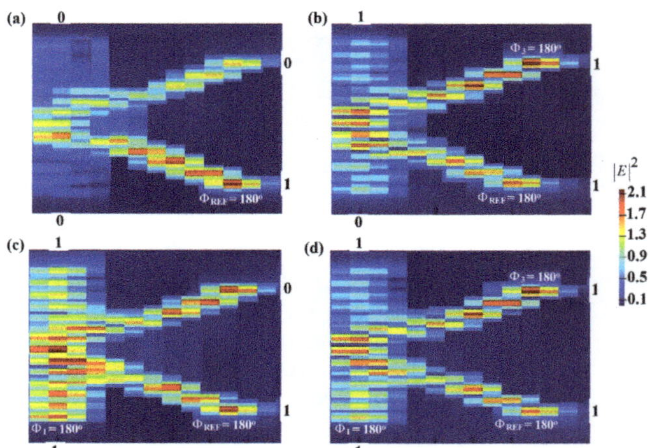


**Figure 7.** OR field intensity distributions, using K-shaped silicon-on-silica waveguide at 1.55 μm: (**a**) '00' input, (**b**) '01' input, (**c**) '10' input, and (**d**) '11' input.

The suggested waveguide obtains a high CR = 33.73 dB due to the significant difference between the mean peak powers of '1' and '0'. Table 4 provides an overview of the outcomes of the OR simulation at 1.55 μm, in terms of T and CR.

**Table 4.** OR simulation outcomes ($T_{th}$ = 0.12).

| $P_{in1}$ | $P_{in3}$ | $P_{in2}$ (REF) | T | $P_{out}$ | CR (dB) |
|---|---|---|---|---|---|
| 0 | 0 | 1 | 0.021 | 0 | 33.73 |
| 0 | 1 | 1 | 0.464 | 1 | |
| 1 | 0 | 1 | 0.852 | 1 | |
| 1 | 1 | 1 | 0.521 | 1 | |

The REF is essential for realizing the XOR, AND, and OR operations. Therefore, using the suggested waveguide at 1.55 μm, we have compared the performance of these three operations in terms of CR in the presence of the REF beam (i.e., REF = '1') and the absence of it, meaning there is no input beam injected into $P_{in2}$. Table 5 indicates the necessity of using the REF to obtain higher CRs.

**Table 5.** CR with and without REF.

| Operation | CR (dB) with REF | CR (dB) without REF |
|---|---|---|
| XOR | 34 | 7.1 |
| AND | 31 | 6.4 |
| OR | 33.73 | 7 |

### 3.4. NOT

To carry out all inverted logic operations, including NOT, NOR, NAND, and XNOR, a clock light (Clk) with an angle of 0° must be sent into the proposed waveguide from $P_{in1}$ of Figure 1. The Clk introduces an additional phase shift on the traveling beams, which changes the waveguide balance and results in an output. One beam is injected into $P_{in3}$ at an angle of 180° to perform the NOT operation. Due to the destructive interference that occurs as a result of the input beams' various phase conditions, when $P_{in3}$ is set to '1', $P_{out}$ produces a logical '0' (i.e., T < $T_{th}$). When $P_{in3}$ is 'OFF', the Clk (all '1's) outputs a logical '1' (i.e., T > $T_{th}$) at $P_{out}$, instead of going through a differencing phase. In this manner, the NOT gate is performed. Using a K-shaped silicon-on-silica waveguide, Figure 8 illustrates the NOT field intensity distributions at 1.55 μm.

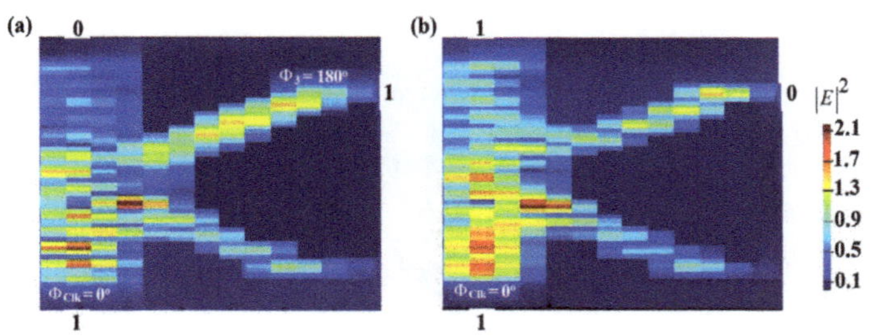


**Figure 8.** NOT field intensity distributions, using K-shaped silicon-on-silica waveguide at 1.55 μm: (**a**) '1' input and (**b**) '0' input.

The suggested waveguide results in a high CR = 30.5 dB for NOT operation. Table 6 provides a summary of the outcomes of the NOT simulation, using the proposed waveguide at 1.55 μm.

**Table 6.** NOT simulation outcomes ($T_{th}$ = 0.12).

| $P_{in1}$ (Clk) | $P_{in3}$ | T | $P_{out}$ | CR (dB) |
|---|---|---|---|---|
| 1 | 1 | 0.032 | 0 | 30.5 |
| 1 | 0 | 0.675 | 1 | |

*3.5. NOR*

Two beams are launched into $P_{in2}$ and $P_{in3}$ to perform the NOR (NOT-OR) operation, and $P_{in1}$ is launched with Clk (all '1's), as shown in Figure 1. When the input beams (01, 10, or 11) are combined and injected at different angles, destructive interference results in a logical '0' at $P_{out}$. If the launched beams' combination is (00), the Clk beam with $\Phi_{Clk}$ = 0° will cancel the phase balance of the three inputs, resulting in a logical '1' at $P_{out}$. Thus, the NOR logic operation is realized, as shown in Figure 9.

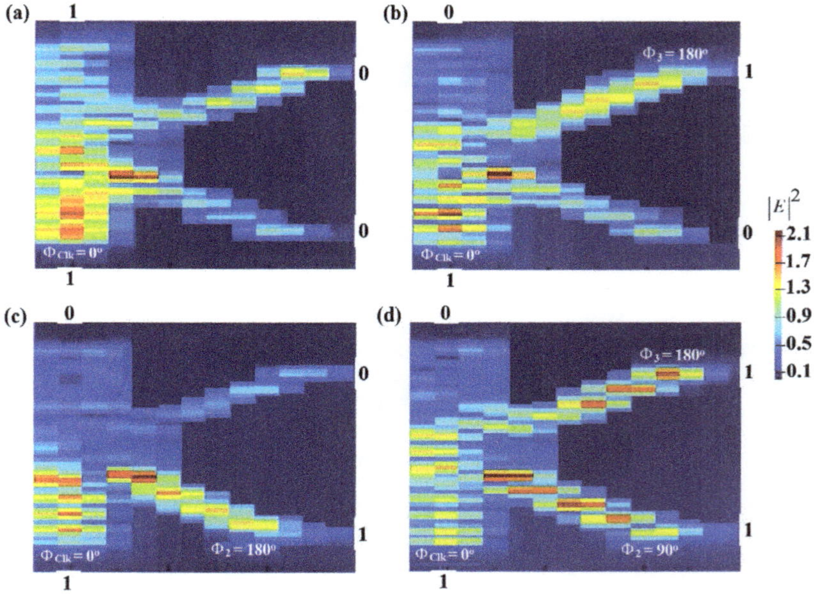


**Figure 9.** NOR field intensity distributions, using K-shaped silicon-on-silica waveguide at 1.55 μm: (**a**) '00' input, (**b**) '01' input, (**c**) '10' input, and (**d**) '11' input.

The suggested waveguide achieves a high CR = 33 dB for the NOR operation, as a result of the large disparity between $P^1_{mean}$ and $P^0_{mean}$. A summary of the simulation outcomes for this logic operation is given in Table 7.

**Table 7.** NOR simulation outcomes ($T_{th}$ = 0.12).

| $P_{in1}$ (Clk) | $P_{in2}$ | $P_{in3}$ | T | $P_{out}$ | CR (dB) |
|---|---|---|---|---|---|
| 1 | 0 | 0 | 0.675 | 1 | 33 |
| 1 | 0 | 1 | 0.032 | 0 | |
| 1 | 1 | 0 | 0.022 | 0 | |
| 1 | 1 | 1 | 0.022 | 0 | |

*3.6. NAND*

The NAND (NOT-AND) can be produced by injecting the Clk into $P_{in1}$ and the other two beams into $P_{in2}$ and $P_{in3}$, respectively. When both $P_{in2}$ and $P_{in3}$ are 'OFF' (i.e., 00), the

Clk with a $\Phi_{Clk} = 0°$ cancels the phase balance of the three inputs, causing $P_{out}$ to become '1'. Constructive interference simply occurs when Clk and (01, 10) are launched at the same angle of 0°, yielding an output of '1'. A '0' output is produced when (11) is launched with Clk at various phases, such as $\Phi_2 = 90°$, $\Phi_3 = 180°$, and $\Phi_{Clk} = 0°$, as illustrated in Figure 10.

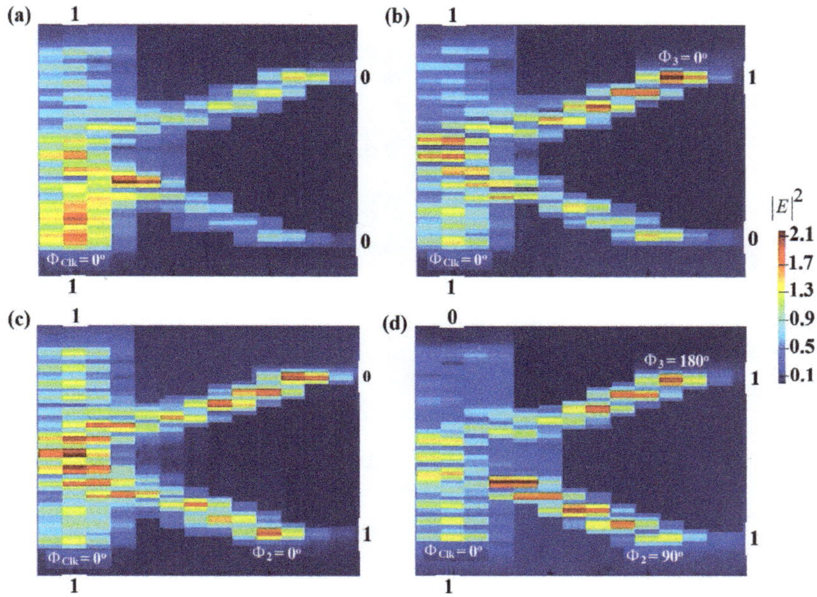


**Figure 10.** NAND field intensity distributions, using K-shaped silicon-on-silica waveguide at 1.55 µm: (**a**) '00' input, (**b**) '01' input, (**c**) '10' input, and (**d**) '11' input.

A summary of the NAND simulation outcomes utilizing the suggested waveguide, which achieves a high CR = 34 dB at 1.55 µm, is shown in Table 8.

**Table 8.** NAND simulation outcomes ($T_{th} = 0.12$).

| $P_{in1}$ (Clk) | $P_{in2}$ | $P_{in3}$ | T | $P_{out}$ | CR (dB) |
|---|---|---|---|---|---|
| 1 | 0 | 0 | 0.675 | 1 | |
| 1 | 0 | 1 | 0.464 | 1 | 34 |
| 1 | 1 | 0 | 0.852 | 1 | |
| 1 | 1 | 1 | 0.022 | 0 | |

### 3.7. XNOR

Similar to the NOR and NAND operations, the Clk enters $P_{in1}$ to create the XNOR (exclusive-XOR) logic function, while the other two beams are injected from $P_{in2}$ and $P_{in3}$, respectively. Constructive interference causes $P_{out}$ to emit a '1' when the combination of the input beams (11) is introduced with the Clk at the same phase of 0°. In contrast, $P_{out}$ produces a '0' when the input beams' combinations, (01) or (10), are inserted with a different phase, as depicted in Figure 11.

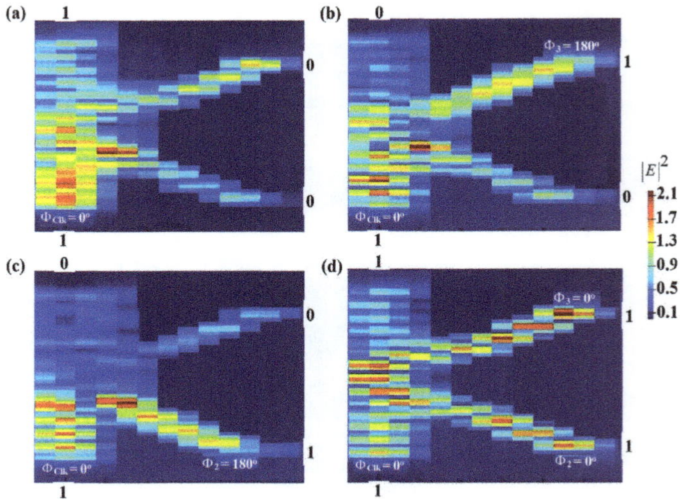


**Figure 11.** XNOR field intensity distributions, using K-shaped silicon-on-silica waveguide at 1.55 µm: (**a**) '00' input, (**b**) '01' input, (**c**) '10' input, and (**d**) '11' input.

Table 9 summarizes the XNOR simulation outcomes with a high CR = 31 dB, using the suggested waveguide.

**Table 9.** XNOR simulation outcomes ($T_{th}$ = 0.12).

| $P_{in1}$ (Clk) | $P_{in2}$ | $P_{in3}$ | T | $P_{out}$ | CR (dB) |
|---|---|---|---|---|---|
| 1 | 0 | 0 | 0.675 | 1 | |
| 1 | 0 | 1 | 0.032 | 0 | 31 |
| 1 | 1 | 0 | 0.022 | 0 | |
| 1 | 1 | 1 | 0.521 | 1 | |

The Nyquist formula gives the speed of a transmission system as 2B $\log_2$[M] [16], where M is the total number of signal levels, and B is the optical bandwidth, which is defined as $B = (c/\lambda^2)\Delta\lambda$, where c is the speed of light in vacuum, λ = 1.55 µm is the optical carrier wavelength, and Δλ is the signal's spectral width. Note $\log_2$[M] is in a binary form, i.e., $\log_2$[M] = log[M]/log[2]. This means that in our case, where B = 30 GHz and for four signal levels (00, 01, 10, and 11), the predicted speed is 120 Gb/s.

Silicon and silica components are readily available, making it easier and more affordable to build the suggested waveguide. As a result, assuming that the necessary technology is available and that the major outcomes of this simulation are valid, the experimental verification of the suggested waveguide may be completed. This is a technology problem that can be resolved in practice, so it is not a crucial obstacle. Several logic operations, on the other hand, have been experimentally implemented based on various optical waveguides and components [31–38].

Table 10 compares the functionality of the considered waveguide, for realizing the intended logic operations at various wavelengths, to that of several waveguides reported on the same topic. This table suggests that the proposed waveguide can achieve faster logic operations with higher CRs than the other listed schemes.

**Table 10.** At various wavelengths, a comparison of our design and other waveguide-based logic function designs.

| Operations | Design | Wavelength (nm) | CR (dB) | Ref. |
|---|---|---|---|---|
| XOR, AND, OR, NOR, NAND, XNOR | Dielectric-loaded waveguides | 471 | 24.41–33.39 | [16] |
| OR, NOT, AND, XOR | Metallic waveguide arrays | 632.8 | 9.3–20 | [17] |
| NOT, XOR, AND, OR, NOR, NAND, XNOR | Nanoring insulator–metal–insulator waveguides | 1550 | −1.1–18.75 | [18] |
| NOT, XOR, AND, OR, NOR, NAND, XNOR | Dielectric–metal–dielectric design | 900 and 1330 | 5.37–22 | [19] |
| AND, OR, NAND, NOR, XOR, Fan-Out, Half adder, Full adder | Photonic crystal circiuts | 1550 | 5.54–11.56 | [20] |
| XOR, AND, OR, NOT, NOR, XNOR, NAND | K-shaped silicon waveguides | 1550 | 30.5–34 | This work |

## 4. Conclusions

Using K-shaped silicon-on-silica waveguides, seven fundamental logic operations, including XOR, AND, OR, NOT, NOR, NAND, and XNOR, were simulated at the 1.55 μm telecommunications wavelength. These operations were simulated by means of FDTD solutions obtained in commercially available software. The correct execution of these logic operations relies on the constructive and destructive interferences that are caused by the suitable phase difference of the launched optical input beams. Compared to other waveguides reported for the same purpose, the suggested K-shaped waveguide achieves logic operations with a higher contrast ratio and operating speed.

**Author Contributions:** Conceptualization, A.K.; data curation, A.K.; formal analysis, A.K.; funding acquisition, A.K.; investigation, A.K.; methodology, A.K.; project administration, A.K.; resources, A.K.; software, A.K.; supervision, K.E.Z.; writing—original draft, A.K.; writing—review and editing, K.E.Z. All authors have read and agreed to the published version of the manuscript.

**Funding:** This research received no external funding.

**Data Availability Statement:** Not applicable.

**Acknowledgments:** A.K. thanks the Chinese Academy of Sciences President's International Fellowship Initiative (Grant No. 2022VMB0013) for the support of this work.

**Conflicts of Interest:** The authors declare no conflict of interest.

## References

1. Houbavlis, C.; Zoiros, K.E.; Kalyvas, M.; Theophilopoulos, G.; Bintjas, C.; Yiannopoulos, K.; Pleros, N.; Vlachos, K.; Avramopoulos, H.; Schares, L.; et al. All-optical signal processing and applications within the Esprit project DO ALL. *J. Lightwave Technol.* **2005**, *23*, 781–801. [CrossRef]
2. Clavero, R.; Mart´ınez, J.M.; Ramos, F.; Mart, J. All-optical packet routing scheme for optical label-swapping networks. *Opt. Express* **2004**, *12*, 4326–4332. [CrossRef] [PubMed]
3. Ji, W.; Zhang, M.; Ye, P. All-optical-packet header and payload separation for unslotted optical-packet-switched networks. *J. Lightwave Technol.* **2007**, *25*, 703–709. [CrossRef]
4. Ma, S.; Sun, H.; Chen, Z.; Dutta, N.K. High-speed all-optical PRBS generation based on quantum-dot semiconductor optical amplifiers. *Opt. Express* **2009**, *17*, 18469–18477. [CrossRef]
5. Singh, S.; Lovkesh; Ye, X.; Kaler, R.S. Design of ultrafast encryption and decryption circuits for secured optical networks. *IEEE J. Quantum Electron.* **2012**, *48*, 1547–1553. [CrossRef]
6. Aikawa, Y.; Shimizu, S.; Uenohara, H. Demonstration of all-optical divider circuit using SOA-MZI-type XOR gate and feedback loop for forward error detection. *J. Lightwave Technol.* **2011**, *29*, 2259–2266. [CrossRef]
7. Kim, S.H.; Kim, J.H.; Choi, J.W.; Son, C.W.; Byun, Y.T.; Jhon, Y.M.; Lee, S.; Woo, D.H.; Kim, S.H. All-optical half adder using cross gain modulation in semiconductor optical amplifiers. *Opt. Express* **2006**, *14*, 10693–10698. [CrossRef]
8. Gayen, D.K.; Bhattachryya, A.; Chattopadhyay, T.; Roy, J.N. Ultrafast all-optical half-adder using quantum-dot semiconductor optical amplifier-based Mach–Zehnder interferometer. *J. Lightwave Technol.* **2012**, *30*, 3387–3393. [CrossRef]

9. Berrettini, G.; Nguyen, A.T.; Lazzeri, E.; Meloni, G.; Scaffardi, M.; Pot, L.; Bogoni, A. All-optical digital circuits exploiting SOA-based loop memories. *IEEE J. Sel. Top. Quantum Electron.* **2012**, *18*, 847–858. [CrossRef]
10. Wang, Y.; Zhang, X.; Dong, J.; Huang, D. Simultaneous demonstration on all-optical digital encoder and comparator at 40 Gb/s with semiconductor optical amplifiers. *Opt. Express* **2007**, *15*, 15080–15085. [CrossRef]
11. Scaffardi, M.; Ghelfi, P.; Lazzeri, E.; Pot, L.; Bogoni, A. Photonic processing for digital comparison and full addition based on semiconductor optical amplifiers. *IEEE J. Sel. Top. Quantum Electron.* **2008**, *14*, 826–833. [CrossRef]
12. Zoiros, K.E.; Houbavlis, T.; Kalyvas, M. Ultra-high speed all-optical shift registers and their applications in OTDM networks. *Opt. Quantum Electron.* **2004**, *36*, 1005–1053. [CrossRef]
13. Kumar, S.; Willner, A.E. Simultaneous four-wave mixing and cross-gain modulation for implementing an all-optical XNOR logic gate using a single SOA. *Opt. Express* **2006**, *14*, 5092–5097. [CrossRef] [PubMed]
14. Jung, Y.J.; Son, C.W.; Jhon, Y.M.; Lee, S.; Park, N. One-level simplification method for all-optical combinational logic circuits. *IEEE Photon. Technol. Lett.* **2008**, *20*, 800–802. [CrossRef]
15. Mashanovich, G.Z.; Milošević, M.M.; Nedeljkovic, M.; Owens, N.; Xiong, B.; Teo, E.J.; Hu, Y. Low loss silicon waveguides for the mid-infrared. *Opt. Express* **2011**, *19*, 7112–7119. [CrossRef]
16. Yao, C.; Kotb, A.; Wang, B.; Singh, S.; Guo, C. All-optical logic gates using dielectric-loaded waveguides with quasi-rhombus metasurfaces. *Opt. Lett.* **2020**, *45*, 3769–3772. [CrossRef]
17. Yanga, W.; Shi, X.; Xing, H.; Chen, X. All-optical logic gates based on metallic waveguide arrays. *Res. Phys.* **2018**, *11*, 837–841. [CrossRef]
18. Abdulnabi, S.H.; Abbas, M.N. All-optical logic gates based on nanoring insulator–metal–insulator plasmonic waveguides at optical communications band. *J. Nanophotonics* **2019**, *13*, 016009. [CrossRef]
19. Al-Musawi, H.K.; Al-Janabi, A.K.; Al-Abassi, S.A.W.; Abusiba, N.A.A.; Al-Fatlawi, N.A.Q. Plasmonic logic gates based on dielectric-metal-dielectric design with two optical communication bands. *Optik* **2020**, *223*, 165416. [CrossRef]
20. Caballero, L.P.; Povinelli, M.L.; Ramirez, J.C.; Guimarães, P.S.S.; Neto, O.P.V. Photonic crystal integrated logic gates and circuits. *Opt. Express* **2022**, *30*, 1976. [CrossRef]
21. Asakawa, K.; Sugimoto, Y.; Nakamura, S. Silicon photonics for telecom and data-com applications. *Opto-Electron. Adv.* **2020**, *3*, 200011. [CrossRef]
22. Sun, C.; Yu, Y.; Zhang, X. Ultra-compact waveguide crossing for a mode-division multiplexing optical network. *Opt. Lett.* **2017**, *42*, 4913–4916. [CrossRef] [PubMed]
23. Green, W.M.J.; Rooks, M.J.; Sekaric, L.; Vlasov, Y.A. Ultra-compact, low RF power, 10 Gb/s silicon Mach-Zehnder modulator. *Opt. Express* **2007**, *15*, 17106–17113. [CrossRef] [PubMed]
24. Dimitropoulos, D.; Raghunathan, V.; Claps, R.; Jalali, B. Phase-matching and nonlinear optical processes in silicon waveguides. *Opt. Express* **2004**, *12*, 149–160. [CrossRef]
25. Kotb, A.; Guo, C. 100 Gb/s all-optical multifunctional AND, XOR, NOR, OR, XNOR, and NAND logic gates in a single compact scheme based on semiconductor optical amplifiers. *Opt. Laser Technol.* **2021**, *137*, 106828. [CrossRef]
26. Zoiros, K.E.; Papadopoulos; Houbavlis, T.; Kanellos, G.T. Theoretical analysis and performance investigation of ultrafast all-optical Boolean XOR gate with semiconductor optical amplifier-assisted Sagnac interferometer. *Opt. Commun.* **2006**, *258*, 114–134. [CrossRef]
27. Passaro, V.M.N.; Notte, M.L. Optimizing SOI slot waveguide fabrication tolerances and strip-slot coupling for very efficient optical sensing. *Sensors* **2012**, *12*, 2436–2455. [CrossRef]
28. Prinzen, A.; Waldow, M.; Kurz, H. Fabrication tolerances of SOI based directional couplers and ring resonators. *Opt. Express* **2013**, *21*, 17212–17220. [CrossRef]
29. Available online: https://www.edmundoptics.com/p/1550nm-0-250mw-fiber-coupled-laser/12219/ (accessed on 13 November 2022).
30. Al-Hetara, A.M.; Shamsan, Z.A. Optical wavelength and dimensions tolerance criterion for multimode interference couplers. *WSEAS Trans. Commun.* **2014**, *13*, 567–571.
31. Kita, S.; Nozaki, K.; Takata, K.; Shinya, A.; Notomi, M. Ultrashort low-loss Ψ gates for linear optical logic on Si photonics platform. *Commun. Phys.* **2020**, *3*, 33. [CrossRef]
32. Donzella, V.; Sherwali, A.; Flueckiger, J.; Grist, S.M.; Fard, S.T.; Chrostowski, L. Design and fabrication of SOI micro-ring resonators based on sub-wavelength grating waveguides. *Opt. Express* **2015**, *23*, 4791–4803. [CrossRef] [PubMed]
33. Pan, D.; Wei, H.; Xu, H. Optical interferometric logic gates based on metal slot waveguide network realizing whole fundamental logic operations. *Opt. Express* **2013**, *21*, 9556. [CrossRef] [PubMed]
34. Gao, L.; Chen, L.; Wei, H.; Xu, H. Lithographically fabricated gold nanowire waveguides for plasmonic routers and logic gates. *Nanoscale* **2018**, *10*, 14771. [CrossRef] [PubMed]
35. Fu, Y.; Hu, X.; Lu, C.; Yue, S.; Yang, H.; Gong, Q. All-optical logic gates based on nanoscale plasmonic slot waveguides. *Nano Lett.* **2012**, *12*, 5784–5790. [CrossRef] [PubMed]
36. Hou, Z.; Sun, Y.; Li, Q.; Xudong, F.; Cheng, R. Smart bio-gel optofluidic Mach–Zehnder interferometers multiphoton-lithographically customized with chemo-mechanical-opto transduction and bio-triggered degradation. *Lab Chip* **2020**, *20*, 3815–3823. [CrossRef]

37. Liang, D.; Zhang, X.; Li, M.; Lin, Z.; Dai, H.; Wu, Z.; Pu, J. Visually Adjusting Coupling Conditions in Light-Emitting Micro-Components. *IEEE Photon. Technol. Lett.* **2019**, *31*, 1425–1428. [CrossRef]
38. Kowsari, A.; Saghaei, H. Resonantly enhanced all-optical switching in microfibre Mach–Zehnder interferometers. *Electron. Lett.* **2018**, *54*, 229–231. [CrossRef]

*electronics*

*Article*

# Investigation of Barrier Layer Effect on Switching Uniformity and Synaptic Plasticity of AlN Based Conductive Bridge Random Access Memory

**Srikant Kumar Mohanty [1], Kuppam Poshan Kumar Reddy [1], Chien-Hung Wu [2,*], Po-Tsung Lee [1], Kow-Ming Chang [3], Prabhakar Busa [4] and Yaswanth Kuthati [4,*]**

1 UST-IPPP, College of Electrical and Computer Engineering, National Yang Ming Chiao Tung University, Hsinchu 30010, Taiwan
2 Department of Optoelectronics and Materials Engineering, Chung Hua University, Hsinchu 30012, Taiwan
3 Institute of Electronics Engineering, National Yang Ming Chiao Tung University, Hsinchu 30010, Taiwan
4 Department of Anesthesiology, Cathy General Hospital, Taipei 10630, Taiwan
* Correspondence: rossiwu@chu.edu.tw (C.-H.W.); yaswanthk1987@gmail.com (Y.K.)

**Abstract:** In this work, we investigated the effect of the tungsten nitride (WNx) diffusion barrier layer on the resistive switching operation of the aluminum nitride (AlN) based conductive bridge random access memory. The WNx barrier layer limits the diffusion of Cu ions in the AlN switching layer, hence controlling the formation of metallic conductive filament in the host layer. The device operated at a very low operating voltage with a $V_{set}$ of 0.6 V and a $V_{reset}$ of 0.4 V. The spatial and temporal switching variability were reduced significantly by inserting a barrier layer. The worst-case coefficient of variations ($\sigma/\mu$) for HRS and LRS are 33% and 18%, respectively, when barrier layer devices are deployed, compared to 167% and 33% when the barrier layer is not present. With a barrier layer, the device exhibits data retention behavior for more than $10^4$ s at 120 °C, whereas without a barrier layer, the device fails after $10^3$ s. The device demonstrated synaptic behavior with long-term potentiation/depression (LTP/LTD) for 30 epochs by stimulating with a train of identical optimized pulses of 1 μs duration.

**Keywords:** CBRAM; synaptic device; AlN; barrier layer; potentiation; depression

**Citation:** Mohanty, S.K.; Reddy, K.P.K.; Wu, C.-H.; Lee, P.-T.; Chang, K.-M.; Busa, P.; Kuthati, Y. Investigation of Barrier Layer Effect on Switching Uniformity and Synaptic Plasticity of AlN Based Conductive Bridge Random Access Memory. *Electronics* **2022**, *11*, 3432. https://doi.org/10.3390/electronics11213432

Academic Editor: Lucas Lamata

Received: 29 September 2022
Accepted: 20 October 2022
Published: 23 October 2022



## 1. Introduction

The conductive bridge random access memory (CBRAM) has demonstrated enormous potential for energy-efficient non-volatile storage with cognitive computing ability due to its high scalability, large dynamic memory window, good analog conductance switching, and low-off state current compared to valence change memory [1–4]. The CBRAM is comprised of a simple metal-insulator-metal (MIM) structure with a top electrode made of an electrochemically active metal (e.g., Cu, Ag), a bottom electrode made of an inert metal (Pt, W), and a sandwiched insulating layer that allows the formation and dissolution of nanoscale metallic conductive filaments [5–9]. The cation ($Ag^+$, $Cu^{2+}$) migration process in filament evolution is functionally equivalent to $Ca^{2+}$ dynamics of biological synapse and it was previously reported by demonstrating synaptic behaviors including short-term plasticity (STP), long-term plasticity (LTP), and spike-time dependent plasticity (STDP) characteristics [10–15]. Despite various attempts to mimic the human brain by utilizing resistive switching architecture, endeavors to match device performance to biological counterparts continue. To justify the suitability for neuromorphic computing, a memristive device must exhibit uniform spatial and temporal switching, long retention, and linear conductance updating capabilities. The defects in the insulating layer and electrode-insulator interface induce uncertainty in the behavior of the conductive filament. Additionally, the inadequate controllability of metal-ion injection and the creation of numerous filaments are implicated in the spatial/temporal switching variability [16–18].

As a switching layer in CRAM devices, a variety of chalcogenides (Ag2S, Cu2S), oxides (HfOx, TaOx, and SiOx), and nitride (SiNx and AlNx) based materials have been employed. For these reasons, AlN is a potential contender among these storage media because of its large energy bandgap (6.0 eV), high thermal conductivity (285 W/m.K), and greater thermal stability, all of which result in enhanced switching behavior. It is also much more compatible with nitride electrodes (TiN and TaN), which are common in semiconductor production foundries. Moreover, AlN-based memory could be easily integrated with power electronics devices, including nitride compound semiconductor-based high electron mobility transistors (HEMTs) [19] that can drive high-speed memristors. The anion migration-based AlN RRAM has shown energy-efficient ultra-high switching speeds and high-density multilevel storage capabilities [20,21]. Recently, the high-performance cation migration-based resistive switching operation of the AlN CBRAM device has been reported for non-volatile storage applications. The switching instability remains the main problem in CBRAM devices due to uncontrolled drift and diffusion of metal-ion species from the active electrode to the switching layer during switching cycles.

Several techniques such as the insertion of a metal layer [22], employing a composite electrode [23], embedding a metal nano-island array [24], and the alloying of conduction channels [25] were deployed to improve the switching stability of the CBRAM device. Compared to other techniques, the insertion of the barrier layer is a cost-effective and easy-to-fabricate method. Transition metals such as titanium, tantalum, and tungsten have been extensively studied as diffusion barrier materials due to their immiscible characteristics with copper [26–28]. Metal nitrides act as better barrier layers compared to their metal parts as nitridation improves chemical stability [29]. The diffusion barrier properties of $WN_x$ in copper interconnect and microelectronic circuits have already been reported [30,31]. WNx has shown better barrier properties for the copper electrode as nitrogen does not react with copper and smaller-size nitrogen molecules fill the void, reducing the diffusion of copper [32]. In this work, we propose the insertion of the $WN_x$ barrier layer to enhance the switching uniformity by limiting Cu diffusion into the switching layer. The reliability characteristics, such as endurance and data retention characteristics, improved significantly with barrier layer insertion. To further demonstrate the suitability of the CBRAM device for neuromorphic applications, the long-term potentiation and depression synaptic activity are illustrated with superior linearity.

## 2. Materials and Methods

The schematic device processing started with the deposition of 500 nm thermally grown $SiO_2$ on a RCA cleaned Si wafer as an isolation layer. For the bottom electrode, a 20 nm Ti adhesion layer with a Pt layer of 60 nm thickness was deposited by a DC sputtering system. Then an AlN switching layer of 10 nm thickness was deposited at room temperature in a mixed Ar and $N_2$ ambient atmosphere using DC sputtering at a working pressure of 7.6 mTorr. Prior to patterning the top electrode, a 7 and 10 nm thick $WN_x$ diffusion barrier layer was deposited by DC sputtering at room temperature by maintaining the flow rate of Ar and $N_2$ at 24 sccm and 12 sccm, respectively. Subsequently, an 80 nm thick Cu layer was deposited as the top electrode using DC sputtering at room temperature to complete the stacked device structure. Both the barrier and the Cu top electrode were patterned using a circular metal shadow mask to a diameter of 120 μm. The electrical DC and pulse characteristics were performed with the Agilent B1500/B1530 semiconductor parameter analyzer. The Cu top electrode was always biased with DC and pulse signals, while the Pt bottom electrode was grounded in the electrical measurement setup. The structural information of the device was taken by using a JOEL JEM-2010 transmission electron microscope (TEM).

## 3. Results and Discussion

Figure 1a depicts the schematic layout of the proposed $Cu/WN_x/AlN/Pt$ CBRAM device. The cross-sectional TEM image of the device shown in Figure 1b clearly indicates the multiple layers in the stacked structure. The device switches from a high-resistance pristine state to a low-resistance state called the forming process. The forming behavior of without barrier layer (WOBL) and with barrier layer (WBL) devices shown in Figure 1c confirms a larger leakage current with low forming voltage in WOBL devices due to the diffusion of copper into the switching layer during top electrode deposition [33,34]. The insertion of the $WN_x$ barrier layer, which prevents copper diffusion, reduces the leakage current and increases the forming voltage. The result shows with a thicker $WN_x$ barrier layer (10 nm), forming voltage increases undesirably and during resistive switching operation, it shows volatile operation. Hence, a WBL device with a barrier layer thickness of 7 nm was compared with the WOBL device for further investigation. The forming voltage distribution for 10 devices is shown in Figure 1d. The average forming voltage of the WOBL device is 1.2 V, whereas it is 2.5 V for the WBL device. The device-to-device variation of $V_{forming}$ is larger in the case of the WOBL device and is mainly due to uneven spatial copper diffusion into the AlN switching layer.

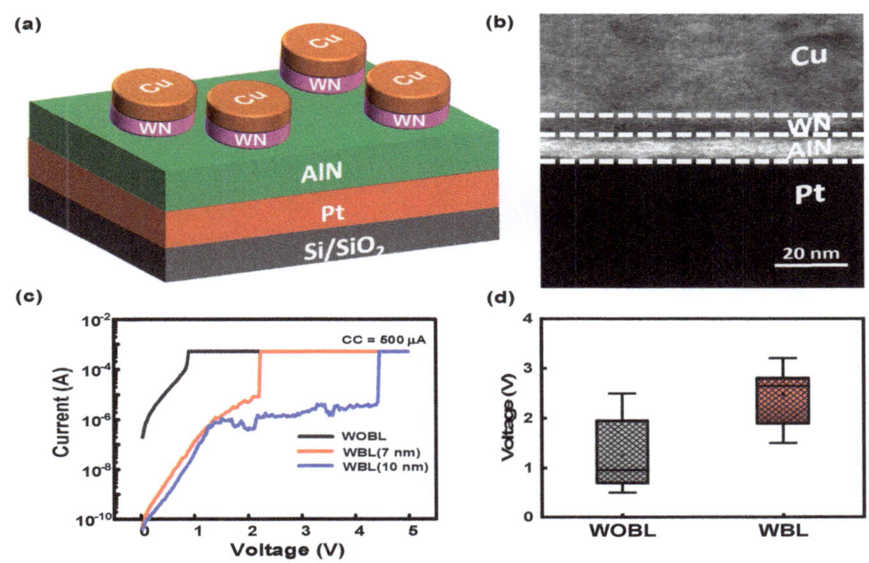


**Figure 1.** (**a**) Schematic structure of the Pt/AlN/WN/Cu RRAM device. (**b**) The TEM cross-sectional image of the device. (**c**) The forming characteristics of WOBL and WBL devices. (**d**) Forming voltage distribution for memristor without/with barrier layer for 10 devices.

Figure 2a,b show the typical DC current-voltage characteristics of the WOBL and WBL devices. Both devices exhibit bipolar switching behavior with a positive bias sweep of 1.2 V at a compliance current (CC) of 1 mA for set operation and a negative sweep of −0.8 V for reset operation, respectively. From the measured 100 DC switching cycles, we observed that the WOBL device shows larger cycle-to-cycle variation in HRS/LRS compared to the WBL device. Without the barrier layer, multiple incomplete conductive paths are formed by the diffusion of copper into the AlN layer during the forming process, which subsequently causes variation in set/reset switching cycles. The device-to-device and cycle-to-cycle on/off resistance distribution is plotted in Figure 2c,d with each device tested for 30 DC cycles. The on-state resistance, LRS, of WOBL devices is significantly lower compared to WBL devices. This indicates filament formation through a large amount of Cu ion migration. The WOBL devices show the worst-case coefficient of variances ($\sigma/\mu$) of 167% and 33% for HRS and LRS, respectively. In the case of the WBL, the variations improved significantly to 38% and 18%, respectively. This result confirms the switching stability with the insertion of the barrier layer. The resistance at LRS increases with the increase of temperature, which confirms the metallic nature of the conductive filament shown in Figure 2f. From the resistance-temperature plot, the temperature resistance coefficient is calculated by using Equation (1) [35]:

$$R(T) = R_0[1 + \alpha(T - T_0)] \quad (1)$$

where $R_0$ is the resistance at temperature $T_0$ and $\alpha$ is the temperature coefficient of resistance. The estimated $\alpha$ = 0.003 confirms the formation of a Cu filament during the on state.

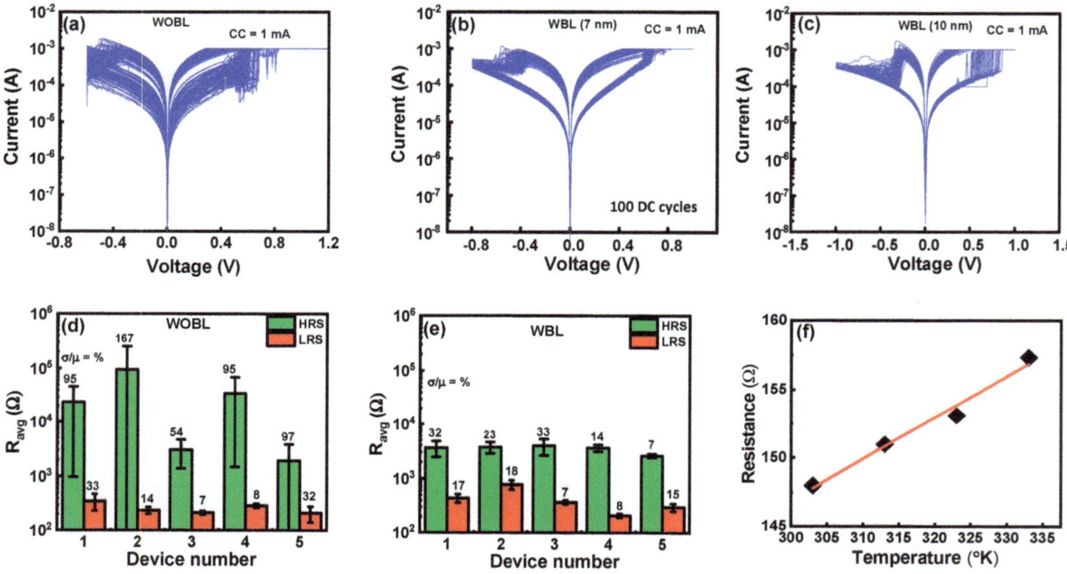


**Figure 2.** Typical DC I-V characteristics for 100 cycles; (**a**) without barrier layer; (**b**) with barrier layer (7 nm); and (**c**) with barrier layer (10 nm). The statistical distribution of LRS and HRS for memristor; (**d**) without barrier layer; (**e**) with barrier layer; and (**f**) temperature-dependent resistance in LRS.

The conduction mechanism of the Cu/$WN_x$/AlN/Pt CBRAM cell was investigated by analyzing the ln(I)~ln(V) graph in different regions as shown in Figure 3. The I-V curve between 0 and 0.25 V with a slope of 1.2 is well-fitted with the ionic conduction Equation (2) given by [36,37]:

$$\ln(j) \propto \ln(V) - (\ln T + a/T) \quad (2)$$

where $J$, $V$, $a$, and $T$ are current density, bias voltage, constant, and temperature. This equation explains that ionic conduction occurs due to the oxidation of electrochemically active metal Cu into Cu ions and then the migration of Cu ions to the Pt bottom electrode with the application of positive bias to the top Cu electrode. The Cu ions are reduced at the bottom Pt electrode and the metallic conductive filaments grow from the bottom electrode to the top electrode, making the connection between the electrodes. As the applied bias voltage increases, the diameter of the filament becomes thicker, resulting in higher slopes of 2.1 and 3.7 between 0.25 to 0.58 V and 0.58 to 0.65 V, respectively. Due to the presence of the $WN_x$ barrier layer, there is a gradual increment in slope. During biasing from 0.55 to 0 V, the current plot fitted well with the ohmic conduction mechanism which can be expressed by Equation (3):

$$\ln(j) \propto \ln(V) - a/T \quad (3)$$

where $J$, $V$, $a$, and $T$ are current density, bias voltage, constant, and temperature. The current conduction from 0 to −0.47 V follows Ohm's law as −0.47 V was not sufficient to rapture the Cu filament. During reset operation with bias, from −0.48 to −0.74 V, there is a slope of −2.2 which indicates the breaking of the Cu filament, while from −0.74 to 0 V there is a positive slope of 1.3, showing the drift of Cu ions from the filament side to the Cu electrode. Similar conduction mechanisms were demonstrated using an Ag-based conductive bridge random access memory system in a prior study [36].

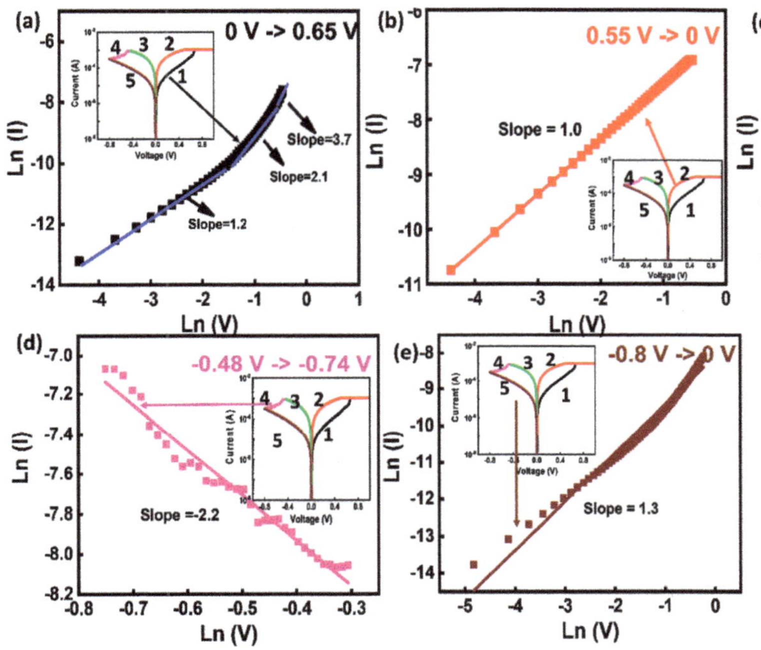


**Figure 3.** Conduction mechanism of WBL device in; (**a**) set operation HRS region from 0 to 0.65 V; (**b**) LRS region from 0.55 to 0 V, (**c**,**e**) reset operation from 0 to −0.47 V and from −0.47 to −0.74 V (**d**) reset operation HRS region from 0.74 to 0 V.

The schematic of the resistive switching mechanism for WOBL and WBL CBRAM devices is depicted in Figure 4, explaining switching stability. Because there is no barrier layer between the Cu atoms and the AlN switching layer, more Cu atoms diffuse into the AlN switching layer for the WOBL device than the WBL sample in the pristine state, as shown in Figure 4a,b. During positive bias, Cu ionization begins at the top electrode and progresses towards the Pt electrode, where it is reduced back to Cu atoms to form the conductive bridge in the CBRAM device. After the complete formation of the conductive path between the electrodes, the device switches to the LRS state. The WOBL layer inherits multiple incomplete conductive filaments with one or a few complete conductive filaments. During the reset process, the electrochemical dissolution of the filament happens, and the Cu ions drift back to the top electrode. The switching variability in the WOBL device is illustrated by taking two competing filamentary paths into account. Assuming the CF1 is an initially formed conductive path which undergoes continuous structural change with several set-reset switching cycles. The gap formed in the filament during the reset event is g1. The selected conductive path will be CF1 only if g1 < g2. The CF2 conductive path undergoes minimal structural change as very little current flows. After several set-reset cycles, the CF2 becomes the selected conductive path with g2 < g1, which leads to a new LRS/HRS value. The change in conductive path is the main source of switching instability, and it is more pronounced with the presence of multiple incomplete conductive filaments. The presence of the $WN_x$ barrier layer limits the diffusion of Cu atoms into the switching layer, preventing the occurrence of multiple conductive paths as shown in Figure 4c. The lower number of filamentary paths improves switching stability, which is an important requirement for synaptic devices.

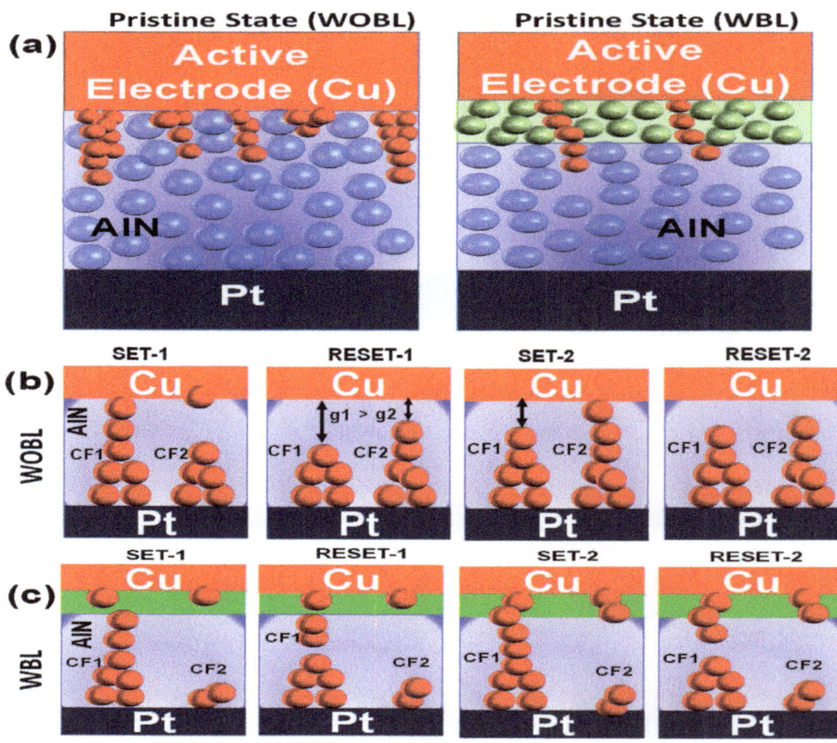


**Figure 4.** Schematic of (**a**) WOBL and WBL devices in a pristine state. Set and reset the switching mechanism of (**b**) WOBL and (**c**) WBL devices.

The endurance plots for both devices are shown in Figure 5a,b for 500 DC cycles. The HRS/LRS were evaluated at a read voltage of 0.2 V. The WBL device has shown a stable and consistent memory window as compared to the WOBL device. To further evaluate the nonvolatile operation of RRAM, the data retention characteristic behavior at HRS/LRS was evaluated at room temperature (RT) as well as the elevated temperature of 120 °C as displayed in Figure 5c,d. Both WOBL and WBL devices exhibited high retention of $10^4$ s at RT. Nonetheless, at 120 °C, the HRS of the WOBL device degrades after $10^3$ s due to temperature-assisted metal diffusion into the switching layer.

The gradual conductance switching rather than abrupt switching upon application of a pulse is an important requirement for the memristive device to behave as an artificial synapse in neuromorphic hardware. To evaluate the synaptic characteristics of the WBL device, 50 identical positive pulses of 0.8 V/1 µs for potentiation and 50 similar negative pulses of −0.8 V/1 µs for depression were applied to the device as shown in Figure 6a. The conductance increased gradually from 400 to 800 µS during the potentiation period, and similar progressive decrementing behavior was observed during the depression period. Figure 6b shows the normalized potentiation/depression curves from which the non-linearity factor can be calculated by using the following Equations (4)–(6) [38]:

$$G_P = B\{1 - e^{\left(\frac{P}{A}\right)}\} + G_{\min} \tag{4}$$

$$G_D = -B\left\{1 - e^{\left(\frac{P-P_{max}}{A}\right)}\right\} + G_{max} \tag{5}$$

$$B = \frac{G_{max} - G_{min}}{1 - e^{\frac{-P_{max}}{A}}} \tag{6}$$

where $G$, $G_{max}$, $G_{min}$, and $P_{max}$ denote the conductance, maximum conductance, minimum conductance, and maximum number of pulses, respectively.

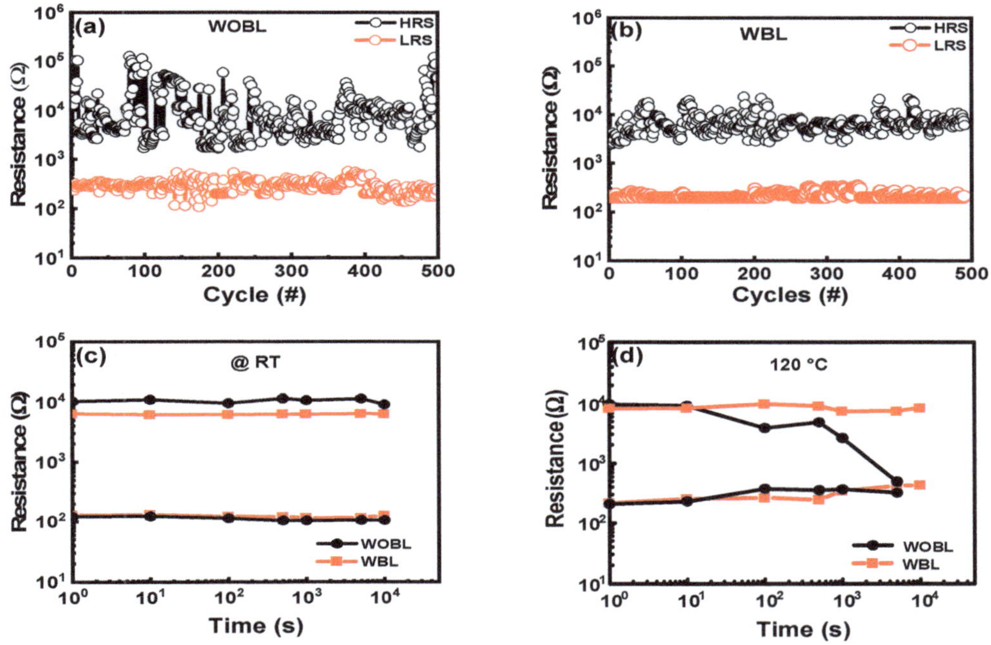


**Figure 5.** **DC** endurance behavior of (**a**) WOBL and (**b**) WBL devices. Data retention characteristics of WOBL and WBL devices were measured at (**c**) room temperature and (**d**) 120 °C.

The non-linearities calculated for potentiation and depression are 2.2 and 3.5, respectively. Figure 6c displays the long-term potentiation/depression (LTP/LTD) for 30 epochs. When switching between successive pulses, the device displays smooth and monotonic transitions of conductance, which is desirable for enhancing neuromorphic systems' learning accuracy. The LTP/LTD characteristics of the device are suitable for pattern recognition applications of the neural network. The neurons transmit signals through the synapse, and the learning algorithm is used to update the weight value of the synaptic device. The WBL device shows a multilevel conductance state with better linearity and could achieve pattern recognition accuracy quickly. The above results confirm the AlN CBRAM with $WN_x$ barrier layer is a potential synaptic device for neuromorphic computing.

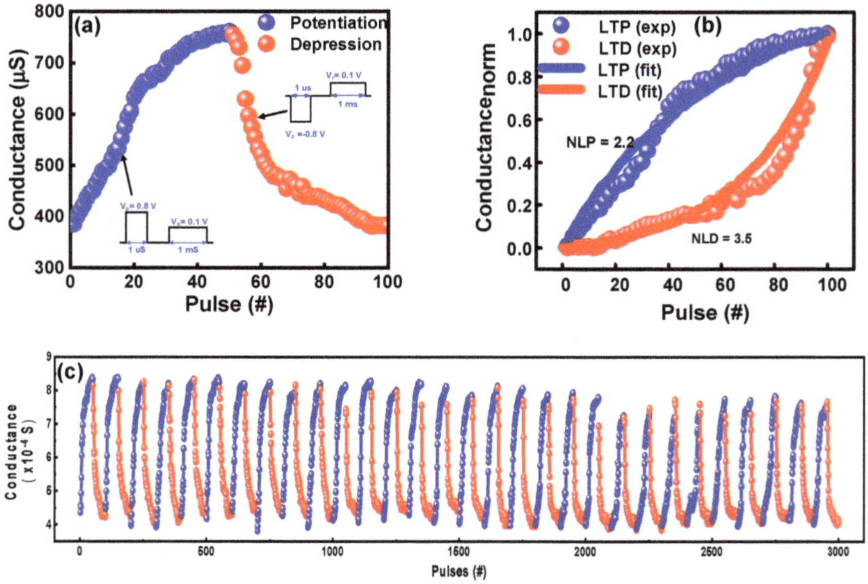


**Figure 6.** (**a**) Synaptic potentiation/depression characteristics measured by applying 50 identical 0.8 V/1 μs potentiation pulses and 50 identical −0.8 V/1 μs depression pulses. (**b**) Normalized conductance with nonlinearity measurement. (**c**) Thirty repeated potentiation/depression cycles of the WBL device.

## 4. Conclusions

In this study, we have demonstrated improved switching stability and excellent data retention along with highly linear 30 epoch LTP/LTD synaptic functions by inserting the $WN_x$ barrier layer in the AlN memristor. The result could be attributed to the insertion of a 7 nm thick $WN_x$ barrier layer, which prevents the formation of multiple conductive bridges by limiting the diffusion of Cu into the switching layer. The cycle-to-cycle and device-to-device switching uniformity improved substantially by regulating Cu migration in the host layer. The CBRAM device demonstrated longer DC endurance for 500 cycles and stable data retention for $10^3$ s at 120 °C. The ohmic and ionic conduction mechanisms of the non-volatile CBRAM cell during set/reset operation were explained by fitting the current transport in the switching layer. Finally, the repeated LTP and LTD synaptic functions with a fixed pulse scheme are demonstrated.

**Author Contributions:** Conceptualization, S.K.M. and C.-H.W.; writing and data curation, S.K.M. and K.P.K.R.; fabrication and measurements, S.K.M., P.B.; formal analysis, review, and investigation, C.-H.W. and Y.K.; project administration, P.-T.L., P.B. and K.-M.C. All authors have read and agreed to the published version of the manuscript.

**Funding:** This work was supported by the Ministry of Science and Technology under the contract MOST 109-2221-E-216-002.

**Conflicts of Interest:** The authors declare no conflict of interest.

## References

1. Waser, R.; Dittmann, R.; Staikov, G.; Szot, K. Redox-Based Resistive Switching Memories—Nanoionic Mechanisms, Prospects, and Challenges. *Adv. Mater.* **2009**, *21*, 2632–2663. [CrossRef]
2. Wang, Z.; Wu, H.; Burr, G.W.; Hwang, C.S.; Wang, K.L.; Xia, Q.; Yang, J.J. Resistive switching materials for information processing. *Nat. Rev. Mater.* **2020**, *5*, 173–195. [CrossRef]
3. Cha, J.-H.; Yang, S.Y.; Oh, J.; Choi, S.; Park, S.; Jang, B.C.; Ahn, W.; Choi, S.-Y. Conductive-bridging random-access memories for emerging neuromorphic computing. *Nanoscale* **2020**, *12*, 14339–14368. [CrossRef]

4. Choi, S.; Tan, S.H.; Li, Z.; Kim, Y.; Choi, C.; Chen, P.-Y.; Yeon, H.; Yu, S.; Kim, J. SiGe epitaxial memory for neuromorphic computing with reproducible high performance based on engineered dislocations. *Nat. Mater.* **2018**, *17*, 335–340. [CrossRef] [PubMed]
5. Liu, Y.; Gao, J.; Wu, F.; Tian, H.; Ren, T.L. The Origin of CBRAM With High Linearity, On/Off Ratio, and State Number for Neuromorphic Computing. *IEEE Trans. Electron. Devices* **2021**, *68*, 2568–2571. [CrossRef]
6. Pan, F.; Gao, S.; Chen, C.; Song, C.; Zeng, F. Recent progress in resistive random access memories: Materials, switching mechanisms, and performance. *Mater. Sci. Eng. R Rep.* **2014**, *83*, 1–59. [CrossRef]
7. Valov, I.; Waser, R.; Jameson, J.R.; Kozicki, M.N. Electrochemical metallization memories—Fundamentals, applications, prospects. *Nanotechnology* **2011**, *22*, 254003. [CrossRef] [PubMed]
8. Sun, W.; Gao, B.; Chi, M.; Xia, Q.; Yang, J.J.; Qian, H.; Wu, H. Understanding memristive switching via in situ characterization and device modeling. *Nat. Commun.* **2019**, *10*, 3453. [CrossRef]
9. Yu, S.; Wong, H.S.P. Compact Modeling of Conducting-Bridge Random-Access Memory (CBRAM). *IEEE Trans. Electron Devices* **2011**, *58*, 1352–1360. [CrossRef]
10. Ilyas, N.; Li, D.; Li, C.; Jiang, X.; Jiang, Y.; Li, W. Analog Switching and Artificial Synaptic Behavior of Ag/SiOx:Ag/TiOx/p++-Si Memristor Device. *Nanoscale Res. Lett.* **2020**, *15*, 30. [CrossRef]
11. Ohno, T.; Hasegawa, T.; Tsuruoka, T.; Terabe, K.; Gimzewski, J.K.; Aono, M. Short-term plasticity and long-term potentiation mimicked in single inorganic synapses. *Nat. Mater.* **2011**, *10*, 591–595. [CrossRef] [PubMed]
12. Yan, X.; Qin, C.; Lu, C.; Zhao, J.; Zhao, R.; Ren, D.; Zhou, Z.; Wang, H.; Wang, J.; Zhang, L.; et al. Robust Ag/ZrO2/WS2/Pt Memristor for Neuromorphic Computing. *ACS Appl. Mater. Interfaces* **2019**, *11*, 48029–48038. [CrossRef] [PubMed]
13. Lim, S.; Sung, C.; Kim, H.; Kim, T.; Song, J.; Kim, J.J.; Hwang, H. Improved Synapse Device With MLC and Conductance Linearity Using Quantized Conduction for Neuromorphic Systems. *IEEE Electron Device Lett.* **2018**, *39*, 312–315. [CrossRef]
14. Sun, B.; Guo, T.; Zhou, G.; Ranjan, S.; Jiao, Y.; Wei, L.; Zhou, Y.N.; Wu, Y.A. Synaptic devices based neuromorphic computing applications in artificial intelligence. *Mater. Today Phys.* **2021**, *18*, 100393. [CrossRef]
15. Sun, B.; Zhou, G.; Yu, T.; Chen, Y.; Yang, F.; Zhao, Y. Multi-factor-controlled ReRAM devices and their applications. *J. Mater. Chem. C* **2022**, *10*, 8895–8921. [CrossRef]
16. Lu, W.; Jeong, D.S.; Kozicki, M.; Waser, R. Electrochemical metallization cells—Blending nanoionics into nanoelectronics? *MRS Bull.* **2012**, *37*, 124–130. [CrossRef]
17. Moon, K.; Lim, S.; Park, J.; Sung, C.; Oh, S.; Woo, J.; Lee, J.; Hwang, H. RRAM-based synapse devices for neuromorphic systems. *Faraday Discuss.* **2019**, *213*, 421–451. [CrossRef]
18. Celano, U.; Goux, L.; Belmonte, A.; Opsomer, K.; Detavernier, C.; Jurczak, M.; Vandervorst, W. Conductive filaments multiplicity as a variability factor in CBRAM. In Proceedings of the 2015 IEEE International Reliability Physics Symposium, Monterey, CA, USA, 19–23 April 2015; pp. MY.11.11–MY.11.13.
19. Mohanty, S.K.; Chen, Y.-Y.; Yeh, P.-H.; Horng, R.-H. Thermal Management of GaN-on-Si High Electron Mobility Transistor by Copper Filled Micro-Trench Structure. *Sci. Rep.* **2019**, *9*, 19691. [CrossRef]
20. Choi, B.J.; Torrezan, A.C.; Strachan, J.P.; Kotula, P.G.; Lohn, A.J.; Marinella, M.J.; Li, Z.; Williams, R.S.; Yang, J.J. High-Speed and Low-Energy Nitride Memristors. *Adv. Funct. Mater.* **2016**, *26*, 5290–5296. [CrossRef]
21. Mohanty, S.K.; Reddy, P.K.; Prasad, O.K.; Wu, C.H.; Chang, K.M.; Lin, J.C. Interface Engineering for 3-Bit per Cell Multilevel Resistive Switching in AlN Based Memristor. *IEEE Electron Device Lett.* **2021**, *42*, 1770–1773. [CrossRef]
22. Belmonte, A.; Kim, W.; Chan, B.T.; Heylen, N.; Fantini, A.; Houssa, M.; Jurczak, M.; Goux, L. A Thermally Stable and High-Performance 90-nm $Al_2O_3$\Cu-Based 1T1R CBRAM Cell. *IEEE Trans. Electron Devices* **2013**, *60*, 3690–3695. [CrossRef]
23. Kuo, C.C.; Chen, I.C.; Shih, C.C.; Chang, K.C.; Huang, C.H.; Chen, P.H.; Chang, T.C.; Tsai, T.M.; Chang, J.S.; Huang, J.C. Galvanic Effect of Au–Ag Electrodes for Conductive Bridging Resistive Switching Memory. *IEEE Electron. Device Lett.* **2015**, *36*, 1321–1324. [CrossRef]
24. Wang, J.; Li, L.; Huyan, H.; Pan, X.; Nonnenmann, S.S. Highly Uniform Resistive Switching in HfO2 Films Embedded with Ordered Metal Nanoisland Arrays. *Adv. Funct. Mater.* **2019**, *29*, 1808430. [CrossRef]
25. Yeon, H.; Lin, P.; Choi, C.; Tan, S.H.; Park, Y.; Lee, D.; Lee, J.; Xu, F.; Gao, B.; Wu, H.; et al. Alloying conducting channels for reliable neuromorphic computing. *Nat. Nanotechnol.* **2020**, *15*, 574–579. [CrossRef]
26. Holloway, K.; Fryer, P.M.; Cabral, C.; Harper, J.M.E.; Bailey, P.J.; Kelleher, K.H. Tantalum as a diffusion barrier between copper and silicon: Failure mechanism and effect of nitrogen additions. *J. Appl. Phys.* **1992**, *71*, 5433–5444. [CrossRef]
27. Ting, C.Y.; Wittmer, M. The use of titanium-based contact barrier layers in silicon technology. *Thin Solid Film.* **1982**, *96*, 327–345. [CrossRef]
28. Farahani, M.M.; Turner, T.E.; Barnes, J.J. Evaluation of Titanium as a Diffusion Barrier Between Aluminum and Silicon for 1.2 μm CMOS Integrated Circuits. *J. Electrochem. Soc.* **1987**, *134*, 2835–2845. [CrossRef]
29. Ono, H.; Nakano, T.; Ohta, T. Diffusion Barrier Effects Against Cu of W-N Layer Formed by Electron Cyclotron Resonance Plasma Nitridation on W Layer. *Jpn. J. Appl. Phys.* **1995**, *34*, 1827–1830. [CrossRef]
30. Hsieh, T.-E.; Lin, Y.-C.; Li, F.-M.; Shi, W.-C.; Huang, Y.-X.; Lan, W.-C.; Chin, P.-C.; Chang, E.Y. GaN High-Electron-Mobility Transistor with WN x /Cu Gate for High-Power Applications. *J. Electron. Mater.* **2015**, *44*, 4700–4705. [CrossRef]
31. Takeyama, M.; Noya, A. Preparation of $WN_x$ Films and Their Diffusion Barrier Properties in Cu/Si Contact Systems. *Jpn. J. Appl. Phys.* **1997**, *36*, 2261–2266. [CrossRef]

32. Asgary, S.; Hantehzadeh, M.R.; Ghoranneviss, M.; Boochani, A. Thermal stability of amorphous tungsten/tungsten nitride synthesis using HFCVD as a diffusion barrier for copper. *Appl. Phys. A* **2016**, *122*, 518. [CrossRef]
33. Panda, D.; Simanjuntak, F.M.; Chandrasekaran, S.; Pattanayak, B.; Singh, P.; Tseng, T.Y. Barrier Layer Induced Switching Stability in Ga:ZnO Nanorods Based Electrochemical Metallization Memory. *IEEE Trans. Nanotechnol.* **2020**, *19*, 764–768. [CrossRef]
34. Iraji-zad, A.; Ahadian, M.M.; Vashaei, Z. Room temperature diffusion of Cu in vanadium pentoxide thin films. *J. Phys. D Appl. Phys.* **2002**, *35*, 1176–1182. [CrossRef]
35. Kang, Y.; Liu, T.; Potnis, T.; Orlowski, M.K. Composite Cu/VO and VO/Cu Nanofilaments in Cu/Ta2O5/Pt Devices. *ECS Solid State Lett.* **2013**, *2*, Q54–Q57. [CrossRef]
36. Kwon, K.-C.; Song, M.-J.; Kwon, K.-H.; Jeoung, H.-V.; Kim, D.-W.; Lee, G.-S.; Hong, J.-P.; Park, J.-G. Nanoscale CuO solid-electrolyte-based conductive-bridging-random-access-memory cell operating multi-level-cell and 1selector1resistor. *J. Mater. Chem. C* **2015**, *3*, 9540–9550. [CrossRef]
37. Seung, H.-M.; Kwon, K.-C.; Lee, G.-S.; Park, J.-G. Flexible conductive-bridging random-access-memory cell vertically stacked with top Ag electrode, PEO, PVK, and bottom Pt electrode. *Nanotechnology* **2014**, *25*, 435204. [CrossRef]
38. Wu, W.; Wu, H.; Gao, B.; Yao, P.; Zhang, X.; Peng, X.; Yu, S.; Qian, H. A Methodology to Improve Linearity of Analog RRAM for Neuromorphic Computing. In Proceedings of the 2018 IEEE Symposium on VLSI Technology, Honolulu, HI, USA, 18–22 June 2018; pp. 103–104.

*electronics*

MDPI

*Article*

# An Analysis of Some Properties and the Use of the Twist Map for the Finite Frenkel–Kontorova Model

**Wolfgang Quapp [1,*] and Josep Maria Bofill [2,3]**

1 Mathematisches Institut, Universität Leipzig, D-04009 Leipzig, Germany
2 Departament de Química Orgánica, Universitat de Barcelona, 08028 Barcelona, Spain
3 Institut de Química Teórica i Computacional, Universitat de Barcelona, (IQTCUB), Martí i Franqués, 1, 08028 Barcelona, Spain
* Correspondence: quapp@math.uni-leipzig.de or quapp@uni-leipzig.de

**Abstract:** We discuss the twist map, with a special interest in its use for the finite Frenkel–Kontorova model. We explain the meaning of the tensile force in some proposed models. We demonstrate that the application of the twist map for the finite FK model is not correct, because the procedure ignores the necessary boundary conditions.

**Keywords:** Frenkel–Kontorova model; average distance; finite chain; tensile force; twist map

**Citation:** Quapp, W.; Bofill, J.M. An Analysis of Some Properties and the Use of the Twist Map for the Finite Frenkel–Kontorova Model. *Electronics* **2022**, *11*, 3295. https://doi.org/10.3390/electronics11203295

Academic Editor: Lucas Lamata

Received: 20 September 2022
Accepted: 10 October 2022
Published: 13 October 2022



## 1. Introduction

The Frenkel–Kontorova model (FK) was proposed in 1938 to discuss dislocations in a chain of particles [1]. In the last century, many aspects of solid-state physics have been discussed using the FK model as an example. One divides a set of particles into a one-dimensional subsystem of interacting elements and a remainder as a substrate. The latter acts on the extracted subsystem by a potential. Of special interest may be electronic applications [2–7] for Wigner electrons or Josephson junctions. Further models are charge-density wave conductors [8–10], charge transport in solids and on crystal surfaces [11], magnetic or ferro- and antiferromagnetic domain walls [12], magnetic superlattices [13], superconductivity [14,15], and vortex matter [16–18], to name a few.

The FK model for a chain of $N$ particles (called atoms) at positions $u_i$ on an axis has the energy [1,19]

$$U(\mathbf{u}) = \sum_{i=1}^{N} \lambda\, V(u_i) + \sum_{i=1}^{N-1} \frac{k}{2}(u_{i+1} - u_i - a_o)^2 \,. \tag{1}$$

The substrate potential function is the usual one

$$V(u_i) = 1 - \cos\Big(\frac{2\,\pi}{a_s} u_i\Big)\,.$$

The parameter $a_s$ is the period of this potential, and $\lambda$ is its amplitude. The second term is the spring term for the nearest neighbor forces between the atoms. Parameter $k$ is its amount. The equilibrium distance, $a_o$, of the chain would hold without the potential $V$, thus for $\lambda = 0$. Usually, the two periods $a_o$ and $a_s$ are different [20].

Sometimes the equilibrium distance, $a_o$, of the chain is missing [21–23]. At first, one has to assume that $a_o = 0$ is used. One treats the FK model as a different one with the pure harmonic potential of the spring forces between the nearest neighbors but often with an additional linear tensile force of amount $\mu$

$$\sum_{i=1}^{N-1} \left[\frac{k}{2}(u_{i+1} - u_i)^2 - \mu\,(u_{i+1} - u_i)\right]. \tag{2}$$

Note that the alternating tensile force of form (2) reduces to two summands

$$-\mu\,(u_N - u_1)\,.$$

If $\mu = 0$, then putting all $u_i = 0$ (or putting all $u_i = n\,a_s$ with an integer $n$) gives zero energy [24–26]. This is the global minimum, but it is a trivial one. If one starts with any finite chain with no zero distances between the atoms, and if one uses the values of the parameters of ref. [21] for $\lambda \leq k$, and $\mu = 0$, then a minimization would return this trivial result.

In Appendix A, we remark that for $\lambda \gg k$, further stationary states can exist for $a_o = 0$ and for $\mu = 0$. However, these states are not global minima.

The paper mainly discusses the application of the twist map on a finite FK chain. In Section 2, we recall the formulas for an equilibrium structure, and we argue against using these formulas to apply the twist map. In Section 3, we explain the meaning of the tensile force. In Sections 4 and 5, we discuss the boundary conditions of a finite FK chain and explain with examples the incorrect results of the twist map. Further small points are the energy of the FK chain and the putative problem of the irrational relations of the two periodicities of the FK model. Some aspects of our treatments include connections to an infinite FK chain [21].

## 2. Equations for Stationary States

At first, it seems that the gap of form (2) does not matter, because for the minimum search for the inner atoms, $u_i$, holds an equation, where the $a_o$ disappears

$$\frac{\partial U(\mathbf{u})}{\partial u_i} = \frac{2\pi\lambda}{a_s}\sin\Big(\frac{2\,\pi u_i}{a_s}\Big) + k[2\,u_i - u_{i+1} - u_{i-1}] = 0,\; i = 2,..,N-1. \tag{3}$$

Note that the linear tensile force with amount $\mu$ also disappears [27]. However, the $a_o$ as well as the $\mu$ do not disappear at all; they emerge in the two boundary equations for the first and the last atom of the chain, $u_1$ and $u_N$ [19,28]

$$\frac{\partial U(\mathbf{u})}{\partial u_1} = \frac{2\pi\lambda}{a_s}\sin\Big(\frac{2\,\pi u_1}{a_s}\Big) - k[u_2 - u_1 - a_o] + \mu = 0 \tag{4}$$

and

$$\frac{\partial U(\mathbf{u})}{\partial u_N} = \frac{2\pi\lambda}{a_s}\sin\Big(\frac{2\,\pi u_N}{a_s}\Big) + k[u_N - u_{N-1} - a_o] - \mu = 0\,. \tag{5}$$

Equations (3)–(5), for all atoms $i = 1, \ldots N$, form a system of coupled equations, where we have to include the boundary conditions Equations (4) and (5). To select a single 'solution' of one or a small number of them will usually not give a correct stationary state of the FK chain.

In contrast to this simple remark, many researchers propose to use the 'twist map', which successively calculates one $u_{i+1}$ if the former $u_i$ and $u_{i-1}$ are known—an assumption that we cannot make. If one does not know the solution, one also does not know the location of the two special atoms at the start of the map [19]. In real calculations, the twist map is started at an arbitrary pair of two points in a finite region [21]. Of course, to start the twist map at arbitrary point pairs (see the $M_i$ in Figure 1 of ref. [21]) provides nice pictures, including the excursion to chaos. We claim that many calculations of twist maps do not use the correct initial values; for example, see Example 1 and Section 6 below.

**Example 1.** *A simple example is a chain with N = 3 atoms, with the parameters $a_s = 2\,\pi$, $k = 2.5$, $\lambda = 1$, and $a_o = 2a_s/3$. The equilibrium structures of a minimum are obtained by the NMinimize procedure of Mathematica 13.0, starting with a chain with $u_i = a_o\,(i-1)$ for $i = 1,..,3$, as shown on the left of Figure 1.*

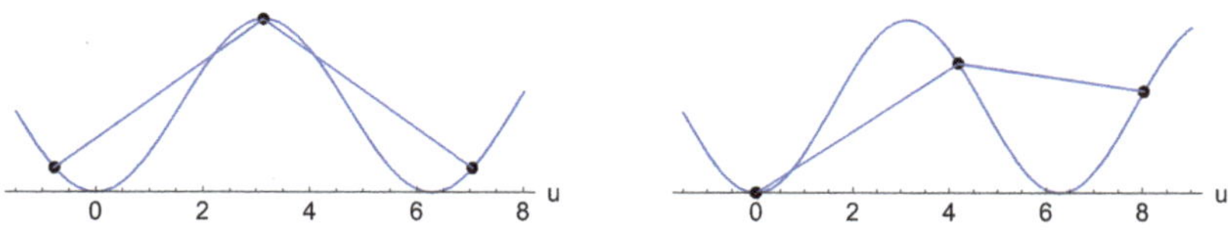


**Figure 1.** (**Left**): Minimum solution for an FK chain with $N = 3$ atoms. (**Right**): With an arbitrary start of $u_1 = 0$, we obtain $u_2 = a_0 = 2a_s/3$, and we obtain with Equation (3) the shown point $u_3$. However, the structure is not a minimum, see text. Note that the atoms are shifted on the site-up potential to help the imagination.

Using the twist map of Equations (4) and (3) starting at an arbitrary point, for example, with $u_1 = 0$, we obtain with Equation (4) $u_2 = a_0 = 2a_s/3$, and we obtain with Equation (3) the structure on the right hand side of Figure 1. It does not look like a minimum. The left minimum has an energy of 2.756 units, but this special twist map 'solution' has an energy of 2.826 units. Its gradient $(0, 0, 0.118)^T$ is not a zero vector.

The reason for the deviation of the gradient from zero is that the third Equation (5) for $i = N = 3$ is not fulfilled for the obtained $u_3$. Only when one begins with $u_1$ of a stationary structure is the third equation fulfilled. Further arguments to understand the result are given in ref. [19]. We conclude with the following theorem.

**Theorem 1.** *Using the twist map at an arbitrary start point is useless. In the general case, it will not provide a stationary structure of the FK chain. The reason is that Equation (5) is not fulfilled, in the general case. This is not restricted to the* $N = 3$ *chain.*

The twist map fulfills the demand of equilibrium only from step to step. Only when the start pair belongs to the correct minimum do the results (under numerical problems, see below) build a correct minimum structure. In the system, one can fulfill some equations—with false values, which vice versa compensate—however, one cannot assume that the other equations obtain the correct solution.

## 3. A Remark on the Length of the Chain

We treat the energy of the chain structure (1). When the spring distance of the chain, $a_o$, and the period of the site-up potential, $a_s$, are equal, and if $\mu = 0$, or if $a_0 = 0$ and $\mu/k = a_s$, or if $a_0 + \mu/k = a_s$, then an equilibrium chain exists in its ground state with all particles at the bottom of the site-up potential, with energy zero for every particle, and thus with zero energy for the chain in its ground state. (This case was originally treated [1] for an infinite chain.)

In any other case, an infinite chain needs an infinite energy in the sum, which is impossible. Strangely, the authors of ref. [29] stated that "... one does not hope that the sum of the energies converges." Nevertheless, they worked with the divergent sum. Tong et al. [30] treated a finite segment of the infinite incommensurate FK chain without the boundary conditions. Without further explanation, the problem was circumscribed by the word 'formal sum' [31,32].

A way out, avoiding the infinite energy of an infinite chain, was proposed [33] by dividing the energy of the chain by $N$, the number of atoms. So, one has a finite energy per atom [34,35], and one can apply the optimization per atom at $u_i$, discussed in Section 2 [36].

## 4. The Meaning of the Tensile Force

If $\lambda = 0$, we see with Equations (4) and (5) that the equilibrium separation of the chain becomes

$$a_0 + \mu/k\,. \tag{6}$$

In addition, if one uses the ansatz of form (2) with $a_0 = 0$, then the value of the tensile force divided by the value of the spring force, $\mu/k$, becomes the placeholder of the $a_0$ in the original FK model. One can put [37,38]

$$l_o = \mu/k \ \rightarrow \ \text{new} \ \ a_o \, .$$

The meaning of the tensile force in (2) is to include an equilibrium separation of the chain, $l_o$ for $a_0$. One can ask why one needs the tensile force.

On the other hand, one can ask for the action of the tensile force, if its amount $\mu$ increases. It has the form

$$-\mu \, \mathbf{f}^T \, \mathbf{u}$$

with the special form of the force vector

$$\mathbf{f}^T \ = \ (-1, 0, \ldots, 0, 1) \, .$$

At $\mu = 0$, we may be in an original minimum (with $a_0 = 0$ or not). Increasing the amount of $\mu$, we can optimize another minimum. The curve, which describes the consecutive series of minima, is a Newton trajectory; it was named the curve of the force-displaced stationary points [19,39–44]. Newton trajectories connect stationary points of an index difference of one [45] on the original potential energy surface. It is a quite interesting property that the special form of the **f** vector acts as a change in the orginal distance, $a_0$, of the atoms of the FK chain, see value (6), and not only as the tilding of the potential of the first and the last atom of the chain. The FK chain is similar to an accordion. If one pulls the ends, all of the bellows relax. Newton trajectories with force **f** can be used to calculate many kinds of solitones of the FK chain and, thus, intermediate minima of the potential energy surface and saddle points with an increasing index [3,6,19,46–48].

**5. The Meaning of the Free Boundary Conditions**

Now, we discuss the boundary conditions (BCs). If one has free BCs, then we cannot start with an arbitrary 'left' BC, because the minimization will probably result in different BC in the end. So, the steps of an arbitrary twist map start in a nebula. The developments by some researchers [21,49–51] (to name a few), which use the twist map, ignore that the chain will find another minimum structure at the boundary, in comparison to an arbitrary twist map result [19].

Of course, if one starts the twist map with the correct $u_1$, $u_2$ of a minimum structure or of another stationary state of the FK chain, then one can regain this stationary structure with the result of the twist map, at least for the first section of the structure; see the example below. However, we do not know where the stationary structure of an FK chain begins if the BCs are free.

**Example 2.** *The example is again the chain with parameters* $a_s = 2\pi$, $k = 2.5$, $\lambda = 1$, *and* $a_o = 2a_s/3$, *for* $N = 3, 4, 8–10, 20, 47, 74, 99, 500, 1000$, *and* $N = 5000$ *atoms. The equilibrium structures, probably minima, are obtained by the NMinimize procedure of Mathematica 13.0 starting with a chain with* $u_i = a_o\,(i-1)$ *for* $i = 1, \ldots, N$. *The first atom of the chain at* $u_1$ *changes its place with* $N$, *as well as the average distance*

$$\tilde{a} = \frac{u_N - u_1}{N-1} \,, \tag{7}$$

*see Table 1. The numbers in Table 1 are rational numbers, because they are the result of a computer calculation. We do not know whether the character of the limit for* $N$ *to* $\infty$ *is rational or irrational. The two values possibly converge for large* $N$ *against defined values around* $u_1 = 0.369$ *and* $a_o$. *However, we do not know the exact limit of* $\tilde{a}$.

**Table 1.** Position of the first atom of a minimum of an FK chain and its average distance.

| N | $u_1$ | $\tilde{a}/2\pi$ | N | $u_1$ | $\tilde{a}/2\pi$ |
|---|---|---|---|---|---|
| 3 | −0.769 | 0.6224 | 47 | 0.390 | 0.6700 |
| 4 | 0.117 | 0.6542 | 74 | 0.370 | 0.6688 |
| 8 | 0.751 | 0.6801 | 99 | 0.358 | 0.6621 |
| 9 | −0.341 | 0.6386 | 500 | 0.368 | 0.6670 |
| 10 | 0.239 | 0.6582 | 1000 | 0.369 | 0.6665 |
| 20 | 0.632 | 0.6736 | 5000 | 0.369 | 0.6667 |

We show in Figure 2 a minimum of an FK chain with $N = 500$ atoms. The central region represents the winding number, the ratio 2/3 of $a_o/a_s$; however, the two boundary regions show the large influence of the free BCs. A swing in at the left boundary needs ≈50 atoms, and an analogous transient process needs ≈50 atoms at the right boundary. One could imagine that an infinite chain with the 2/3 ratio could be represented by the central piece of Figure 2. However, a cut anywhere relaxes the strain at the border [20], and the shown boundary pattern will emerge again. It also holds for $N = 1000$ and $N = 5000$, where we obtain similar pictures.

The structures of the FK chain at the reported minima are mirror symmetrical up to a length of $N = 46$; see the left panel of Figure 1. Beginning with $N = 47$, the symmetry is broken, and the different form at the two boundaries emerges, as in Figure 2.

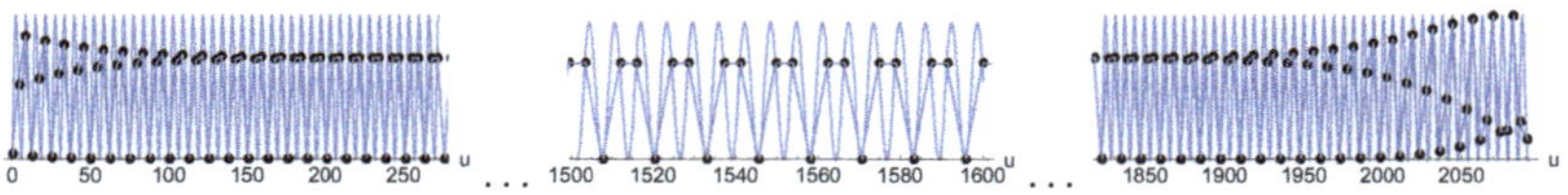

**Figure 2.** Minimum solution for a fixed $a_o$, see text. FK chain with $N = 500$ atoms, with the first 66 atoms on the left and the last 66 atoms on the right. The center is an enlarged piece with the regular pattern representing the rational relation to $a_s$ sorting three atoms in two wells.

In Figure 3, we show the corresponding twist map representation of the first 66 atoms for $N = 500$. There, we used the map $\Phi_i = u_i(\text{mod } a_s)$ for the minimization result. Consecutive pairs of atoms are shown, $(\Phi_i, \Phi_{i+1})$, $(\Phi_{i+1}, \Phi_{i+2})$, $(\Phi_{i+2}, \Phi_{i+3})$, … and so on. The dashed triangle is the start set of points, and the red triangle shows the periodic stable cycle of the central region of the FK chain. The swing in at the left boundary of the chain is the path from the dashed to the red triangle. Note that the points of the inner cycle form a kind of a unit cell defined by the used winding number, 2/3, but they are not free of tension. This is demonstrated by the relaxation at the boundaries.

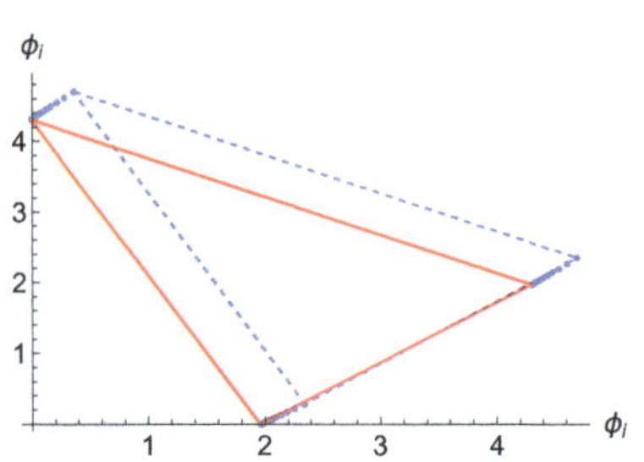

**Figure 3.** Twist map representation of the first 66 atoms of a long chain. The dashed line is the start triangle of points, and the red line is the quasi-stable cycle of three atoms in one cell of two troughs, as in the central region of Figure 2. (The calculation was a minimization.)

## 6. The Twist Map for a Finite FK Chain

In the right side of Figure 4, we depict the result of the twist map used with the two correct initial points, $u_1 = 0.369$ and $u_2 = 4.701$, of the optimized structure. The calculation was performed with Mathematica 13.0 with the usual accuracy. The result fits the true structure up to $N \approx 300$ atoms. Then, the twist map 'breaks out', and at $N = 353$ the former right BC appears. Its way out emerges by numerical effects, which can accumulate over the consecutive iterations.

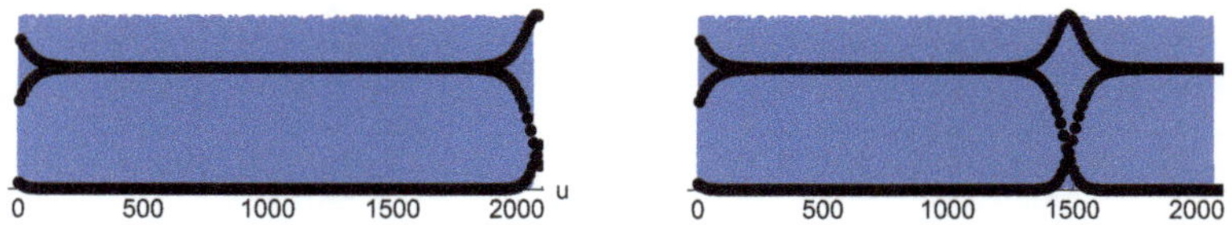


**Figure 4.** (**Left**): FK chain, optimized minimum with $N = 500$ atoms. (**Right**): The result of the twist map with the two points of the minimum, $u_1$ and $u_2$, at the start. It repeats the left BC; however, it 'finishes' at $N = 353$ with a right BC; then, the calculation is reflected.

The twist map must include the handling of the tension of the central quasi-unit cells. The $N = 4$ minimum shown on the left of Figure 5, with a $u_1$ value of 0.117, is two times 'shorter' than the putative unit cell of $4\pi/3$, see Table 1. The distances of the minimized structure also change from atom to atom in the central part. It only looks like a unit cell. In reality, a unit cell does not exist.

The result for an infinite chain [21] does not hold here, where one has a unit cell if $a_o/a_s$ is a rational number. There is no period of the kind

$$u_{i+3} = u_i + 2\,a_s\,, \quad \text{but it holds} \quad u_{i+3} = u_i + 2\,a_s + \epsilon_i$$

with variable values of $\epsilon_i$. The smallest $\epsilon_i$ we find at the central region of the chain with $\epsilon_{244} \approx 2 \times 10^{-11}$.

The fact that even the twist map calculation by Equation (3) finds (though at a false place) a correct upper boundary structure, as the direct optimization does, is very interesting. One has to consider that the quasi-perfect cyclic behavior in the central region is not the behavior of a limit cycle, in contrast to the propositions of ref. [36].

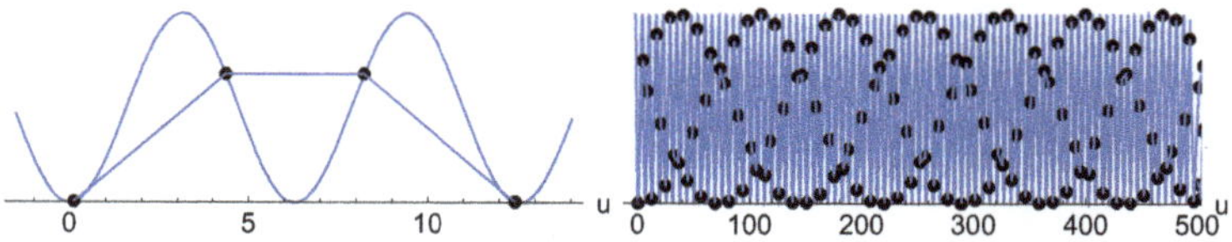


**Figure 5.** (**Left**): FK chain minimum for $N = 4$ atoms; it is not the unit cell of Figure 2, center; however, it is shorter, see Table 1. (**Right**) The result of a twist map calculation with the two points, $u_1 = 0$ and $u_2 = 2a_s/3$, of a putative unit cell at the start. It does not result in the quasi-unit cells of Figure 2.

Next, we show a twist map calculation in the right panel of Figure 5 and in Figure 6. The start is the putative unit cell with $u_1 = 0$ and $u_2 = 4\,\pi/3$. However, the twist map does not find the structure of consecutive unit cells. It finds that the 3-cycle of Figures 2 and 3 rotates and does not form a correct stable cycle, at least not for three atoms. See the twist map representation in Figure 6.

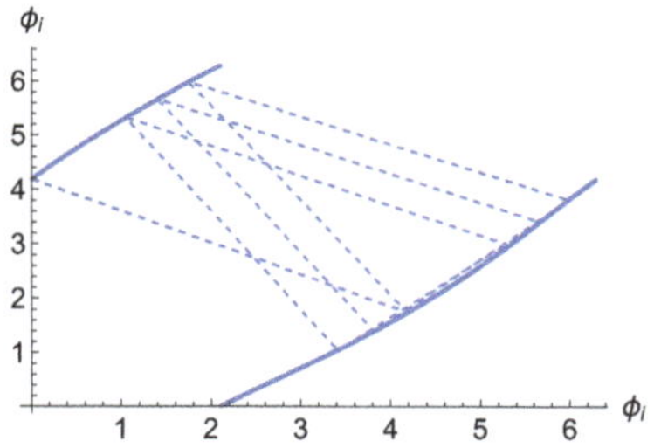


**Figure 6.** Twist map representation of a twist map calculation with $u_1 = 0$ and $u_2 = 4\pi/3$. The dashed line represents the start sequence of points.

## 7. The Disappearence of Incommensurabilities

For a given length of a finite FK chain with $N$ atoms for $\lambda = 0$,

$$L_o = (N-1)\, a_o,$$

one can determine an integer $M$ such that the chain of length $L_o$ covers up $M$ troughs of the site potential. If $\lambda$ is set correctly, the FK chain will fit into the $M$ troughs forming a structure of a minimum with an average separation $\tilde{a}$, with Equation (7). It holds independently of whether the numbers, $L_o, a_o, \tilde{a}$, and $a_s$, are rational or irrational. One can cite ref. [52] "...a finite chain on a periodic substrate will always be locked because of its free ends." The relation of $N$ to $M$ is in every case a commensurable ratio between two integers [53]; see Example 3.

**Example 3.** *We treat an example of a chain with a putative 'irrational' winding number [54] of $a_o = 0.873a_s$ and the other parameters $a_s = 2\pi$, $k = 2.5$, and $\lambda = 1$, for $N = 500$. The equilibrium structure, probably a minimum, is obtained by the NMinimize procedure of Mathematica starting with a chain with $u_i = a_o\,(i-1)$ for $i = 1,\ldots,N$. Figure 7 shows the structure and the corresponding twist map representation.*

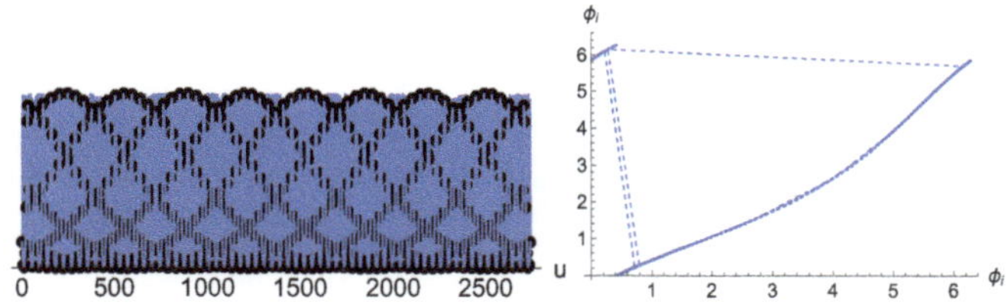


**Figure 7.** (**Left**): Optimized FK chain minimum for $N = 500$ atoms with winding number 0.873. (**Right**): Twist map representation of the result. The dashed line is the start sequence of nine points plus the first step of the next cycle. It is a quasi-cyclic result.

The structure seems not to be very erratic. The twist map shows a quasi-cycle of nine points. However, the winding number 0.873 is not fully irrational.

## 8. Remark on the Aubry Transformation of a Finite Chain

There should be a value $\lambda_C$, where a qualitative change in the behavior of the infinite FK chain occurs [21,55,56]. For a finite chain, there is no such value. If $\lambda > 0$, and $k > 0$, then the action of the site-up potential will fix the chain in a minimum, if we start a minimization for the 'relaxed' chain with distances $a_0$; see Section 7.

The putative Aubry-phase transition demonstrated by Figure 6 in ref. [54] is an incorrect interpretation. The first eigenvalue of the second derivatives of the potential of the chain is the frequency for a collective movement of the chain. For an unpinned chain, it has to be zero; thus, the potential has to be flat. However, in the example used in [54], the first eigenvalue is greater than zero. Thus, the potential is a minimum in both treated cases, $\lambda > \lambda_C$ and $\lambda < \lambda_C$, and the chain is pinned. Of course, the pinning is small if the $\lambda$ is small. The chain can collectively vibrate; however, it cannot move.

For the interpretation of the experiment [57] with $N$ = 1–5 atoms, we assume that the reported effect is for other reasons.

## 9. Discussion

We understand the long history of the twist map in the following way:

For a finite FK chain, the parameter $a_0$ is given; however, the localization of the minimum is unknown. We do not know the values of $u_1$, $u_N$, and $\tilde{a}$. Thus, we cannot start the twist map.

To salvage this nevertheless, one assumes a limit $l$ of the average separation and starts with this $l$ anywhere with assumed points $u_1$ and $u_2$ for the initial values. One postulates that every arbitrary pair of the start values is possible.

It is clear that the result is not helpful for an original finite FK chain, see Figure 1.

We note that many researchers have treated finite chains [9,33,35,54,57–65] (to name a few); however, they use a useless contrast of rational with irrational numbers in the finite FK model.

In a positive contrast, the treatment in [66] sorts the FK chain in a 'commensurable' way into the site-up potential.

## 10. Conclusions

We are surprised by researchers' attention to the twist map in the field of the FK model. For a finite FK chain, the twist map theory can only be a supplement; otherwise, important physics conclusions of the model are lost.

**Author Contributions:** Conceptualization, W.Q. and J.M.B.; software, W.Q.; validation, J.M.B.; writing—original draft preparation, W.Q.; writing—review and editing, J.M.B.; funding acquisition, J.M.B. All authors have read and agreed to the published version of the manuscript.

**Funding:** This research was funded by the Spanish Ministerio de Economıa y Competitividad, Project No. PID2019-109518GB-I00, and the Spanish Structures of Excellence Maria de Maeztu program through Grant MDM-2017-0767.

**Institutional Review Board Statement:** Not applicable.

**Informed Consent Statement:** Not applicable.

**Data Availability Statement:** All structure data of FK chains, and Mathematica files are available by e-mail from W.Q.

**Conflicts of Interest:** The authors declare no conflict of interest.

## Abbreviations

The following abbreviations are used in this manuscript:

| | |
|---|---|
| BC | Boundary condition |
| FK | Frenkel–Kontorova |

## Appendix A

For $k = 0$ and $\mu = 0$, the FK model shortens to the pure site-up potential without spring forces. Any arbitrary distribution of the atoms at the bottom of the siteup potential is possible.

For $\lambda \gg \lambda_c > k > 0$ in the FK model, further minimum structures are possible under $a_0 = 0$, aside from the zero solution discussed in Section 1. $\lambda_c$ depicts a critical value. Note that every small $k > 0$ prevents the length of the FK chain from becoming too large.

In Figure A1, we depict a case for $N = 8$, $\lambda = 10$, $k = 1$, and $\mu = 0$. The structure has two 'slightly stretched' springs, three 'stretched' springs, and two 'overstretched' springs, which are possible because of the much smaller $k$ parameter. Note that most of the atoms are not at the minima of the site-up potential, as claimed for large $\lambda$ [34].

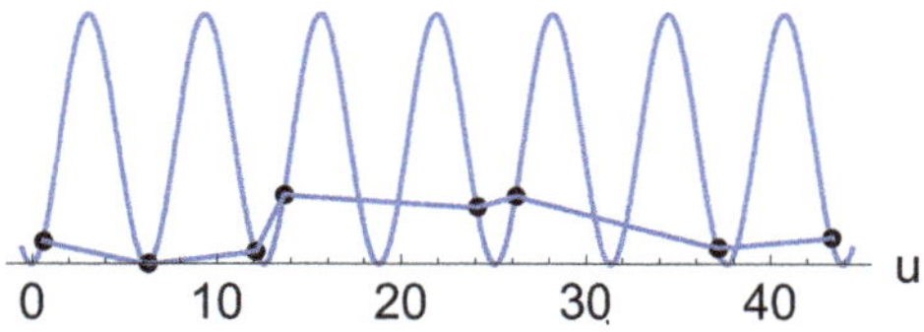


**Figure A1.** A nonzero FK minimum to potential (2).

## References


1. Kontorova, T.A.; Frenkel, Y.I. To the theory of plastic deformations and twinnings I. *Zh. Eksp. Teor. Fis.* **1938**, *8*, 89–95. (In Russian)
2. Lin, J.Y.; Smorodin, A.V.; Badrutdinov, A.O.; Konstantinov, D. Transport Properties of a Quasi-1D Wigner Solid on Liquid Helium Confined in a Microchannel with Periodic Potential. *J. Low Temp. Phys.* **2019**, *195*, 289–299. [CrossRef]
3. Quapp, W.; Lin, J.Y.; Bofill, J.M. The movement of a one-dimensional Wigner solid explained by a modified Frenkel-Kontorova model. *Eur. Phys. J. B* **2020**, *93*, 227. [CrossRef]
4. Duwel, A.E.; Watanabe, S.; Triás, E.; Orlando, T.P.; van der Zant, H.S.J.; Strogatz, S.H. Discreteness-induced resonances and ac voltage amplitudes in long one-dimensional Josephson junction arrays. *J. Appl. Phys.* **1997**, *82*, 4661–4668. [CrossRef]
5. Ustinov, A.V. Solitons in Josephson-junction arrays. *Phys. D* **1998**, *123*, 315–329. [CrossRef]
6. Quapp, W.; Bofill, J.M. Description of Shapiro steps on the potential energy surface of a Frenkel-Kontorova model Part II: Free boundaries. *Eur. Phys. J. B* **2021**, *94*, 64. [CrossRef]
7. Quapp, W.; Bofill, J.M. Description of zero field steps on the potential energy surface of a Frenkel-Kontorova model for annular Josephson junction arrays. *Eur. Phys. J. B* **2021**, *94*, 105. [CrossRef]
8. Grüner, G. The dynamics of charge-density waves. *Rev. Mod. Phys.* **1988**, *60*, 1129–1181. [CrossRef]
9. Garcia-Mata, I.; Zhirov, O.V.; Shepelyansky, D.L. The Frenkel-Kontorova model with cold trapped ions. *Eur. Phys. J. D* **2007**, *41*, 325–330. [CrossRef]
10. Pouget, J.P. The Peierls instability and charge density wave in one-dimensional electronic conductors. *Comptes Rendus Phys.* **2016**, *17*, 332–356. [CrossRef]
11. Braun, O.; Dauxois, T.; Paliy, M.V.; Peyrard, M. Mobility and diffusivity in a generalized Frenkel-Kontorova model. *Phys. Rev. B* **1996**, *54*, 321. [CrossRef]
12. Ekomasov, E.G.; Murtazin, R.R.; Bogomazova, O.B.; Gumerov, A.M. One-dimensional dynamics of domain walls in two-layer ferromagnet structure with different parameters of magnetic anisotropy and exchange. *J. Magn. Magn. Mater.* **2013**, *339*, 133–137. [CrossRef]
13. Trallori, L. Magnetic superlattices, classical spin chains, and the Frenkel-Kontorova model. *Phys. Rev. B* **1998**, *57*, 5923. [CrossRef]
14. Que, W.; Walker, M.B. Generalized Frenkel-Kontorova model for structural modulations in bismuth high-Tc superconductors and related compounds. *Phys. Rev. B* **1992**, *46*, 14772. [CrossRef] [PubMed]
15. Dinsmore, R.C., III; Bae, M.H.; Bezryadin, A. Fractional order Shapiro steps in superconducting nanowires. *Appl. Phys. Lett.* **2008**, *93*, 192505. [CrossRef]
16. Besseling, R.; Niggebrugge, R.; Kes, P.H. Transport properties of vortices in easy flow channels: A Frenkel-Kontorova study. *Phys. Rev. Lett.* **1999**, *82*, 3144–3147. [CrossRef]
17. Kokubo, N.; Besseling, R.; Vinokur, V.M.; Kes, P.H. Mode Locking of Vortex Matter Driven through Mesoscopic Channels. *Phys. Rev. Lett.* **2002**, *88*, 247004. [CrossRef]
18. Iaroshenko, O.; Rybalko, V.; Vinokur, V.M.; Berlyand, L. Vortex phase separation in mesoscopic superconductors. *Sci. Rep.* **2013**, *3*, 1758. [CrossRef]
19. Quapp, W.; Bofill, J.M. Newton Trajectories for the Frenkel-Kontorova model. *Mol. Phys.* **2019**, *117*, 1541–1558. [CrossRef]
20. Kochendörfer, A.; Seeger, A. Theorie der Versetzungen in eindimensionalen Atomreihen. I. Periodisch angeordnete Versetzungen. *Zeitschr. Phys.* **1950**, *127*, 533–550. [CrossRef]
21. Aubry, S. The twist map, the extended Frenkel-Kontorova model and the devil's staircase. *Phys. D* **1983**, *7*, 240–258. [CrossRef]
22. Hobart, R.; Celli, V. A Solution to the Frenkel-Kontorova Dislocation Model. *J. Appl. Phys.* **1962**, *33*, 60–62. [CrossRef]
23. Hu, B.; Shi, J. Nonanalytic twist maps and Frenkel-Kontorova models. *Phys. D* **1994**, *71*, 23–38. [CrossRef]
24. Aubry, S. Defectibility and frustration in incommensurate structures: The devil's stair case transformation. *Ferroelectrics* **1980**, *24*, 53–60. [CrossRef]
25. Guo, Y.; Qu, Z.; Zhang, Z. Lyapunov stability and precise control of the frictional dynamics of a one-dimensional particle array. *Phys. Rev. B* **2006**, *73*, 094118. [CrossRef]
26. Theodorakopoulos, N.; Peyrard, M.; MacKay, R.S. Nonlinear Structures and Thermodynamic Instabilities in a One-Dimensional Lattice System. *Phys. Rev. Lett.* **2004**, *93*, 258101. [CrossRef]
27. Aubry, S. Devils staircase and order without periodicity in classical condensed matter. *J. Phys. Fr.* **1983**, *44*, 147–162. [CrossRef]

28. Frank, F.C.; van der Merwe, J.H. One-Dimensional Dislocations. I. Static Theory. *Proc. R. Soc. Lond. A* **1949**, *198*, 205–216.
29. Blass, T.; de la Llave, R. The Analyticity Breakdown for Frenkel-Kontorova Models in Quasi-periodic Media: Numerical Explorations. *J. Stat. Phys.* **2013**, *150*, 1183–1200. [CrossRef]
30. Tong, P.; Li, B.; Hu, B. Wave transmission, phonon localization, and heat conduction of a one-dimensional Frenkel-Kontorova chain. *Phys. Rev. B* **1999**, *59*, 8639. [CrossRef]
31. Aubry, S.; MacKay, R.S.; Baesens, C. Equivalence of uniform hyperbolicity for symplectic twist maps and phonon gap for Frenkel-Kontorova models. *Phys. D* **1992**, *56*, 123–134. [CrossRef]
32. Salahshoor, H.; de la Llave, R. A Numerical Investigation of the Pinning Phenomenon in Quasi-Periodic Frenkel-Kontorova Model Under an External Force. *J. Stat. Phys.* **2018**, *173*, 398–410. [CrossRef]
33. Hu, B.; Li, B.; Zhao, H. Mode-locking of incommensurate phase by quantum zero point energy in the Frenkel-Kontorova model. *Europhys. Lett.* **2001**, *53*, 342. [CrossRef]
34. Griffiths, R.B.; Chou, W. Effective Potentials: A New Approach and New Results for One-Dimensional Systems with Competing Length Scales. *Phys. Rev. Lett.* **1986**, *56*, 1929. [CrossRef] [PubMed]
35. Sharma, S.R.; Bergersen, B.; Joos, B. Aubry transition in a finite modulated chain. *Phys. Rev. B* **1984**, *29*, 6335–6340. [CrossRef]
36. Baesens, C.; MacKay, R.S. Gradient dynamics of tilted Frenkel-Kontorova models. *Nonlinearity* **1998**, *11*, 949–964. [CrossRef]
37. Aubry, S.; Abramovici, G. Chaotic trajectories in the standard map. The concept of anti-intergability. *Phys. D* **1990**, *43*, 199–219. [CrossRef]
38. Aubry, S.; Le Daeron, P.Y. The diskrete Frenkel-Kontorova model and its extensions I. Exact results for the ground-states. *Phys. D* **1983**, *8*, 381. [CrossRef]
39. Avdoshenko, S.M.; Makarov, D.E. Reaction Coordinates and Pathways of Mechanochemical Transformations. *J. Phys. Chem. B* **2015**, *120*, 1537–1545. [CrossRef]
40. Bofill, J.M.; Ribas-Ariño, J.; García, S.P.; Quapp, W. An Algorithm to Locate Optimal Bond Breaking Points on a Potential Energy Surface. *J. Chem. Phys.* **2017**, *147*, 152710. [CrossRef]
41. Quapp, W. Can We Understand the Branching of Reaction Valleys for more than two Degrees of Freedom? *J. Math. Chem.* **2015**, *54*, 137–148. [CrossRef]
42. Quapp, W.; Bofill, J.M. A contribution to a theory of mechanochemical pathways by means of Newton trajectories. *Theor. Chem. Acc.* **2016**, *135*, 113. [CrossRef]
43. Quapp, W.; Bofill, J.M. Mechanochemistry on the Müller-Brown Surface by Newton Trajectories. *Int. J. Quant. Chem.* **2018**, *118*, e25522. [CrossRef]
44. Quapp, W.; Bofill, J.M.; Ribas-Ariño, J. Towards a Theory of Mechanochemistry, Simple Models from the Very Beginnings. *Int. J. Quant. Chem.* **2018**, *118*, e25775. [CrossRef]
45. Hirsch, M.; Quapp, W. Reaction Channels of the Potential Energy Surface: Application of Newton Trajectories. *J. Mol. Struct. Theochem* **2004**, *683*, 1–13. [CrossRef]
46. Quapp, W.; Bofill, J.M. A Model for a Driven Frenkel-Kontorova Chain. *Eur. Phys. J. B* **2019**, *92*, 95–117. [CrossRef]
47. Quapp, W.; Bofill, J.M. Sliding Paths for Series of Frenkel-Kontorova Models—A Contribution to the Concept of 1D-superlubricity. *Eur. Phys. J. B* **2019**, *92*, 193. [CrossRef]
48. Quapp, W.; Bofill, J.M. Description of Shapiro steps on the potential energy surface of a Frenkel-Kontorova model Part I: Chain in a box. *Eur. Phys. J. B* **2021**, *94*, 66. [CrossRef]
49. Bichoutskaia, E.; Heggie, M.I.; Lozovik, Y.E.; Popov, A.M. Multi-Walled Nanotubes: Commensurate-Incommensurate Phase Transition and NEMS Applications. *Fuller. Nanotubes Carbon Nanostruct.* **2006**, *14*, 131–140. [CrossRef]
50. Black, R.C.; Satija, I.I. Universal phase diagram in the generalized Frenkel-Kontorova model. *Phys. Rev. B* **1991**, *44*, 4089–4096. [CrossRef]
51. Vanossi, A.; Bechinger, C.; Urbakh, M. Structural lubricity in soft and hard matter systems. *Nat. Commun.* **2020**, *11*, 4657. [CrossRef] [PubMed]
52. Stoyanov, S.; Müller-Krumbhaar, H. Resonance-induced Cluster-mobility: Dynemics of a finite Frenkel-Kontorova model. *Surf. Sci.* **1985**, *159*, 49–68. [CrossRef]
53. Quapp, W.; Bofill, J.M. Comment on 'Out-of-equilibrium Frenkel-Kontorova model' (A. Imparato, 2021, J. Stat. Mech. 013214). *J. Stat. Mech.* **2022**, *2022*, 013204. [CrossRef]
54. Muñoz, J.M.; Sawant, R.; Maffei, A.; Wang, X.; Barontini, G. Realizing the Frenkel-Kontorova model with Rydberg-dressed atoms. *Phys. Rev. A* **2020**, *102*, 043308. [CrossRef]
55. Axel, F.; Aubry, S. Polarisation and transition by breaking of analyticity in a one-dimensional model for incommensurate structures in an electric field. *J. Phys. A* **1987**, *20*, 487. [CrossRef]
56. Peyrard, M.; Aubry, S. Critical behaviour at the transition by breaking of analyticity in the discrete Frenkel-Kontorova model. *J. Phys. C Solid State Phys.* **1983**, *16*, 1593–1608. [CrossRef]
57. Bylinskii, A.; Gangloff, D.; Counts, I.; Vuletić, V. Observation of Aubry-type transition in finite atom chains via friction. *Nat. Mater.* **2016**, *15*, 717–721. [CrossRef]
58. Babushkin, A.Y.; Abkaryan, A.K.; Dobronets, B.S.; Krasikov, V.S.; Filonov, A.N. The Ground State of the Frenkel-Kontorova Model. *Phys. Solid State* **2016**, *58*, 1834–1845. [CrossRef]
59. Ma, Y.; Wang, J.; Xu, X.; Wei, Q.; Kais, S. A density-matrix renormalization group Study of one-dimensional incommensurate quantum Frenkel-Kontorova model. *J. Phys. Soc. Jpn.* **2014**, *83*, 094605. [CrossRef]

60. Mikheikin, I.D.; Kuznetsov, M.Y.; Makhonina, E.V.; Pervov, V.S. Defects in Inorganic Suprastructures with Incommensurate Structural Elements: The Static Frenkel-Kontorova Model for Finite Systems. *Dokl. Phys. Chem.* **2001**, *376*, 52–55. [CrossRef]
61. Nasilowski, R. An unorthodox analysis of the Frenkel-Kontorova model. *Phys. A* **1996**, *230*, 266–284. [CrossRef]
62. Novak, I.; Truskinovsky, L. Segmentation in cohesive systems constrained by elastic environments. *Phil. Trans. R. Soc. A* **2017**, *375*, 20160160. [CrossRef] [PubMed]
63. Wang, C.L.; Tekić, J.; Duan, W.S.; Shao, Z.G.; Yang, L.P. Existence and stability of the resonant phenomena in the dc- and ac-driven overdamped Frenkel-Kontorova model with the incommensurate structure. *Phys. Rev. E* **2011**, *84*, 046603. [CrossRef] [PubMed]
64. Zhirov, O.V.; Lages, J.; Shepelyansky, D.L. Thermoelectricity of cold ions in optical lattices. *Eur. Phys. J. D* **2019**, *73*, 149. [CrossRef]
65. Zhirov, O.V.; Lages, J.; Shepelyansky, D.L. Thermoelectricity Modeling with Cold Dipole Atoms in Aubry Phase of Optical Lattice. *Appl. Sci.* **2020**, *10*, 2090. [CrossRef]
66. Thomas, A.; Leoni, T.; Siri, O.; Becker, C.; Unzog, M.; Kern, C.; Puschnig, P.; Zeppenfeld, P. A one-dimensional high-order commensurate phase of tilted molecules. *Phys. Chem. Chem. Phys.* **2022**, *24*, 9118. [CrossRef] [PubMed]

electronics

MDPI

*Review*

# Recent Progress of Non-Cadmium and Organic Quantum Dots for Optoelectronic Applications with a Focus on Photodetector Devices

**Hasan Shabbir and Marek Wojnicki ***

Faculty of Non–Ferrous Metals, AGH University of Science and Technology, Mickiewicza Ave. 30, 30-059 Krakow, Poland; shabbir@agh.edu.pl
* Correspondence: marekw@agh.edu.pl

**Abstract:** Quantum dots (QDs) are zero-dimensional (0D) nanomaterials with charge confinement in all directions that significantly impact various applications. Metal-free organic quantum dots have fascinating properties such as size-dependent bandgap tunability, good optical absorption coefficient, tunability of absorption and emission wavelength, and low-cost synthesis. Due to the extremely small scale of the materials, these characteristics originated from the quantum confinement of electrons. This review will briefly discuss the use of QDs in solar cells and quantum dots lasers, followed by a more in-depth discussion of QD application in photodetectors. Various types of metallic materials, such as lead sulfide and indium arsenide, as well as nonmetallic materials, such as graphene and carbon nanotubes, will be discussed, along with the detection mechanism.

**Keywords:** quantum dot; confinement; tunable photoexcitation; solar cell; quantum dot laser; carbon dot; quantum dots photodetector

## 1. Nanomaterials

Nanomaterials have garnered much attention in recent years due to their fascinating and distinctive characteristics that surpass the limitations of bulk materials. Any substance with at least one dimension under 100 nm can be categorized as a nanomaterial [1,2]. Materials with all dimensions larger than nanoscale are called 3D materials or bulk materials, while researchers use a few classifications based on the dimensions of nanomaterials. If two dimensions are larger than the nanoscale, they are called 2D nanomaterials, e.g., quantum wells and nanolayers, a heterostructure composed of two barrier layers sandwiching a single thin well layer. This layer, which contains electrons and holes, is just about 100 Å thick; as a result, the electrons and holes start behaving like waves. In practice, standing waves in a direction perpendicular to the layers correspond to the permissible states in this structure. The system is quantized because only specific waves are standing waves. Many quantum mechanical processes can be observed and controlled in thin quantum well layered semiconductor structures. Modern epitaxial crystal growth techniques can produce them with great precision.

If one dimension of a material is in the nanoscale range and two dimensions are larger than nanoscale, it is categorized as a 1D nanomaterial, e.g., nanorods and nanowires [3,4]. 1D nanomaterials are used in solid-state electronics and as diagnostic tools in medical sciences. They can conduct at the level of one degree of freedom. This enables the utilization of nanowires in applications where electrical conduction is required [5].

If all dimensions of a material are smaller than 10 nm, it is called a 0D material, e.g., nanoparticles and quantum dots (QDs).

As the size limits approach the zero dimension and the region becomes extremely confined, it is also referred to as a dot. QDs are produced when material sizes become smaller and smaller in each dimension, which causes them to behave like groups of atoms

**Citation:** Shabbir, H.; Wojnicki, M. Recent Progress of Non-Cadmium and Organic Quantum Dots for Optoelectronic Applications with a Focus on Photodetector Devices. *Electronics* **2023**, *12*, 1327. https://doi.org/10.3390/electronics12061327

Academic Editor: Lucas Lamata

Received: 30 January 2023
Revised: 3 March 2023
Accepted: 6 March 2023
Published: 10 March 2023

and exhibit fascinating features. The electron–hole pair is also produced by some energy input, such as UV radiation, and the pair is kept together by Coulombic forces [6]. QDs' size ranges from 2 to 10 nm, or between 10 and 50 atoms, and they have 100 to 1000 electrons. Almost all material systems, including metal, insulators, and semiconductors, exhibit size-dependent electrical or optical characteristics in the quantum size domain. Due to its fundamental and technological significance, the semiconductor's energy band gap alteration is the most appealing [7]. The materials for the next generation of flat panel displays, photovoltaic, and optoelectronic devices are semiconductors with extensively tunable energy band gaps [8].

Quantum dots (QDs) are tiny semiconductor particles that can be utilized in photodetectors and other applications. They are typically only a few nanometers in size. QDs can be employed as a light-sensitive substance, also referred to as a photoconductive substance, in a photodetector. One example of a photodetector using QDs is a QD infrared photodetector (QDIP) [9,10]. QDIPs use a layer of QDs as the active material, which absorbs the incoming infrared light and generates an electrical current [11]. It has been demonstrated that they have good stability, quick response times, and high sensitivity.

Another example is a QD-based ultraviolet photodetector. The photoactive material employed in the detector is QDs, and it has a high sensitivity and quick response time for detecting UV light [12,13]. Figure 1 shows the different types of materials based on dimensions.

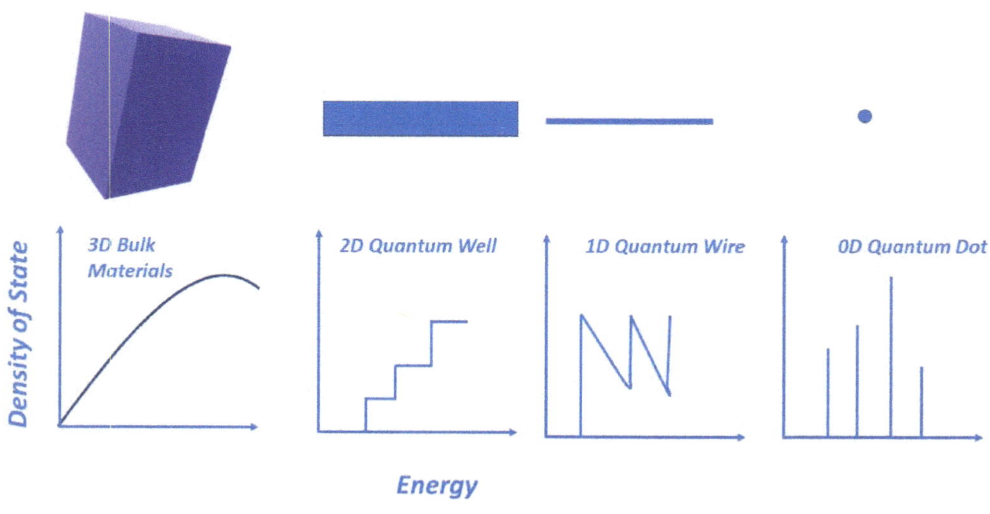


**Figure 1.** Schematic of nanomaterials based on types.

A third example is a QD-based hybrid photodetector, which combines the benefits of QDs and conventional semiconductor materials, such as silicon [14,15]. These devices combine silicon and QDs to produce a sensitive and effective photodetector. Overall, due to their distinct optical and electrical characteristics and their potential to enhance the performance of these devices, quantum dots are a promising material for use in photodetectors [16].

Although the discovery of QDs is familiar due to the toxic nature of initially observed QDs such as cadmium, their application is limited [17]. The semiconductor quantum dots used to make quantum dots (QDs) photodetectors include cadmium sulfide (CdS), lead sulfide (PbS), indium arsenide (InAs), and organic quantum dots [18]. Due to their distinct optical and electrical characteristics, these QDs have potential applications in photodetector sensors. The most extensively researched QD material for photodetector sensors is cadmium sulfide (CdS) [19]. They can efficiently convert light into an electrical signal due to their high quantum yield and absorption coefficient, allowing them to absorb much light. These characteristics make CdS QDs ideal for ultraviolet and infrared photodetectors [20].

Nevertheless, CdS have many disadvantages. Cadmium is a poisonous heavy metal that can cause serious health issues if inhaled, and proper precautions are needed to dispose of devices made up of cadmium. Quantum dots based on CdS are not biodegradable and can affect the ecosystem [21]. CdS are also more expensive than other quantum dot materials, so researchers are working on different QD materials useful for photodetection applications [22].

The second most commonly used QDs for photodetection applications are lead sulfide (PbS) QDs. They are especially beneficial for infrared photodetectors because they feature a band gap that makes them sensitive to infrared light [23]. Quantum dots made of indium arsenide (InAs), which have a high near-infrared absorption coefficient and are, therefore, particularly effective for near-infrared photodetectors, are another potential material for use in photodetector sensors [24,25]. Overall, the wavelength range of the light that needs to be detected and the specific specifications of the device will determine which QD material is best [26].

This review is focused on non-cadmium nanomaterials for electronic applications. As light-emitting diodes also work with the mechanism of electron–hole recombination, QDs also have potential applications in this area [27]. In terms of applications, quantum dots are used in various fields, including electronics, biomedicine, and renewable energy. For example, quantum dots can be used in electronics to make highly efficient solar cells, LED lights, and other devices. In biomedicine, quantum dots are used as imaging agents to study cells and tissues in the body [28]. In renewable energy technologies, quantum dots are used to produce hydrogen fuel from water using sunlight [29].

## 2. Optical and Electrical Properties of Quantum Dots

There are different energies for the electrons in quantum dots. When QDs absorb energy above their band gap, an exciton (electron–hole pair) is created. The electron and hole pair is confined if the diameter of the nanomaterial is smaller than its exciton Bohr radius, often known as the average physical distance between the electron and hole [30]. This electron and hole pair confinement is called the quantum confinement effect, which causes a discrete packet of energy and cannot be considered continuous [31,32]. One of the fascinating phenomena that QDs exhibit is photoluminescence. In this phenomenon, QDs absorb higher-energy photons and release low-energy photons. The electron moves into an excited state upon absorption. When it recombines with a pair, it emits lower energy light. This process takes place in femtoseconds [33].

Photoluminescence generally depends not on the type of materials but on the size [34]. As the quantum dot size decreases, the band gap of the quantum dot increases, causing the emission wavelength to shift toward a lower wavelength, and this is known as the blue shift, which is emission in the visible range. However, the emission can also occur in other regions of the electromagnetic spectrum, such as ultraviolet or infrared. QDs have exciting optical properties such as better photostability, high molar extinction, and high quantum yield (ratio of photons emitted vs. photons absorbed). QDs can be observed using confocal microscopy, fluorescence spectroscopy, total internal reflection microscopy, etc. [35,36].

QDs also have fascinating electrical properties, such as when sufficient energy is absorbed by the QD electron excited from the valence band to the conduction band while leaving the hole (empty state) in the valence band. This hole can be considered a positive mobile charge in the valence band. QDs possess a discrete energy state like an electron of a single atom [37]. Typically, a photon can only form one exciton and excite one electron across the band gap of fluorescent materials, with extra energy being lost as heat. However, multiple exciton generation is possible in QDs, which can be produced simultaneously, increasing the efficiency with which the nanocrystals convert energy [38].

Quantum yield measures the efficiency with which a system converts absorbed light into a different form of energy, such as heat or electricity. In the case of quantum dots, it refers to the efficiency with which the dots convert absorbed light into emitted light [39,40].

The equation for quantum yield is typically expressed as the ratio of the number of photons emitted by the quantum dots to the number of photons absorbed by the dots. This ratio can be expressed as follows:

$$\text{Quantum yield} = \frac{\text{Number of photons emitted}}{\text{Number of photons absorbed}}$$

The quantum yield plays an important role in the properties of QDs. The confinement behaviors produced by QDs' core/shell structures are closely connected to the core and shell materials [41]. For instance, the electron–hole pair is confined within the core when the energy band gap of the shell material is more significant than that of the core material, and when the energy band gap of core materials is more significant, the electron–hole pair is confined in the shell [42]. QDs, like carbon quantum dots, are extremely sensitive to the presence of extra charges, such as electrons or holes, on their surfaces and in their environment, which can change the photoluminescence and absorption wavelength of QDs. The functional group present on the surface of carbon quantum dots can absorb and trap the photon, which can change the emission wavelength of carbon QDs [43]. Due to these fascinating properties, quantum dots are used in many applications, such as solar cells [44], lasers [45], and photodetectors [46].

## 3. Quantum Dot Lasers

Semiconductor spherical nanocrystals (colloidal quantum dots), monolayers, and nanowires have found usage in the laser industry. CDs offer wide electromagnetic spectrum emissions from infrared to the visible range due to their size tunability, optical transition, and solution processibility. Compared to conventional semiconductors, QD-based semiconductors are easier to manipulate in size and shape [47,48]. Due to the unique absorption and emission properties of QDs, it is possible to produce lasing from a single exciton. A QD laser provides even spacing, low threshold amplification, and continuous wave operation. The shell absorbs the energy and acts as an antenna when the core–shell QDs are excited by photons having energies above the energy band gap. These QD outpace Auger recombination, so the threshold for lasing operation decreases [45].

## 4. Quantum Dots for Solar Cells

For more than two decades, semiconductor quantum dots (QDs) have attracted much attention due to the optoelectronic properties based on zero dimensions. Through the QD band energy level, colloidal quantum dots and self-assembled quantum dots improve the photogeneration of carriers for photovoltaic applications [49]. By expanding the range of photoexcitation, theoretically, it is possible to achieve maximum thermodynamic efficiency although the mechanism of carrier transport and carrier collection also contribute to its efficiency. Self-assembled and colloidal quantum dots are the two most-used types of QDs in solar cell applications [50]. The growth of solar cell wafers is also crucial for their efficiency, including surface passivation and well-controlled heterointerface [51].

Future risks include rising temperatures caused by global warming, shrinking fossil fuel supplies, and ever-increasing energy needs, which will require scientists and researchers to continuously work on developing efficient and affordable renewable energy alternatives [52]. Sunlight, which is abundantly available worldwide and can be converted into electricity or other forms of energy without polluting the environment, is one of the most commonly used renewable energy sources. Photovoltaic systems are based on converting sunlight into electricity, another valuable form of energy [53]. Photovoltaic device usage to produce electrical energy has increased, but obstacles still need to be overcome before these devices can significantly contribute to satisfying the world's energy needs [54].

Generally, there are three types of solar cells: traditionally, silicon-based monocrystalline and polycrystalline solar cells have been used [55]. The second type is thin-film solar cells, which are less expensive due to the low cost of absorbing material compared to silicon. These include copper gallium indium diselenide, cadmium telluride, and organic solar panels [56]. The third type, perovskite solar cells, uses hybrid organic–inorganic materials as the light-harvesting active layer. Perovskite materials, such as all-inorganic cesium lead halides and methylammonium lead halides, are cheap and straightforward to manufacture. Carrier mobility, doping density, trap density, and diffusion length in films are crucial factors that affect the performance of photovoltaic devices [57].

Junwei Yang et al. [58] reported synthesizing CdTe core QD-based solar cells. They used the aqueous route to synthesize QDs. CdTe core QDs were first synthesized and then deposited on a $TiO_2$ photoanode. An increase in absorption spectrum is observed for CdTe core QDs compared to CdTe cells. The power conversion efficiency of QD-based solar cells is increased up to 22% compared to CdTe/CdS plain solar cells. The suppressed charge recombination is also confirmed in QD-based solar cells. The authors also reported that thickness plays an important role in quantum efficiency. With optimization of $CdSe_xS_{1-x}$ shell thickness, about 7.24% conversion efficiency can be achieved, while Jun Du reported that Zn–Cu–In–Se alloyed QDs increase the absorption efficiency when deposited on $TiO_2$ film electrode, and reported conversion efficiency up to 11.66% [44].

## 5. Nanomaterials for Photodetection

Nanomaterials have properties that are suitable for photodetection, especially two-dimensional nanomaterials Two-dimensional (2D) materials have emerged as promising candidates for photodetectors due to their unique electrical and optical properties. In particular, their ultra-thin structure allows for efficient light absorption and carrier transport, leading to high sensitivity and fast response times. Two examples of 2D materials that are commonly used for photodetectors are:

Graphene is a single layer of carbon atoms arranged in a honeycomb lattice. It has excellent electrical conductivity and high carrier mobility, which makes it an attractive material for photodetectors [59]. When light is absorbed by graphene, it generates electron–hole pairs, which the graphene's high carrier mobility can efficiently collect. Graphene photodetectors have shown high responsivity and fast response times, making them useful for optical communications and imaging applications, while graphene hybrid composite materials with metallic nanomaterials and inorganic nanomaterials can increase photosensitivity [60].

Transition metal dichalcogenides (TMDs): TMDs are a family of 2D materials consisting of a transition metal atom sandwiched between two chalcogen atoms. They have a direct bandgap, which allows for efficient light absorption and emission [61]. TMDs have shown high sensitivity to light in the visible and near-infrared regions, making them useful for applications such as biomedical imaging and environmental monitoring. In addition, TMDs can be easily integrated with other electronic components, making them promising for on-chip optoelectronics [62]. 2D materials have shown great potential for photodetector applications due to their unique electrical and optical properties. Graphene and TMDs are just two examples of 2D materials that are being actively researched for their photodetection capabilities. As research in this area continues, it is likely that even more promising 2D materials will be discovered and developed for photodetector applications.

Infrared light detection is essential for biological sensing, spectroscopy, and in-depth 3D imaging. At the same time, semiconductors such as silicon, Cds, HgCdTe, and HgS are used for photodetectors. However, these materials have a few drawbacks, such as toxicity, low responsivity, and high cost of fabrication. The imaging performance of the quantum well photodetector, which is comparable to that of the mercury cadmium telluride (HgCdTe) detector, has recently attracted a lot of interest [63].

Following progress in the nanomaterials industry, it has been found that they have some advantages for photodetection over conventional materials, and the most-used nanomaterials are quantum well.

Quantum well infrared photodetectors (QWIP) absorb electrons through electronic intersubband polariton in a quantum well [64]. To be used for infrared detection, the energy difference between the first and second quantized states of the quantum well infrared photodetector's quantum wells must match the energy of the incoming infrared photons. III-V direct band gap semiconductor gallium arsenide (GaAs) is usually used to make QWIPs [65].

Therefore, research is also being carried out to create high-performance photodetectors using these nanomaterials. This review article focuses on quantum dot photodetectors [66]. They have been used in applications such as gas sensors and medical images, etc. Quantum well infrared detectors also have some limitations, such as the intersubband transition in the conduction band due to infrared photon polarization not being allowed in quantum wells [67,68]. This limitation can be overcome by using grating. However, they have a few more drawbacks, such as higher dark current, low quantum yield, lower responsivity, and detection level, and require cryogenic temperature to operate, so these limitations can be fulfilled by quantum dot photodetectors that have similar properties to quantum wells and also have some advantages due to the confinement of electrons in three dimensions [69]. QDs have reduced dependence on temperature for carrier transport, and carrier lifetime is also 10–100 times longer, which means that the electron remains excited for a more extended period, which causes a reduction in the dark current and increases sensitivity. QD-based detectors can absorb normal incident light without grating [70].

Similar to a quantum well, a quantum dot exhibits detection through the confinement-induced intraband photoexcitation of electrons. A photocurrent is generated by attracting the electrons in the collector with an electric field generated by the applied bias.

Colloidal QDs are low-cost materials, due to solution-based fabrication into semiconductor wafers and tunable band gap material, that are used as light emitters in flat panel displays. QDs are a cheap alternative to III–V group-based materials for imaging and sensing [71]. Research is being already carried out to make it possible to place QDs in a complementary metal–oxide–semiconductor (CMOS) circuit and increase the responsivity of QDs while decreasing the low noise equivalent power. If QDs can emit electrons across a wide range of the electromagnetic spectrum, they will be more practically efficient [72].

Infrared detectors are mainly categorized based on the material used for detection. Some of these are already extensively used in industry, including direct band gap semiconductor alloys such as HgCdTe and InSb, majority carrier Type-I superlattices, e.g., GaAs/AlGaAs QWIPs, and extrinsic semiconductors, e.g., Si: Ga, Ge: Hg. The fourth type is silicon Schottky barriers such as PtSi, and IrS. At the same time, the other two types include high-temperature superconductors and QD-based photodetectors, e.g., InAs/GaAs-based detectors [73,74]. The QD-based photodetector will be discussed in this review. Examples of quantum dots for photodetection are given below.

## 6. Lead(II) Sulfide (PbS) Quantum Dots

Lead(II) sulfide (PbS) QDs are special semiconductors with distinctive optoelectronic characteristics. Their strong absorption coefficient and band gap of 0.35 eV make them suitable for photovoltaic, photoconductor, and light-emitting diode applications [75,76]. PbS quantum dots are advantageous for sensing and imaging applications because they also have a high quantum yield and a narrow emission spectrum. Furthermore, their band gap can be modified by varying the quantum dot size, enabling further improvement of their optoelectronic characteristics [60,77].

One of these characteristics is their high absorption coefficient, which enables effective light absorption over a broad range of wavelengths. This is because PbS quantum dots have a significant quantum confinement effect, which confines the electrons to a limited area and causes them to have a high absorption coefficient [78].

The high responsivity of PbS quantum dots is a crucial characteristic that makes them effective as photodetectors which means more transformation of absorbed photons into an electrical signal [59,79]. Due to their high electron–hole recombination rate and rapid electron mobility, PbS quantum dots have a high responsiveness that enables the effective conversion of absorbed photons into electrical current. PbS quantum dots are also ideally suited for applications with high sensitivity and low noise because of their excellent signal-to-noise ratio and low dark current [80,81]. Therefore, lead sulfide (PbS) quantum dots (QDs) are considered potential material options for next-generation light, affordable, and flexible photodetectors due to their wide tunable band gaps, high absorption coefficients, and simple solution synthesis methods.

Alberto Maulu et al. synthesized PbS QDs and then modified their surface properties by using 3-mercaptopropionic acid (MPA) and tetrabutylammonium iodide (TBAI) [82]. It was observed that modification with this ligand significantly alters the properties of QDs. MPA-based QDs showed a decrease in dark current and low noise ratio, which causes the enhancement of detective and photosensitivity compared to TBAI-based QDs. MPA-modified QDs have functional group (Pb–OH bonds), while TBAI has a functional group (Pb–I bond), which is the primary reason for the Pb enhancement properties.

QDs used the trap sensitization method to show photoconductivity, increasing the responsivity but decreasing the response time for low optical power. At medium power incidents, light, when traps are filled with both types of photodetector, shows similar types of properties.

Urvashi Bothra et al. investigated the effect of ligands on PbS properties. They showed that a $PbX_2$ [X = I, Br])-based photodetector decreases the trap density compared to a PbS-based device, which increases the performance of (PbS) $PbX_2$-based devices [83].

Jin Beom Kwon et al. fabricated short-wave infrared (SWIR) photodetectors based on PbS QDs [84]. These infrared photodetectors minimize the harmful effect caused by the low wavelength band of SWIR. They fabricated a PbS photodetector with an absorbance peak at 1410 nm and used poly(3-hexylthiophene) (P3HT) as a conductive polymer with PbS QDs. Measurements were made on the properties of the synthesized PbS QDs, and the current–voltage (I–V) features of the PbS SWIR photodetectors. The maximal responsiveness of the PbS SWIR photodetector with P3HT optimization was found to be 2.26 times greater than that of the PbS SWIR photodetector without P3HT. Moreover, the former demonstrated a reduced operating voltage due to the high hole mobility and an adequate highest occupied molecular orbital level of P3HT. Figure 2 shows the I–V characteristics of the PbS-based photodetector.

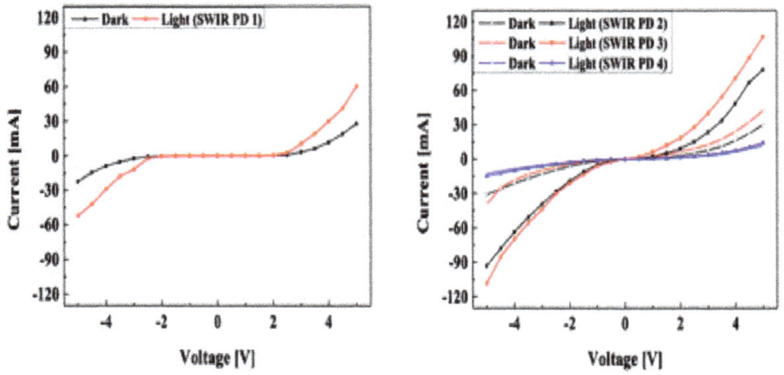


(a) (b)

**Figure 2.** I–V characteristics of (**a**) PbS SWIR photodetector without P3HT polymer and (**b**) PbS SWIR photodetector with P3HT polymer [84].

To compare their performance, they measured the I–V characteristics and responsiveness of the manufactured PbS SWIR photodetectors. The electron–hole pairs produced in the photoactive layer of the PbS SWIR photodetectors were removed from the electrodes by an external electric field when exposed to IR light. When the IR light source was turned off, the dark current was measured, and the light current was measured when the light source was turned on. The output power of the IR light source was 1 $W/m^2$, and the voltage sweep range was between −5 V and 5 V. The highest current difference between the dark current and light current and responsivity was recorded at −5 V in the case of the PbS SWIR photodetector without a P3HT layer. The largest current difference was 23.72 mA, and when this was divided by the input voltage, the corresponding responsivity was calculated as 2635.6 A/W. The PbS SWIR photodetector with a P3HT layer exhibited the best properties when applying 20 mg/mL of P3HT. The greatest current difference was confirmed to be 5891.1 A/W, which is 2.26 times greater than SWIRs, and the maximum current difference was 53.02 mA at −5V.

The drawbacks of current single-layer PbS-QDs photodetectors include excessive dark currents, slow light responses, and poor on–off ratios. It has been demonstrated that combining high-conducting graphene with organic materials and metal nanoparticles to create hybrid PbS-QDs devices can increase photoresponsivity. However, these hybrid devices bring other problems, and Zhenwei Ren et al. [80] synthesized the bilayer QDs device, which can be easily integrated into a flexible polyimide substrate. It increases the responsivity, signal-to-noise ratio, linear dynamic, and detection.

Moreover, this improvement in the device's properties is attributable to QDs' significant role. For example, the QDs junction accelerates carrier separation when the light is on. When the light is off, the QDs support the recombination of accumulated carriers in two films via the interface. Junction control is more dominant than trap control, which is the main determining factor in single-layer devices. They fabricated the bilayer photodetector using tetrabutylammonium iodide (TBAI)- and 1,2-ethanedithiol (EDT)-treated QDs. They compared it with a single-layer detector and observed that the bilayer photodetector enhanced optoelectronic properties. The bilayer PbS- TBAI/PbS-EDT devices have higher light current and detection efficiency, while the responsivity of PbS- TBAI/PbS-EDT is comparable to PbS-EDT devices due to the very close photocurrent.

It can be concluded that surface functional chemistry and modification of PbS with ligands improve the performance, and hybrid structure with modifications from different ligands can also improve the performance.

## 7. Indium Arsenide (InAs) Quantum Dots

Indium arsenide (InAs) quantum dots (QDs) are also potential materials for photodetector sensors due to their distinct optical and electrical characteristics. These QDs are especially beneficial for NIR photodetectors due to their high near-infrared (NIR) absorption coefficient [85]. NIR imaging is one possible use for InAs QD-based photodetectors. Due to InAs QDs' band gap, developing highly sensitive NIR imaging systems with them is possible. These gadgets are used in telecommunications for sensing, night vision, and medical imaging [86]. Deep-tissue imaging, optical communication, and covert illumination are just a few of the optoelectronics and biomedical uses for indium arsenide quantum dots, which typically emit in the near-infrared spectrum. Systems with bigger optical band gaps have yet to be produced, despite theory suggesting that further quantum confinement through size reduction could enable visible light emission.

NIR spectroscopy is another use for InAs QD-based photodetectors. Due to the NIR light detection capability, highly sensitive NIR spectroscopy devices can be made [87]. Additionally, the development of optoelectronic devices such as solar cells, LEDs, and lasers has utilized InAs QD-based photodetectors. These devices can increase the efficiency of solar cells and other optoelectronic devices by effectively converting light into electrical energy. Applications for these devices include bio-photonics, environmental monitoring, and chemical sensing [88,89].

Overall, because of their distinct optical and electrical characteristics, as well as their capacity to detect NIR light, InAs QDs have the potential to be utilized in a variety of photodetector applications. They might find use in spectroscopy, optoelectronic devices, and NIR imaging [90]. InAs quantum dots (QDs) might have a different band gap depending on their size and shape. Due to quantum confinement, the band gap of InAs QDs is typically between 0.35 and 0.4 eV, which is lower than that of bulk InAs (0.36 eV) [91], which is shown in Figure 3.

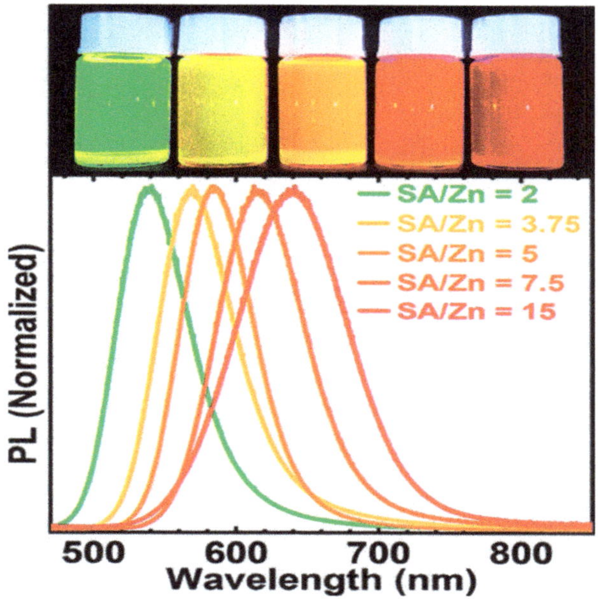


**Figure 3.** Photoluminescence spectra of In(Zn)As/ZnSe/ZnS QDs synthesized with different amounts of stearic acid (SA) ligand relative to the initial zinc precursor under UV illumination (Reproduced from [92], with permission from ACS Publications).

A low-temperature nanocluster synthesis technique was used by Daryl Darwan et al. to create highly luminous, visible-light-emitting In(Zn)As/ZnSe/ZnS QDs [92]. Each QD has an ultra-confined In(Zn)As nanocluster and emits light with a high photoluminescence quantum efficiency of 58% at tunable wavelengths between 538 and 640 nm. Theoretical calculations support the hypothesis that In(Zn)As nanoclusters are responsible for the infrared to visible spectral shift. This suggests that optical tuning in the visible region is also possible by utilizing confined semiconductor systems with a wide range of applications. The schematic is shown in Figure 4. Using an ultra-confinement technique, they were able to extend the spectrum coverage of InAs-based QDs into the visible range.

## 8. Organic Materials Based on Quantum Dots

Organic materials, such as polymers or small organic compounds, are used to create organic quantum dots (OQDs), a particular kind of quantum dot. Due to their special optical and electronic characteristics, they could be used in photodetector sensors. Organic photovoltaics is one potential use for OQD-based photodetectors (OPVs) [93,94]. The active layer in OPVs, which transform light into electrical energy, can be made of OQDs. OQDs may absorb various wavelengths, which can enhance the efficiency of OPVs. They could also be used to make solar cells that are less expensive than ordinary inorganic solar cells. OQD-based photodetectors can also be used in flexible and wearable gadgets. OQDs offer thin, bendable, and solution-processable advantages, making them ideal for application in bendable and flexible products such as flexible screens and wearable technology [95].

Devices for sensing and bioimaging have also been created using OQDs. OQDs have the potential to be used in biosensing and biomedical imaging, as well as optical imaging. They are advantageous in these applications because of their high quantum yield, strong stability, and great sensitivity [96]. OQDs are a promising material for photodetector sensors overall because of their distinct optical and electronic characteristics and their potential for usage in a variety of applications, including solar cells, flexible and wearable devices, and bio-imaging and sensing equipment [97].

## 9. Carbon-Based Materials for Photodetection

Carbon materials, such as carbon nanotubes, graphene, and fullerenes, are used to synthesize carbon-based quantum dots (CQDs), a particular kind of quantum dot. Due to their extraordinary optical and electronic characteristics, they could be used in photodetector sensors [98].

The photodetection of diverse wavelength ranges is one potential use for CQD-based photodetectors. It has been demonstrated that carbon-based quantum dots have a broad absorption spectrum, making them appropriate for use in photodetectors that must detect a variety of wavelengths [99]. They are, therefore, very beneficial for multi-spectral imaging and sensing. High-speed optoelectronics is a further possible use for CQD-based photodetectors. Because of their rapid response times, carbon-based quantum dots can be used in high-speed optoelectronic systems, including optical communications and high-speed data transfer [100]. Devices enabling sensing and bioimaging have also been made with CQDs. The unique optical and electrical characteristics of carbon-based quantum dots make them suitable for biosensing and bioimaging applications, including in vivo imaging. They are helpful in various applications due to their high quantum yield, outstanding stability, and great sensitivity [101]. CQDs are a promising material for photodetector sensors overall because of their distinctive optical and electronic properties and their potential to be employed in a variety of applications, including multi-spectral imaging and sensing, high-speed optoelectronics, and bio-imaging and sensing devices [102].

Both graphene quantum dots (GQD) and carbon dots (CD) are being studied for potential application in photodetector sensors. Small, carbon-based nanoparticles known as CDs have been shown to exhibit robust fluorescence characteristics [103]. On the other hand, GQDs, which are small graphene flakes, have been discovered recently to have fluorescence characteristics similar to those of CDs, with the added advantage of being more stable and simpler to functionalize. CDs and GQDs are promising candidates for photodetector sensors due to their efficiency at absorbing light and transforming it into an electrical signal [104].CDs' performance is negatively influenced by their inefficient near-infrared (NIR-I and NIR-II) excitation and emission. By resolving this, CDs can be utilized for in vivo bioimaging.

## 10. Graphene Quantum Dots (GQDs)

There may be applications for graphene quantum dots (GQDs), tiny graphene sheets, in photodetector sensors. Their unique electrical and optical features make them suitable for sensing technologies. GQDs have been found to possess strong fluorescence properties, effectively absorbing light and turning it into an electrical signal [105]. GQDs are extremely stable and may be functionalized with various chemical groups, enabling precise control over their optical and electrical characteristics. They are beneficial for sensing applications requiring specificity and sensitivity [106].

Graphene, in particular, is a 2D monolayer of $sp^2$ hybrid carbon. At the Dirac point, graphene's conduction band meets with its valence band. This causes a linear dispersion relationship, making it a zero-band gap semiconductor with zero density of electronic states and an incredibly high room temperature carrier mobility [107]. The broad spectrum response and quick response of graphene are perfect photoelectric qualities due to its zero-band gap semiconducting nature. Although graphene quantum dots have lesser absorption and responsiveness, researchers are primarily focusing on hybridizing graphene with high-absorption materials to enhance photodetection capability like quantum dots [108].

GQDs are also ideally suited for photodetectors and other optoelectronic devices due to their outstanding charge transfer characteristics. Different photodetector designs, including p–n junction, Schottky, and photoconductive ones, can incorporate GQDs [109]. They are helpful for various sensing applications because they can detect a wide range of wavelengths, from visible to infrared.

Molahalli Vandana et al. fabricated a UV-based photodetector by utilizing graphene QDs with polypyrrole polymer [110]. They used two different amounts of graphene, 20 and 40 mL in polypyrrole, labeled as PGC2 and PGC4, while single graphene QDs (GQDs) were also utilized. The active layer of ITO electrode was illuminated with UV light of 265 and 355 nm for about 200 s. The graphene QDs are an excellent light absorber and electron donor to increase the carrier concentration. Figure 4 compares the photocurrent of GQDs, PGC2, and PGC4.

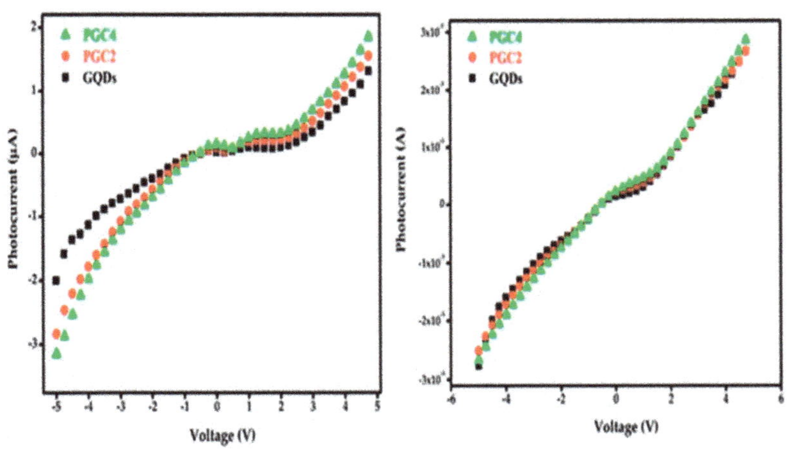


**Figure 4.** Comparison between the photocurrents of three photodetector devices under illumination [110].

PGC2 and PGC4 show responsivity of about 0.33 μA/W and 0.36 μA/W and corresponding photocurrent of 2.65 μA and 2.9 μA when illuminated at wavelength 265 nm, while a responsivity of 1.93 and 2.33 μA/W and photocurrent of 15.5 μA and 18.7 μA was observed for PGC2 and PGC4 when illuminated with UV light of 355 nm. Based on the results, which show that the PGC4 photodiode had the highest 18.7μA photocurrent, it can be concluded that the quantum dots in the polymer boosted electron mobility due to the high electrical properties of the nanocomposite.

## 11. Carbon Dots

Carbon QDs are nanomaterials made of carbon, similar to substances such as nanodiamonds, fullerenes, carbon nanotubes, graphene, and graphene oxide. Small, water-soluble carbon nanoparticles known as carbon dots (CDs) have been shown to have potential uses in photodetector sensors [111]. CDs are a desirable material for use in sensing technologies due to their distinctive electrical and optical characteristics. Strong fluorescence qualities have been discovered in them, making them efficient at absorbing light and transforming it into an electrical signal [112]. CDs' high quantum yield and outstanding stability have been found to make them appropriate for use in photodetectors [113]. The precise control of their optical and electrical properties is made possible by their simplicity in synthesis and functionalization. As a result, they are advantageous for sensing applications that demand specificity and sensitivity [114]. CDs' superior charge transport characteristics have also been discovered, making them suitable for photodetectors and other optoelectronic devices. Many photodetector designs, including p–n junction, Schottky, and photoconductive ones, can incorporate CDs [115]. They are helpful for various sensing applications because they can detect a wide range of wavelengths, from visible to infrared. Additionally, CDs are non-toxic and biocompatible, which makes them advantageous in medicinal applications such as cancer therapy and bioimaging. Additionally, CDs can be functionalized with different biomolecules to enable the detection of certain biomolecules and biological activities [116].

They have gained attention due to their environmentally friendliness, chemical stability, and good conductivity, making them useful in various fields such as biomedical and biotechnological research, solar cells, light-emitting devices, imaging, electrochemical studies, and electrochemiluminescence studies [117]. Carbon dots are a class of nanoparticles composed mainly of carbon and oxygen, while some amino and carboxyl functional groups are also present on the surface [118]. These nanodots have unique optical, electrical, and chemical properties that make them highly versatile and helpful in various applications. Some of the critical properties of carbon dots include: Size: Carbon dots are nanoparticles with diameters ranging from 1 to 10 nanometers. Their small size allows them to display unique properties and quantum confinement effects, which are helpful for a number of applications. These properties make carbon dots valuable for various applications, including sensors, catalysts, and energy storage and conversion. Luminescence: Carbon dots can emit light, making them useful for applications such as biosensors and imaging [119]. Chemical stability: Carbon dots are highly stable in many different chemical environments, including basic and acidic conditions. Their stability makes them useful for sensing and imaging applications. Conductivity: Since CDs have excellent electrical conductivity, carbon dots can be used in photovoltaic and energy storage applications [120]. Biocompatibility: Due to their non-toxicity and biocompatibility, carbon dots can be used in biological systems without damaging healthy cells. Surface functionalization: The surface of carbon dots can be modified with different chemical groups, enabling them to be useful for specific purposes [121]. Optical properties: Carbon dots can absorb light in the ultraviolet and near-infrared parts of the spectrum, making them useful for specific applications. For example, their strong absorbance in these spectrum regions can be utilized in solar cells to absorb and convert sunlight into electricity and in LED lighting to provide energy-efficient and long-lasting light sources [122]. CDs can emit light when excited by an external energy source, such as electricity or heat. The emission wavelength, or the specific range of wavelengths that the light emitted by CDs falls within, can be manipulated by changing the size and composition of the CDs. This can be achieved by using different starting materials or synthesis methods while synthesizing CDs. Solubility: Since they are highly soluble in a wide range of solvents, including water, carbon dots are simple to handle and incorporate into different systems [123]. High surface area: The small size of carbon dots results in a high surface area, which makes them ideal for applications such as catalysis and drug delivery [123].

One crucial property of CDs is their ability to emit light when excited by light, also known as photoluminescence. The photophysical responses of CDs, including their light absorption and emission, are influenced by the isolated network of $sp^2$ carbon bonds in their structure [123]. In contrast, extended networks of carbon bonds, as found in carbon nanotubes, graphite, and graphene, do not interact with light as strongly. This may be due to the rapid recombination of photogenerated electrons and holes without the emission of radiation. CDs typically absorb light over a broad range of wavelengths [124]. They can absorb light over a wide range of wavelengths, but the specific wavelength at which they emit light is determined by the population of a particular energy level or "domain" and the surface properties of the quantum dots. This phenomenon, known as excitation-dependent emission, is a characteristic of CDs arising from these nanomaterials' surface state. Essentially, the surface of the quantum dots plays a crucial role in determining the specific wavelength at which they will emit light when excited by an external energy source [125].

Carbon dots have a high absorption coefficient, which implies they can absorb a lot of light, in addition to their fluorescence and quenching qualities. Due to their ability to transform light into energy that may be utilized to treat a range of medical problems, they are valuable in phototherapy applications [126].

K. Sarkar et al. [127] show the use of carbon dots (CDs) synthesized from Carica Papaya through a facile synthesis method as a potential broadband photodetector in a hybrid (organic–inorganic) heterostructure with a silicon wafer. The engineered CDs' maximum detectivity and responsivity are around 1 A $W^{-1}$ and $2 \times 10^{12}$ Jones, respectively. The device structure ensures a promising photodetector technology comparable to silicon process technology through a cheap and simple fabrication process. Additional optimization effort and defect engineering can speed up switching.

Sergii Kalytchuk et al. demonstrated the synthesis of purple-emissive CDs utilizing folic acid as the only precursor in a simple and reproducible process [128]. The CDs have a high photoluminescence quantum yield of 54.6% and emit light at a wavelength of 390 nm, which is independent of the excitation wavelength and integrates the emission color spectrum of intensely glowing carbon dots to purple. They employed CDs to increase the silicon photodetector's UV range sensitivity. It was found that the photoresponsivity of the silicon photodetector in the UV range of 0.8 to 2.5 mA/W was enhanced by 203.8% with CDs' integration.

Di Li et al. [129], by modifying the surface of CDs with functional groups of sulfoxide/carbonyl, changed the optical band gap of CDs. They proposed a mechanism for improved NIR emission, which can be seen in Figure 5. The CDs have a layered structure, and the outer layers have bonding sites. The interaction between the S = O/C = O functional group and layers increased surface oxidation, which caused the reduction in the lowest unoccupied orbital (LUMO). As a result, these functional groups disrupt CDs' organized structure. Due to the increased surface oxidation, the LUMO orbitals are lower in the outer layers compared to the inner layers and cause the NIR emission of CDs.

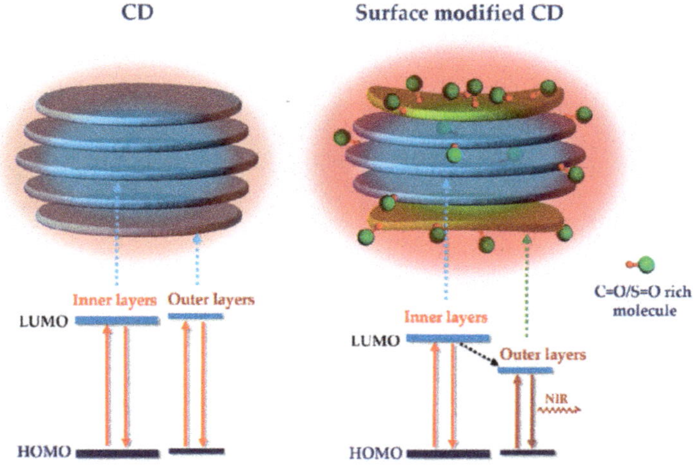


**Figure 5.** Non-treated CDs (left column) and CDs modified with S = O/C = O rich molecules (right column). The red (oxygen atom) and green double-bonded balls represent the C = O/S = O rich molecule (Reproduced from [129], with permission from John Wiley & Sons, Inc.).

## 12. Future Prospects and Conclusions

Several optoelectronics applications that have been utilized as a result of advancements in nanomaterial synthesis were covered in this study. Researchers have used QDs in various devices, including solar cells, lasers, and photodetectors. Quantum dots act differently from bulk materials in light emission and absorption due to the quantum confinement effect, which is favorable for laser and solar panel applications. This review briefly discussed different optoelectronic devices based on QDs.

QDs in photodetectors have a bright future, with the possibility of increased performance, reduced costs, and new applications. Future quantum dot photodetectors are expected to have several benefits over current photodetectors. They could, for instance, offer greater sensitivity, faster response times, and lower noise levels. Since they can be tailored to absorb light over a broad range of wavelengths by altering their size and composition, they are also anticipated to be more flexible.

Factors contributing to the usefulness of QDs as compared to a traditional photodetector that is discussed in this review include:

1. Quantum dot-based photodetectors have the potential to provide greater efficiency than conventional detectors, particularly in the infrared spectrum because they can be engineered to have a very narrow absorption spectrum that fits the wavelength of the incident light, whereas organic QDs have a wide surface area, which enables them to efficiently absorb light and produce a large number of excitons (electron–hole pairs) which are useful for photodetectors.
2. Low-cost, solution-based techniques can be used to make quantum dot-based photodetectors. This could make them more affordable than current photodetectors, particularly for applications involving large areas, while organic QDs can be deposited on flexible substrates, which enables the use of wearable electronics, among many other uses.
3. A faster reaction time is possible because of QDs, which can be used to construct photodetectors with quick response times that are suited for high-speed applications.
4. QDs can be combined with other electronic components to build hybrid devices with cutting-edge features. For instance, these could be combined with complementary metal–oxide–semiconductor technology to produce inexpensive, high-performing bio-imaging systems.
5. Since QDs are more flexible in combining organic and inorganic materials, their properties can be easily changed to suit different needs.

PbS quantum dots have a high electron–hole recombination rate and rapid electron mobility, which increases the responsivity and improves the efficiency of the conversion of photons into electric current. PbS quantum dots' properties can also be modified with ligand reactions. Bilayer PbS quantum dots with different modification ligands, tetrabutylammonium iodide (TBAI), and 1,2-ethanedithiol (EDT) devices have higher light currents and detections.

This review study also discusses indium arsenide (InAs) quantum dots (QDs). It was shown that due to their high infrared absorption coefficient, they make good photodetectors in the region of infrared detection. They are helpful for imaging at infrared wavelengths as well.

This review discusses organic quantum dots such as graphene, particularly carbon quantum dots. CDs' particular electrical and optical properties make them desirable materials for sensing technologies. They have been shown to possess strong fluorescence properties, effectively absorbing light and converting it into an electrical signal. Their ease of synthesis and functionalization allows precise control of their optical and electrical properties. They are, therefore, beneficial for photodetectors that require specificity and sensitivity.

**Author Contributions:** Conceptualization, H.S.; writing—original draft preparation, H.S.; data curation, H.S.; writing—review and editing, H.S. and M.W. All authors have read and agreed to the published version of the manuscript.

**Funding:** This research received financial support from the AGH University of Science and Technology.

**Institutional Review Board Statement:** Not applicable.

**Informed Consent Statement:** Informed consent was obtained from all subjects involved in the study.

**Data Availability Statement:** Not applicable.

**Conflicts of Interest:** Author declares no conflict of interest.

## References


1. Korotcenkov, G. Current trends in nanomaterials for metal oxide-based conductometric gas sensors: Advantages and limitations. part 1: 1D and 2D nanostructures. *Nanomaterials* **2020**, *10*, 1392. [CrossRef] [PubMed]
2. Daulbayev, C.; Sultanov, F.; Bakbolat, B.; Daulbayev, O. 0d 1d 2d Nanomater. Visible Photoelectrochem. Water Splitting. A Review. *Int. J. Hydrog. Energy* **2020**, *45*, 33325–33342. [CrossRef]
3. Fang, L.; Feng, J.J.; Shi, X.; Si, T.; Song, Y.; Jia, H.; Li, Y.; Li, H.-W.; Zhang, Q.J. 0D, 1D and 2D nanomaterials for visible photoelectrochemical water splitting. A review. *Chem. Commun.* **2019**, *55*, 10476–10479. [CrossRef] [PubMed]
4. Yang, T.; Liu, Y.; Wang, H.; Duo, Y.; Zhang, B.; Ge, Y.; Zhang, H.; Chen, W. Recent advances in 0D nanostructure-functionalized low-dimensional nanomaterials for chemiresistive gas sensors. *J. Mater. Chem. C* **2020**, *8*, 7272–7299. [CrossRef]
5. Garnett, E.; Mai, L.; Yang, P. Introduction: 1D nanomaterials/nanowires. *Chem. Rev.* **2019**, *119*, 8955–8957. [CrossRef]
6. Nozik, A.J. Quantum dot solar cells. *Phys. E: Low-Dimens. Syst. Nanostruct.* **2002**, *14*, 115–120. [CrossRef]
7. Bourlinos, A.B.; Stassinopoulos, A.; Anglos, D.; Zboril, R.; Karakassides, M.; Giannelis, E.P. Surface functionalized carbogenic quantum dots. *Small* **2008**, *4*, 455–458. [CrossRef]
8. Liu, X.; Zhang, Y.; Yu, T.; Qiao, X.; Gresback, R.; Pi, X.; Yang, D.J.P.; Characterization, P.S. Optimum quantum yield of the light emission from 2 to 10 nm hydrosilylated silicon quantum dots. *Part. Syst. Charact.* **2016**, *33*, 44–52. [CrossRef]
9. Martyniuk, P.; Rogalski, A. Quantum-dot infrared photodetectors: Status and outlook. *Prog. Quantum Electron.* **2008**, *32*, 89–120. [CrossRef]
10. Stiff-Roberts, A.D. Quantum-dot infrared photodetectors: A review. *J. Nanophotonics* **2009**, *3*, 031607. [CrossRef]
11. Liu, H.; Gao, M.; McCaffrey, J.; Wasilewski, Z.; Fafard, S. Quantum dot infrared photodetectors. *Appl. Phys. Lett.* **2001**, *78*, 79–81. [CrossRef]
12. Li, M.Y.; Yu, M.; Su, D.; Zhang, J.; Jiang, S.; Wu, J.; Wang, Q.; Liu, S. Ultrahigh responsivity UV photodetector based on Cu nanostructure/ZnO QD hybrid architectures. *Small* **2019**, *15*, 1901606. [CrossRef] [PubMed]
13. Praveen, S.; Veeralingam, S.; Badhulika, S. A Flexible Self-Powered UV Photodetector and Optical UV Filter Based on β-Bi2O3/SnO2 Quantum Dots Schottky Heterojunction. *Adv. Mater. Interfaces* **2021**, *8*, 2100373. [CrossRef]
14. Chen, G.; Liang, B.; Liu, X.; Liu, Z.; Yu, G.; Xie, X.; Luo, T.; Chen, D.; Zhu, M.; Shen, G. High-performance hybrid phenyl-C61-butyric acid methyl ester/Cd3P2 nanowire ultraviolet–visible–near infrared photodetectors. *ACS Nano* **2014**, *8*, 787–796. [CrossRef]
15. Dang, V.Q.; Trung, T.Q.; Kim, D.I.; Duy, L.T.; Hwang, B.U.; Lee, D.W.; Kim, B.Y.; Toan, L.D.; Lee, N.E. Ultrahigh responsivity in graphene–ZnO nanorod hybrid UV photodetector. *Small* **2015**, *11*, 3054–3065. [CrossRef]
16. Manga, K.K.; Wang, J.; Lin, M.; Zhang, J.; Nesladek, M.; Nalla, V.; Ji, W.; Loh, K. High-performance broadband photodetector using solution-processible PbSe–$TiO_2$–graphene hybrids. *Adv. Mater.* **2012**, *24*, 1697–1702. [CrossRef]
17. Banyai, L.; Gilliot, P.; Hu, Y.; Koch, S. Surface-polarization instabilities of electron-hole pairs in semiconductor quantum dots. *Phys. Rev. B* **1992**, *45*, 14136. [CrossRef]
18. Gandhi, S.A.; Sutariya, P.G.; Soni, H.N.; Chaudhari, D. Quantum dots: Application in medical science. *Int. J. Nano Dimens.* **2023**, *14*, 29–40.
19. Najm, N.I.; Hassun, H.K.; al-Maiyaly, B.K.H.; Hussein, B.H.; Shaban, A.H. Highly selective CdS: Ag heterojunction for photodetector applications. *AIP Conf. Proc.* **2019**, *2123*, 020031.
20. Hakami, J. Highly induced photosensing behavior of Erbium (Er), Yttrium (Y) and Terbium (Tb) doped nanostructured Cadmium Sulphide (CdS) thin films prepared by nebulizer spray pyrolysis method. *J. Alloys Compd.* **2022**, *924*, 166577. [CrossRef]
21. Habeebu, S.S.; Liu, J. Klaassen CDJT pharmacology, a. Cadmium-induced apoptosis in mouse liver. *Toxicol. Appl. Pharmacol.* **1998**, *149*, 203–209. [CrossRef] [PubMed]
22. Ning, X.; Lu, G. Photocorrosion inhibition of CdS-based catalysts for photocatalytic overall water splitting. *Nanoscale* **2020**, *12*, 1213–1223. [CrossRef] [PubMed]
23. McDonald, S.A.; Konstantatos, G.; Zhang, S.; Cyr, P.W.; Klem, E.J.; Levina, L.; Sargent, E. Solution-processed PbS quantum dot infrared photodetectors and photovoltaics. *Nat. Mater.* **2005**, *4*, 138–142. [CrossRef]
24. Kim, E.-T.; Madhukar, A.; Ye, Z.; Campbell, J. High detectivity InAs quantum dot infrared photodetectors. *Appl. Phys. Lett.* **2004**, *84*, 3277–3279. [CrossRef]
25. Marzin, J.-Y.; Gérard, J.-M.; Izraël, A.; Barrier, D.; Bastard, G. Photoluminescence of single InAs quantum dots obtained by self-organized growth on GaAs. *Phys. Rev. Lett.* **1994**, *73*, 716. [CrossRef] [PubMed]
26. Zhao, N.; Osedach, T.P.; Chang, L.-Y.; Geyer, S.M.; Wanger, D.; Binda, M.T.; Arango, A.C.; Bawendi, M.G.; Bulovic, V. Colloidal PbS quantum dot solar cells with high fill factor. *ACS Nano* **2010**, *4*, 3743–3752. [CrossRef]

27. Bretagnon, T.; Lefebvre, P.; Valvin, P.; Bardoux, R.; Guillet, T.; Taliercio, T.; Gil, B.; Grandjean, N.; Semond, F.; Damilano, B. Radiative lifetime of a single electron-hole pair in Ga N/Al N quantum dots. *Phys. Rev. B* **2006**, *73*, 113304. [CrossRef]
28. Li, X.; Rui, M.; Song, J.; Shen, Z.; Zeng, H. Carbon and graphene quantum dots for optoelectronic and energy devices: A review. *Adv. Funct. Mater.* **2015**, *25*, 4929–4947. [CrossRef]
29. Litvin, A.; Martynenko, I.; Purcell-Milton, F.; Baranov, A.; Fedorov, A.; Gun'Ko, Y. Colloidal quantum dots for optoelectronics. *J. Mater. Chem. A* **2017**, *5*, 13252–13275. [CrossRef]
30. Li, S.-Y.; He, L. Recent progresses of quantum confinement in graphene quantum dots. *Front. Phys.* **2022**, *17*, 33201. [CrossRef]
31. Takagahara, T.; Takeda, K. Theory of the quantum confinement effect on excitons in quantum dots of indirect-gap materials. *Phys. Rev. B* **1992**, *46*, 15578. [CrossRef] [PubMed]
32. Wise, F. Lead salt quantum dots: The limit of strong quantum confinement. *Acc. Chem. Res.* **2000**, *33*, 773–780. [CrossRef] [PubMed]
33. Giessen, H.; Woggon, U.; Fluegel, B.; Mohs, G.; Hu, Y.; Koch, S.; Peyghambarian, N. Femtosecond optical gain in strongly confined quantum dots. *Opt. Lett.* **1996**, *21*, 1043–1045. [CrossRef] [PubMed]
34. Ruedas-Rama, M.J.; Orte, A.; Hall, E.A.; Alvarez-Pez, J.M.; Talavera, E. Quantum dot photoluminescence lifetime-based pH nanosensor. *Chem. Commun.* **2011**, *47*, 2898–2900. [CrossRef] [PubMed]
35. Mintz, K.J.; Zhou, Y.; Leblanc, R.M. Recent development of carbon quantum dots regarding their optical properties, photoluminescence mechanism, and core structure. *Nanoscale* **2019**, *11*, 4634–4652. [CrossRef] [PubMed]
36. Zhu, S.; Song, Y.; Zhao, X.; Shao, J.; Zhang, J.; Yang, B. The photoluminescence mechanism in carbon dots (graphene quantum dots, carbon nanodots, and polymer dots): Current state and future perspective. *Nano Res.* **2015**, *8*, 355–381. [CrossRef]
37. Benyettou, F.; Aissat, A.; Djebari, M.; Vilcot, J.R. Electrical properties of InAsP/Si quantum dot solar cell. *Int. J. Hydrog. Energy* **2017**, *42*, 19512–19517. [CrossRef]
38. Jolley, G.; Fu, L.; Lu, H.F.; Tan, H.H.; Jagadish, C.J. The role of intersubband optical transitions on the electrical properties of InGaAs/GaAs quantum dot solar cells. *Prog. Photovolt. Res. Appl.* **2013**, *21*, 736–746. [CrossRef]
39. Emerson, R. The quantum yield of photosynthesis. *Annu. Rev. Plant Physiol.* **1958**, *9*, 1–24. [CrossRef]
40. Weber, G.; Teale, F. Determination of the absolute quantum yield of fluorescent solutions. *Trans. Faraday Soc.* **1957**, *53*, 646–655. [CrossRef]
41. Hamood, R.; Abd El-sadek, M.; Gadalla, A.J.V. Facile synthesis, structural, electrical and dielectric properties of CdSe/CdS core-shell quantum dots. *Vacuum* **2018**, *157*, 291–298. [CrossRef]
42. Grieve, K.; Mulvaney, P.; Grieser, F. Synthesis and electronic properties of semiconductor nanoparticles/quantum dots. *Curr. Opin. Colloid Interface Sci.* **2000**, *5*, 168–172. [CrossRef]
43. Kouwenhoven, L.P.; Austing, D.; Tarucha, S. Few-electron quantum dots. *Rep. Prog. Phys.* **2001**, *64*, 701. [CrossRef]
44. Du, J.; Du, Z.; Hu, J.-S.; Pan, Z.; Shen, Q.; Sun, J.; Long, D.; Dong, H.; Sun, L.; Zhong, X. Zn–Cu–In–Se quantum dot solar cells with a certified power conversion efficiency of 11.6%. *J. Am. Chem. Soc.* **2016**, *138*, 4201–4209. [CrossRef] [PubMed]
45. Marko, I.P.; Adams, A.R.; Sweeney, S.J.; Mowbray, D.J.; Skolnick, M.S.; Liu, H.Y.; Groom, K. Recombination and loss mechanisms in low-threshold InAs-GaAs 1.3-/spl mu/m quantum-dot lasers. *IEEE J. Sel. Top. Quantum Electron.* **2005**, *11*, 1041–1047. [CrossRef]
46. Park, Y.-S.; Bae, W.K.; Pietryga, J.M.; Klimov, V.I. Auger recombination of biexcitons and negative and positive trions in individual quantum dots. *ACS Nano* **2014**, *8*, 7288–7296. [CrossRef] [PubMed]
47. Geiregat, P.; Van Thourhout, D.; Hens, Z. A bright future for colloidal quantum dot lasers. *NPG Asia Mater.* **2019**, *11*, 41. [CrossRef]
48. Ustinov, V.M.; Zhukov, A.E.; Zhokov, A.E.; Maleev, N.A.; Egorov, A.Y. *Quantum Dot Lasers*; Oxford University Press on Demand: Oxford, UK, 2003; Volume 11.
49. Semonin, O.E.; Luther, J.M.; Choi, S.; Chen, H.-Y.; Gao, J.; Nozik, A.J.; Beard, M.C. Peak external photocurrent quantum efficiency exceeding 100% via MEG in a quantum dot solar cell. *Science* **2011**, *334*, 1530–1533. [CrossRef]
50. Carey, G.H.; Abdelhady, A.L.; Ning, Z.; Thon, S.M.; Bakr, O.M.; Sargent, E.H. Colloidal quantum dot solar cells. *Chem. Rev.* **2015**, *115*, 12732–12763. [CrossRef]
51. Kamat, P.V. Quantum dot solar cells. The next big thing in photovoltaics. *J. Phys. Chem. Lett.* **2013**, *4*, 908–918. [CrossRef]
52. Sawadogo, W.; Abiodun, B.J.; Okogbue, E.C. Impacts of global warming on photovoltaic power generation over West Africa. *Renew. Energy* **2020**, *151*, 263–277. [CrossRef]
53. Green, M. Photovoltaic technology and visions for the future. *Prog. Energy* **2019**, *1*, 013001. [CrossRef]
54. Andreev, V.A.C.M.; Grilikhes, V.A.; Rumyantsev, V.D. *Photovoltaic Conversion of Concentrated Sunlight*; Wiley: New York, NY, USA, 1997.
55. Imamzai, M.; Aghaei, M.; Thayoob, Y.H.M.; Forouzanfar, M. A review on comparison between traditional silicon solar cells and thin-film CdTe solar cells. In Proceedings of the National Graduate Conference Nat-Grad, Kajang, Malaysia, 8–10 November 2012; pp. 1–5.
56. Brendel, R.J. Review of layer transfer processes for crystalline thin-film silicon solar cells. *Jpn. J. Appl. Phys.* **2001**, *40*, 4431. [CrossRef]
57. Rong, Y.; Hu, Y.; Mei, A.; Tan, H.; Saidaminov, M.I.; Seok, S.I.; McGehee, M.D.; Sargent, E.H.; Han, H.J.S. Challenges for commercializing perovskite solar cells. *Science* **2018**, *361*, eaat8235. [CrossRef] [PubMed]

58. Yang, J.; Zhong, X. CdTe based quantum dot sensitized solar cells with efficiency exceeding 7% fabricated from quantum dots prepared in aqueous media. *J. Mater. Chem. A* **2016**, *4*, 16553–16561. [CrossRef]
59. Ge, B.; Nie, C.; Shen, J. A hybrid photodetector of graphene/TiO2/inorganic PbS quantum dots for fast response. *Jpn. J. Appl. Phys.* **2022**, *61*, 040903. [CrossRef]
60. Kundu, B.; Özdemir, O.; Dalmases, M.; Kumar, G.; Konstantatos, G. Hybrid 2D-QD MoS2–PbSe Quantum Dot Broadband Photodetectors with High-Sensitivity and Room-Temperature Operation at 2.5 µm. *Adv. Opt. Mater.* **2021**, *9*, 2101378. [CrossRef]
61. Kagkoura, A.; Skaltsas, T.; Tagmatarchis, N. Transition-Metal Chalcogenide/Graphene Ensembles for Light-Induced Energy Applications. *Chem. A Eur. J.* **2017**, *23*, 12967–12979. [CrossRef]
62. Jaiswal, K.; Girish, Y.R.; De, M.J. Group-VI-Chalcogenide-Based Nanomaterials in Photo/Thermal Organic Transformations. *Acc. Mater. Res.* **2022**, *3*, 1033–1048. [CrossRef]
63. Chatterjee, A.; Pendyala, N.B.; Jagtap, A.; Koteswara Rao, K.S.R. Uncooled Mid-Wave Infrared Focal Plane Array Using Band Gap Engineered Mercury Cadmium Telluride Quantum Dot Coated Silicon ROIC. *E J. Surf. Sci. Nanotechnol.* **2019**, *17*, 95–100. [CrossRef]
64. Levine, B. Quantum-well infrared photodetectors. *J. Appl. Phys.* **1993**, *74*, R1–R81. [CrossRef]
65. Fox, M.; Ispasoiu, R.; Materials, P. *Quantum Wells, Superlattices, and Band-Gap Engineering*; Springer: Cham, Switzerland, 2017.
66. Phillips, J. Evaluation of the fundamental properties of quantum dot infrared detectors. *J. Appl. Phys.* **2002**, *91*, 4590–4594. [CrossRef]
67. Goossen, K.; Lyon, S. Performance aspects of a quantum-well detector. *J. Appl. Phys.* **1988**, *63*, 5149–5153. [CrossRef]
68. Rogalski, A. Quantum well photoconductors in infrared detector technology. *J. Appl. Phys.* **2003**, *93*, 4355–4391. [CrossRef]
69. Bhattacharya, P.; Ghosh, S.; Stiff-Roberts, A. Quantum dot opto-electronic devices. *Annu. Rev. Mater. Res.* **2004**, *34*, 1. [CrossRef]
70. Pan, D.; Towe, E.; Kennerly, S. Photovoltaic quantum-dot infrared detectors. *Appl. Phys. Lett.* **2000**, *76*, 3301–3303. [CrossRef]
71. Liu, M.; Yazdani, N.; Yarema, M.; Jansen, M.; Wood, V.; Sargent, E. Colloidal quantum dot electronics. *Nat. Electron.* **2021**, *4*, 548–558. [CrossRef]
72. Liu, J.; Liu, P.; Chen, D.; Shi, T.; Qu, X.; Chen, L.; Wu, T.; Ke, J.; Xiong, K.; Li, M. A near-infrared colloidal quantum dot imager with monolithically integrated readout circuitry. *Nat. Electron.* **2022**, *5*, 443–451. [CrossRef]
73. Brown, A.E. *Application of Mobility Spectrum Analysis to Modern Multi-Layered IR Device Material*; University of Illinois: Chicago, IL, USA, 2016.
74. Rogalski, A.; Chrzanowski, K.J.M.; Systems, M. Infrared devices and techniques (revision). *Metrol. Meas. Syst.* **2014**, *21*, 565–618. [CrossRef]
75. Hoyer, P.; Könenkamp, R. Photoconduction in porous TiO2 sensitized by PbS quantum dots. *Appl. Phys. Lett.* **1995**, *66*, 349–351. [CrossRef]
76. Wang, H.; Kubo, T.; Nakazaki, J.; Segawa, H. Solution-processed short-wave infrared PbS colloidal quantum Dot/ZnO nanowire solar cells giving high open-circuit voltage. *ACS Energy Lett.* **2017**, *2*, 2110–2117. [CrossRef]
77. Gao, J.; Perkins, C.L.; Luther, J.M.; Hanna, M.C.; Chen, H.-Y.; Semonin, O.E.; Nozik, A.J.; Ellingson, R.J.; Beard, M. N-Type transition metal oxide as a hole extraction layer in PbS quantum dot solar cells. *Nano Lett.* **2011**, *11*, 3263–3266. [CrossRef] [PubMed]
78. Liu, Y.; Wu, H.; Shi, G.; Li, Y.; Gao, Y.; Fang, S.; Tang, H.; Chen, W.; Ma, T.; Khan, I. Merging Passivation in Synthesis Enabling the Lowest Open-Circuit Voltage Loss for PbS Quantum Dot Solar Cells. *Adv. Mater.* **2022**, *35*, 2207293. [CrossRef]
79. Zhang, C.; Lian, L.; Yang, Z.; Zhang, J.; Zhu, H. Quantum confinement-tunable ultrafast charge transfer in a PbS quantum dots/WSe2 0D–2D hybrid structure: Transition from the weak to strong coupling regime. *J. Phys. Chem. Lett.* **2019**, *10*, 7665–7671. [CrossRef]
80. Wei, Y.; Ren, Z.; Zhang, A.; Mao, P.; Li, H.; Zhong, X.; Li, W.; Yang, S.; Wang, J. Hybrid organic/PbS quantum dot bilayer photodetector with low dark current and high detectivity. *Adv. Funct. Mater.* **2018**, *28*, 1706690. [CrossRef]
81. Kufer, D.; Nikitskiy, I.; Lasanta, T.; Navickaite, G.; Koppens, F.H.; Konstantatos, G. Hybrid 2D–0D MoS2–PbS quantum dot photodetectors. *Adv. Mater. Interfaces* **2015**, *27*, 176–180. [CrossRef]
82. Maulu, A.; Navarro-Arenas, J.; Rodríguez-Cantó, P.J.; Sánchez-Royo, J.F.; Abargues, R.; Suárez, I.; Martínez-Pastor, J. Charge transport in trap-sensitized infrared PbS quantum-dot-based photoconductors: Pros and cons. *Nanomaterials* **2018**, *8*, 677. [CrossRef]
83. Bothra, U.; Albaladejo-Siguan, M.; Vaynzof, Y.; Kabra, D. Impact of Ligands on the Performance of PbS Quantum Dot Visible–Near-Infrared Photodetectors. *Adv. Opt. Mater.* **2022**, *11*, 2201897. [CrossRef]
84. Kwon, J.B.; Han, M.; Jung, D.G.; Kong, S.H.; Jung, D.J. High sensitivity shortwave infrared photodetector based on PbS QDs using P3HT. *Nanomaterials* **2021**, *11*, 2683. [CrossRef]
85. Franke, D.; Harris, D.K.; Chen, O.; Bruns, O.T.; Carr, J.A.; Wilson, M.W.; Bawendi, M. Continuous injection synthesis of indium arsenide quantum dots emissive in the short-wavelength infrared. *Nat. Commun.* **2016**, *7*, 12749. [CrossRef]
86. Banin, U.; Cao, Y.; Katz, D.; Millo, O. Identification of atomic-like electronic states in indium arsenide nanocrystal quantum dots. *Nature* **1999**, *400*, 542–544. [CrossRef]
87. Kim, S.-W.; Zimmer, J.P.; Ohnishi, S.; Tracy, J.B.; Frangioni, J.V.; Bawendi, M.G. Engineering InAs x P1-x/InP/ZnSe III−V Alloyed Core/Shell Quantum Dots for the Near-Infrared. *J. Am. Chem. Soc.* **2005**, *127*, 10526–10532. [CrossRef]

88. Jalali, H.B.; De Trizio, L.; Manna, L.; Di Stasio, F. Indium arsenide quantum dots: An alternative to lead-based infrared emitting nanomaterials. *Chem. Soc. Rev.* **2022**, *51*, 9861–9881. [CrossRef]
89. Park, Y.; Jeong, S.; Kim, S. Medically translatable quantum dots for biosensing and imaging. *J. Photochem. Photobiol. C Photochem. Rev.* **2017**, *30*, 51–70. [CrossRef]
90. Fu, Y.; Ferdos, F.; Sadeghi, M.; Zhao, Q.; Wang, S.; Larsson, A.J.S. Strain and optical transitions in InAs quantum dots on (001) GaAs. *Superlattices Microstruct.* **2001**, *30*, 205–213. [CrossRef]
91. Dion, C.; Desjardins, P.; Shtinkov, N.; Robertson, M.; Schiettekatte, F.; Poole, P.; Raymond, S. Intermixing during growth of InAs self-assembled quantum dots in InP: A photoluminescence and tight-binding investigation. *Phys. Rev. B* **2008**, *77*, 075338. [CrossRef]
92. Darwan, D.; Lim, L.J.; Wang, T.; Wijaya, H.; Tan, Z. Ultra-Confined Visible-Light-Emitting Colloidal Indium Arsenide Quantum Dots. *Nano Lett.* **2021**, *21*, 5167–5172. [CrossRef]
93. Nguyen, H.T.; Das, R.; Duong, A.T.; Lee, S. Influence of quantum dot concentration on the opto-electronic properties of colloidal quantum-dots LEDs. *Opt. Mater.* **2020**, *109*, 110251. [CrossRef]
94. Wang, T.; Nie, C.; Ao, Z.; Wang, S.; An, T. Recent progress in gC 3 N 4 quantum dots: Synthesis, properties and applications in photocatalytic degradation of organic pollutants. *J. Mater. Chem. A* **2020**, *8*, 485–502. [CrossRef]
95. Weiss, E.A. Organic molecules as tools to control the growth, surface structure, and redox activity of colloidal quantum dots. *Acc. Chem. Res.* **2013**, *46*, 2607–2615. [CrossRef]
96. Li, J.; Zhu, J.-J. Quantum dots for fluorescent biosensing and bio-imaging applications. *Analyst* **2013**, *138*, 2506–2515. [CrossRef] [PubMed]
97. Vercelli, B.J.C. The role of carbon quantum dots in organic photovoltaics: A short overview. *Coatings* **2021**, *11*, 232. [CrossRef]
98. Alaghmandfard, A.; Sedighi, O.; Rezaei, N.T.; Abedini, A.A.; Khachatourian, A.M.; Toprak, M.S.; Seifalian, A.J. Recent advances in the modification of carbon-based quantum dots for biomedical applications. *Mater. Sci. Eng. C* **2021**, *120*, 111756. [CrossRef] [PubMed]
99. Namdari, P.; Negahdari, B.; Eatemadi, A.J.B. Synthesis, properties and biomedical applications of carbon-based quantum dots: An updated review. *Biomed. Pharmacother.* **2017**, *87*, 209–222. [CrossRef] [PubMed]
100. Xu, A.; Wang, G.; Li, Y.; Dong, H.; Yang, S.; He, P.; Ding, G. Carbon-based quantum dots with solid-state photoluminescent: Mechanism, implementation, and application. *Small* **2020**, *16*, 2004621. [CrossRef] [PubMed]
101. Luo, P.G.; Yang, F.; Yang, S.-T.; Sonkar, S.K.; Yang, L.; Broglie, J.J.; Liu, Y.; Sun, Y.-P. Carbon-based quantum dots for fluorescence imaging of cells and tissues. *RSC Adv.* **2014**, *4*, 10791–10807. [CrossRef]
102. Anilkumar, P.; Wang, X.; Cao, L.; Sahu, S.; Liu, J.-H.; Wang, P.; Korch, K.; Tackett Ii, K.N.; Parenzan, A.; Sun, Y.-P. Toward quantitatively fluorescent carbon-based "quantum" dots. *Nanoscale* **2011**, *3*, 2023–2027. [CrossRef]
103. Walther, B.K.; Dinu, C.Z.; Guldi, D.M.; Sergeyev, V.G.; Creager, S.E.; Cooke, J.P.; Guiseppi-Elie, A.J. Nanobiosensing with graphene and carbon quantum dots: Recent advances. *Mater. Today* **2020**, *39*, 23–46. [CrossRef]
104. Paulo, S.; Palomares, E.; Martinez-Ferrero, E.J.N. Graphene and carbon quantum dot-based materials in photovoltaic devices: From synthesis to applications. *Nanomaterials* **2016**, *6*, 157. [CrossRef]
105. Bacon, M.; Bradley, S.J.; Nann, T.J.P.; Characterization, P.S. Graphene quantum dots. *Part. Syst. Charact.* **2014**, *31*, 415–428. [CrossRef]
106. Tian, P.; Tang, L.; Teng, K.; Lau, S. Graphene quantum dots from chemistry to applications. *Mater. Today Chem.* **2018**, *10*, 221–258. [CrossRef]
107. Zhang, M.K.; Liu, W.D.; Gong, Y.P.; Liu, Q.; Chen, Z. Graphene/Quantum Dot Heterostructure Photodetectors: From Material to Performance. *Adv. Opt. Mater.* **2022**, *10*, 2201889. [CrossRef]
108. De Fazio, D.; Uzlu, B.; Torre, I.; Monasterio-Balcells, C.; Gupta, S.; Khodkov, T.; Bi, Y.; Wang, Z.; Otto, M.; Lemme, M.C. Graphene–quantum dot hybrid photodetectors with low dark-current readout. *ACS Nano* **2020**, *14*, 11897–11905. [CrossRef] [PubMed]
109. Konstantatos, G.; Badioli, M.; Gaudreau, L.; Osmond, J.; Bernechea, M.; De Arquer, F.P.G.; Gatti, F.; Koppens, F.H. Hybrid graphene–quantum dot phototransistors with ultrahigh gain. *Nat. Nanotechnol.* **2012**, *7*, 363–368. [CrossRef] [PubMed]
110. Vandana, M.; Devendrappa, H.; Padova, P.D.; Hegde, G.J.N. Polymer Nanocomposite Graphene Quantum Dots for High-Efficiency Ultraviolet Photodetector. *Nanomaterials* **2022**, *12*, 3175. [CrossRef] [PubMed]
111. Cao, L.; Yang, S.-T.; Wang, X.; Luo, P.G.; Liu, J.-H.; Sahu, S.; Liu, Y.; Sun, Y.-P. Competitive performance of carbon "quantum" dots in optical bioimaging. *Theranostics* **2012**, *2*, 295. [CrossRef]
112. Zhang, X.; Zeng, Q.; Xiong, Y.; Ji, T.; Wang, C.; Shen, X.; Lu, M.; Wang, H.; Wen, S.; Zhang, Y.J. Energy level modification with carbon dot interlayers enables efficient perovskite solar cells and quantum dot based light-emitting diodes. *Adv. Funct. Mater.* **2020**, *30*, 1910530. [CrossRef]
113. Hu, L.; Sun, Y.; Li, S.; Wang, X.; Hu, K.; Wang, L.; Liang, X.-J.; Wu, Y. Multifunctional carbon dots with high quantum yield for imaging and gene delivery. *Carbon* **2014**, *67*, 508–513. [CrossRef]
114. Meziani, M.J.; Dong, X.; Zhu, L.; Jones, L.P.; LeCroy, G.E.; Yang, F.; Wang, S.; Wang, P.; Zhao, Y.; Yang, L.; et al. Visible-light-activated bactericidal functions of carbon "Quantum" dots. *ACS Appl. Mater. Interfaces* **2016**, *8*, 10761–10766. [CrossRef]
115. Chen, X.; Yang, C.; Sun, H.; Ning, S.; Zhou, H.; Zhang, H.; Wang, S.; Feng, G.; Zhou, S. Enhanced photoresponsivity in carbon quantum dots-coupled graphene/silicon Schottky-junction photodetector. *Laser Phys. Lett.* **2019**, *16*, 076201. [CrossRef]

116. da Silva, J.C.E.; Gonçalves, H.M.R. Analytical and bioanalytical applications of carbon dots. *TrAC Trends Anal. Chem.* **2011**, *30*, 1327–1336. [CrossRef]
117. Chu, K.-W.; Lee, S.L.; Chang, C.-J.; Liu, L.J.P. Recent progress of carbon dot precursors and photocatalysis applications. *Polymers* **2019**, *11*, 689. [CrossRef] [PubMed]
118. Zhang, X.; Zhang, Y.; Wang, Y.; Kalytchuk, S.; Kershaw, S.V.; Wang, Y.; Wang, P.; Zhang, T.; Zhao, Y.; Zhang, H. Color-switchable electroluminescence of carbon dot light-emitting diodes. *ACS Nano* **2013**, *7*, 11234–11241. [CrossRef] [PubMed]
119. Kurian, M.; Paul, A. Recent trends in the use of green sources for carbon dot synthesis–A short review. *Carbon Trends* **2021**, *3*, 100032. [CrossRef]
120. Guo, R.; Li, L.; Wang, B.; Xiang, Y.; Zou, G.; Zhu, Y.; Hou, H.; Ji, X. Functionalized carbon dots for advanced batteries. *Energy Storage Mater.* **2021**, *37*, 8–39. [CrossRef]
121. Shabbir, H.; Tokarski, T.; Ungor, D.; Wojnicki, M. Eco Friendly Synthesis of Carbon Dot by Hydrothermal Method for Metal Ions Salt Identification. *Materials* **2021**, *14*, 7604. [CrossRef] [PubMed]
122. Li, X.; Zhao, S.; Li, B.; Yang, K.; Lan, M.; Zeng, L. Advances and perspectives in carbon dot-based fluorescent probes: Mechanism, and application. *Coord. Chem. Rev.* **2021**, *431*, 213686. [CrossRef]
123. De, B.; Karak, N. A green and facile approach for the synthesis of water soluble fluorescent carbon dots from banana juice. *RSC Adv.* **2013**, *3*, 8286–8290. [CrossRef]
124. Tepliakov, N.V.; Kundelev, E.V.; Khavlyuk, P.D.; Xiong, Y.; Leonov, M.Y.; Zhu, W.; Baranov, A.V.; Fedorov, A.V.; Rogach, A.L.; Rukhlenko, I.D. Sp2–sp3-Hybridized atomic domains determine optical features of carbon dots. *ACS Nano* **2019**, *13*, 10737–10744. [CrossRef]
125. Shi, W.; Han, Q.; Wu, J.; Ji, C.; Zhou, Y.; Li, S.; Gao, L.; Leblanc, R.M.; Peng, Z.J. Synthesis Mechanisms, Structural Models, and Photothermal Therapy Applications of Top-Down Carbon Dots from Carbon Powder, Graphite, Graphene, and Carbon Nanotubes. *Int. J. Mol. Sci.* **2022**, *23*, 1456. [CrossRef]
126. Li, W.-K.; Feng, J.-T.; Ma, Z.-Q. Nitrogen, sulfur, boron and flavonoid moiety co-incorporated carbon dots for sensitive fluorescence detection of pesticides. *Carbon* **2020**, *161*, 685–693. [CrossRef]
127. Sarkar, K.; Devi, P.; Lata, A.; Ghosh, R.; Kumar, P. Engineering carbon quantum dots for enhancing the broadband photoresponse in a silicon process-line compatible photodetector. *J. Mater. Chem. C* **2019**, *7*, 13182–13191. [CrossRef]
128. Kalytchuk, S.; Zdražil, L.; Scheibe, M.; Zbořil, R. Purple-emissive carbon dots enhance sensitivity of Si photodetectors to ultraviolet range. *Nanoscale* **2020**, *12*, 8379–8384. [CrossRef] [PubMed]
129. Li, D.; Jing, P.; Sun, L.; An, Y.; Shan, X.; Lu, X.; Zhou, D.; Han, D.; Shen, D.; Zhai, Y. Near-infrared excitation/emission and multiphoton-induced fluorescence of carbon dots. *Adv. Mater.* **2018**, *30*, 1705913. [CrossRef]

*electronics*

*Article*

# Rational Distributed Bragg Reflector Design for Improving Performance of Flip-Chip Micro-LEDs

**Yuechang Sun, Lang Shi, Peng Du, Xiaoyu Zhao and Shengjun Zhou ***

Center for Photonics and Semiconductors, School of Power and Mechanical Engineering, Wuhan University, Wuhan 430072, China
* Correspondence: zhousj@whu.edu.cn; Tel.: +86-027-5085-3293

**Abstract:** The distributed Bragg reflector (DBR) has been widely used in flip-chip micro light-emitting diodes (micro-LEDs) because of its high reflectivity. However, the conventional double-stack DBR has a strong angular dependence and a narrow reflective bandwidth. Here, we propose a wide reflected angle $Ti_3O_5/SiO_2$ DBR (WRA-DBR) for AlGaInP-based red and GaN-based green/blue flip-chip micro-LEDs (RGB flip-chip micro-LEDs) to overcome the drawbacks of the double-stack DBR. The WRA-DBR consisting of six sub-DBRs has high reflectivity within the visible light wavelength region at an incident angle of light ranging from 0° to 60°. Furthermore, the influence of the WRA-DBR and double-stack DBR on performances of RGB flip-chip micro-LEDs is numerically investigated based on the finite-difference time-domain method. Owing to higher reflectivity and less angular dependence of the WRA-DBR, the RGB flip-chip micro-LEDs with the WRA-DBR have a stronger electric field intensity in the top side in comparison with RGB flip-chip micro-LEDs with the double-stack DBR, which indicates that more photons can be extracted from micro-LEDs with the WRA-DBR.

**Keywords:** distributed Bragg reflector; flip-chip; micro-LED; FDTD

**Citation:** Sun, Y.; Shi, L.; Du, P.; Zhao, X.; Zhou, S. Rational Distributed Bragg Reflector Design for Improving Performance of Flip-Chip Micro-LEDs. *Electronics* **2022**, *11*, 3030. https://doi.org/10.3390/electronics11193030

Academic Editor: Lucas Lamata

Received: 3 September 2022
Accepted: 20 September 2022
Published: 23 September 2022



## 1. Introduction

Micro light-emitting diodes (micro-LEDs) have been highlighted as a promising candidate for the realization of next-generation display panels with high dynamic range (HDR) and high resolution because of their numerous advantages such as high luminance, outstanding power efficiency, fast response time, stability, long lifetime, and wide color gamut [1–9]. The full-color micro-LED display panels can be realized by a combination of efficient red, green, and blue micro-LEDs (RGB micro-LEDs) [10–13]. However, to realize the above applications, the external quantum efficiency (EQE) of RGB micro-LEDs should be further improved. A number of methods applied to broad-area LEDs for improving the efficiency could also be applied to micro-LEDs, such as flip-chip technology [6,14–20], surface roughening [21,22], chip shaping [23–25], and patterned sapphire substrate (PSS) [26–29]. Among these methods, flip-chip technology is widely used in micro-LEDs due to its advantage in light extraction efficiency (LEE), heat dissipation, and current spreading.

It is well known that the light generated in the multiple quantum wells (MQWs) active region emits in all directions. In flip-chip micro-LEDs, we expect that light emits to air mainly from the top substrates. However, a myriad of light generated in MQWs travels downward, which does not make any contributions to the LEE. A bottom reflector plays an essential role in improving the LEE by reflecting the light back into the top substrates. The distributed Bragg reflector (DBR) has been extensively used as the bottom reflector owing to its high reflectivity in a specific wavelength region [30–32]. A typical DBR is made by stacking low refractive index dielectric layers (e.g., $SiO_2$) and high refractive index dielectric layers (e.g., $TiO_2$ and $Ti_3O_5$), exhibiting a low absorption in the visible light region. Five pairs of $TiO_2/SiO_2$ single-stack DBR designed for GaN-based green LEDs exhibited that the reflective bandwidth with a high reflectivity (>80%) was larger than

100 nm [31]. A double-stack DBR optimized for two different central wavelengths was designed to provide a higher reflectivity and a larger reflective bandwidth in comparison with a single-stack DBR [33]. Nevertheless, the double-stack DBR still shows severe angular dependence, which hinders further improvement of LEE in flip-chip micro-LEDs. A full-angle DBR composed of 14 sub-DBRs optimized for central wavelength in blue, green, and red wavelength regions exhibited a wide reflective bandwidth and alleviated the angular dependence [34]. However, the reflective bandwidth of the full-angle DBR is limited in the blue and green light wavelength region, which is not suitable for red micro-LEDs. Therefore, to improve the performance of the RGB flip-chip micro-LEDs, a DBR with wider reflective bandwidth covering the whole visible light wavelength region and less angular dependence is required.

To overcome the above-mentioned drawbacks of the DBRs, we propose a wide reflected angle $Ti_3O_5/SiO_2$ DBR (WRA-DBR). The WRA-DBR is composed of six sub-DBRs optimized for different central wavelengths ranging from the blue light wavelength region to the red light wavelength region. The WRA-DBR has a superior reflectivity within the whole visible light region and exhibits less angular dependence. The average reflectivity of the WRA-DBR at a normal incident angle of light within the wavelength region from 400 nm to 700 nm can reach up to 99.73%. Furthermore, the average reflectivity of the WRA-DBR is up to 97.93% when the incident angle of light is 60°. Compared to conventional double-stack DBR, the WRA-DBR exhibits a wider reflective bandwidth and alleviates the angular dependence. To further evaluate the effect of the WRA-DBR on the optical performance of RGB micro-LEDs, a simulation was carried out by using finite difference time domain (FDTD) method. The result demonstrates that the WRA-DBR can act as a potential candidate for realizing high-efficiency RGB flip-chip micro-LEDs.

## 2. Model and Methods

The DBR, based on the thin-film interference effect, is an array of multilayer stacks composed of two kinds of dielectric layers with high and low refractive indices. The dielectric layer thickness can be calculated by the following formula [35]:

$$n_1\, t_1 = n_2\, t_2 = \lambda/4, \tag{1}$$

where the $n_1$ and $n_2$ are refractive indices of the high refractive layer and the low refractive layer, respectively, the $t_1$ and $t_2$ are thicknesses of the high refractive layer and the low refractive layer, respectively, and $\lambda$ is the central wavelength.

Figure 1a shows the schematic illustration of the single-stack DBR. The single-stack DBR consists of $N$ periods of layer 1 and layer 2. The refractive indices of layer 1 and layer 2 are $n_1$ and $n_2$, respectively. The thicknesses of layer 1 and layer 2 are $t_1$ and $t_2$, respectively. We consider the refractive indices of the incident medium and the substrate are $n_0$ and $n_3$, respectively. The reflection coefficient and reflectivity of the single-stack DBR can be calculated based on the transfer matrix method (TMM). From the TMM, the total transfer matrix $M$ can be calculated according to the following equation [36,37]:

$$\boldsymbol{M} = \begin{pmatrix} M_{11} & M_{12} \\ M_{21} & M_{22} \end{pmatrix} = \boldsymbol{D}_0^{-1}\left[\boldsymbol{D}_1\boldsymbol{P}_1\boldsymbol{D}_1^{-1}\boldsymbol{D}_2\boldsymbol{P}_2\boldsymbol{D}_2^{-1}\right]^N\boldsymbol{D}_3 \tag{2}$$

the reflection coefficient can be expressed as

$$r = \frac{M_{21}}{M_{11}}, \tag{3}$$

the dynamical matrix for the medium $\alpha$ can be expressed as [36,37]

$$D_\alpha = \begin{cases} \begin{pmatrix} 1 & 1 \\ n_\alpha \cos\theta_\alpha & -n_\alpha \cos\theta_\alpha \end{pmatrix} \text{for TE wave} \\ \begin{pmatrix} \cos\theta_\alpha & \cos\theta_\alpha \\ n_\alpha & -n_\alpha \end{pmatrix} \text{for TM wave} \end{cases}, \ \alpha = 0,\ 1,\ 2 \text{ and } 3 \quad (4)$$

the propagation matrix is given by

$$P_\alpha = \begin{pmatrix} \exp(i\phi_\alpha) & 0 \\ 0 & \exp(-i\phi_\alpha) \end{pmatrix},\ \alpha = 1,\ 2 \quad (5)$$

and

$$\phi_\alpha = \frac{2\pi}{\lambda} n_\alpha t_\alpha \cos\theta_\alpha,\ \alpha = 1,\ 2 \quad (6)$$

where $\lambda$ is the wavelength of incident light, and $\theta_\alpha$ is the incident angle of light in the dielectric layer calculated by Snell's law of refraction:

$$n_0 \sin\theta_0 = n_1 \sin\theta_1 = n_2 \sin\theta_2 = n_3 \sin\theta_3 \quad (7)$$

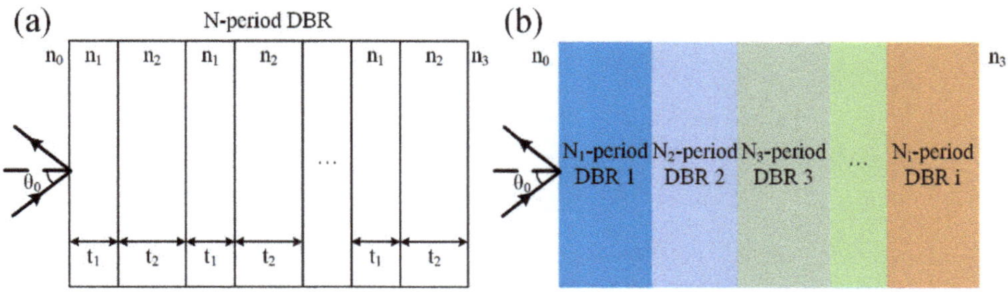


**Figure 1.** (**a**) Schematic illustration of a single-stack DBR consisting of N periods of layer 1 and layer 2. (**b**) Schematic illustration of DBR consisting of several sub-DBRs. Ni (i = 1, 2, 3 . . . ) is the periods of the sub-DBRs.

Then, the reflectivity of the single-stack DBR can be calculated by the following equation:

$$R = r^2 \quad (8)$$

Figure 1b shows the schematic illustration of the DBR consisting of several single-stack DBRs (sub-DBRs). $N_i$ (i = 1, 2, 3 . . . ) is the periods of the sub-DBRs. The total transfer matrix of the DBR is

$$M = \begin{pmatrix} M_{11} & M_{12} \\ M_{21} & M_{22} \end{pmatrix} = D_0^{-1}\left[D_1P_1D_1^{-1}D_2P_2D_2^{-1}\right]^{N1} \cdots \left[D_1P_1D_1^{-1}D_2P_2D_2^{-1}\right]^{Ni} D_3 \quad (9)$$

Then, the reflectivity of the DBR can be calculated from the above-mentioned equations.

## 3. Results and Discussion

Based on TMM, we investigate the angular dependence and reflective bandwidth of the double-stack $Ti_3O_5/SiO_2$ DBR. The double-stack $Ti_3O_5/SiO_2$ DBR is designed and modeled by TFCalc software. The double-stack DBR consists of two sub-DBRs, which are composed of 12 periods of $Ti_3O_5/SiO_2$. The refractive indices of $Ti_3O_5$ and $SiO_2$ in the simulation are fixed at 2.37 and 1.46, respectively. The central wavelengths of the two sub-DBRs are optimized for 465 nm and 630 nm, respectively. Therefore, the thicknesses of $Ti_3O_5/SiO_2$ dielectric layers are 48.97 nm/79.68 nm and 66.34 nm/108.00 nm, respectively.

Figure 2a shows the schematic illustration of the double-stack DBR structure. Figure 2b shows the reflectance spectra of the double-stack DBR. The electroluminescent (EL) spectra of red, green, and blue micro-LEDs are also shown in Figure 2b. The peak wavelengths of red, green, and blue micro-LEDs are 630 nm, 520 nm, and 465 nm, respectively. It is clearly seen in Figure 2b that the reflectivity of the double-stack DBR is high within the whole visible light region when the incident angle of light is 0°. However, as the incident angle increases, the reflective bands exhibit a blueshift and the reflectivity decreases sharply in blue and green light wavelength regions. This result indicates that the double-stack DBR has a strong angular dependence.

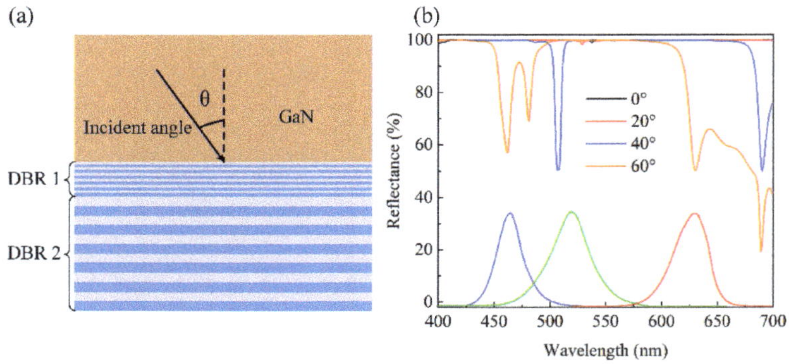


**Figure 2.** (**a**) Schematic illustration of double-stack DBR structure. (**b**) Reflectance spectra of the double-stack DBR as a function of incident angle.

To suppress the sharp decrease in the reflectivity in blue and green light wavelength regions, DBR III is designed by inserting a sub-DBR stack into the double-stack DBR. The inserted sub-DBR is optimized for a central wavelength at 520 nm, and the thickness of $Ti_3O_5/SiO_2$ of the inserted sub-DBR is 55.81 nm/90.82 nm. The total periods of DBR III are the same as that of double-stack DBR, which means that DBR III is composed of 24 periods of $Ti_3O_5/SiO_2$. Therefore, each sub-DBR of DBR III is made of eight periods of $Ti_3O_5/SiO_2$. Figure 3b shows the reflectance spectra of DBR III as a function of incident angle of light. After adding a sub-DBR, the sharp decrease of reflectivity of the double-stack DBR in the blue and green light wavelength region is alleviated, revealing that the added sub-DBR can broaden the reflective bandwidth. However, the reflectivity of DBR III in the red light wavelength region at a large incident angle of light remains relatively low.

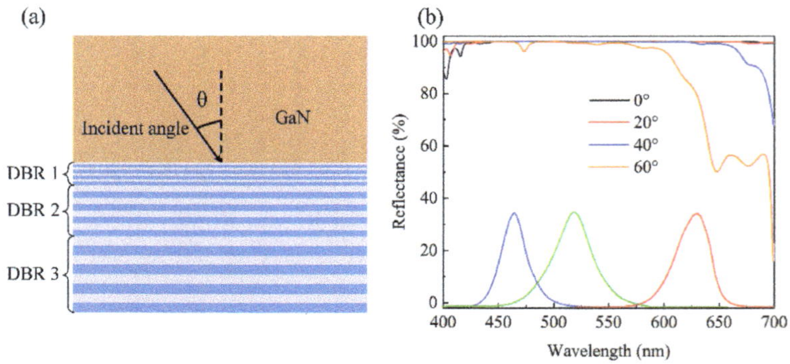


**Figure 3.** (**a**) Schematic illustration of DBR III structure. (**b**) Reflectance spectra of the DBR III as a function of incident angle.

To further improve the reflectivity of the DBR in the red light wavelength region, DBR IV with a combination of DBR III and another sub-DBR optimized for the central wavelength at 650 nm is designed, and the thickness of $Ti_3O_5/SiO_2$ of the inserted sub-DBR is 68.45 nm/111.38 nm. Each sub-DBR of DBR IV is composed of six periods of $Ti_3O_5/SiO_2$. Figure 4a shows the schematic illustration of the DBR IV structure. Figure 4b shows the reflectance of DBR IV. As shown in Figure 4b, the reflectivity is high when the incident angle of light is below 40°. However, the reflectivity of DBR IV at a large incident angle of light is not greatly improved in comparison with DBR III. To further enhance the performance of DBR IV, we investigate the influence of the sub-DBR optimized for different central wavelengths on the reflectivity. Figure 4c shows the reflectance spectra of DBR IV with the added sub-DBR optimized for different central wavelengths when the incident angle of light is 60°. The average reflectivity increases as the central wavelength of the sub-DBR increases from 650 nm to 730 nm. When the central wavelength of the inserted sub-DBR is 730 nm, the average reflectivity of DBR IV is larger than 90%. Therefore, we adopted the inserted sub-DBR optimized for the central wavelength at 730 nm.

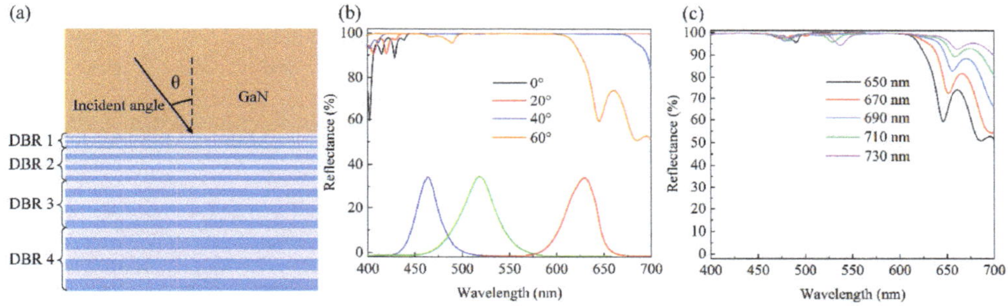


**Figure 4.** (**a**) Schematic illustration of DBR IV structure; (**b**) reflectance spectra of the DBR IV as a function of incident angle; (**c**) reflectance spectra of DBR IV with an added sub-DBR optimized for different central wavelengths at an incident angle of 60°.

To overcome the drawbacks outlined above, the WRA-DBR structure with a combination of DBR IV and another two sub-DBRs is designed. The thicknesses of $Ti_3O_5/SiO_2$ of the two sub-DBRs are 46.34 nm/75.39 nm and 78.98 nm/128.51 nm, respectively. Meanwhile, the central wavelengths of sub-DBRs and number of periods of $Ti_3O_5/SiO_2$ in the WRA-DBR are 440 nm/4 periods, 465 nm/4 periods, 520 nm/4 periods, 630 nm/5 periods, 730 nm/3 periods, and 750 nm/4 periods, respectively. Figure 5a shows the schematic of the WRA-DBR structure. Figure 5b shows the reflectance spectra of the WRA-DBR at different incident angles of light. Among all the mentioned DBRs, the WRA-DBR possesses the widest reflective bandwidth. It should be mentioned that the average reflectivity of the WRA-DBR at a normal incident angle of light is 99.73% within the wavelength region from 400 nm to 700 nm. The reflectivity of the WRA-DBR can reach up to 97.93% within the wavelength region from 400 nm to 700 nm when the incident angle of light is 60°, indicating that the WRA-DBR has alleviated the angular dependence.

We investigate the influence of the double-stack DBR and the WRA-DBR on the performance of RGB flip-chip micro-LEDs using the finite-difference time-domain (FDTD) method. Figure 6 shows the simulation models of the RGB flip-chip micro-LEDs with the double-stack DBR and the WRA-DBR. The optical parameters and the thicknesses of each layer of the RGB flip-chip micro-LEDs are shown in Table 1. The AlGaInP-based red micro-LED consists of ITO, p-GaP, p-AlInP, MQWs, n-AlGaInP, and n-GaAs. The GaN-based green and blue micro-LEDs are composed of ITO, p-GaN, MQWs, and n-GaN. The sidewalls of the RGB micro-LEDs are inclined and the inclination angle is 70° [38]. For simplicity, the metal electrode layer of the micro-LEDs is neglected. Considering the memory of the computer and computing time, the size of micro-LEDs is set to be 8 μm × 8 μm. The perfect matched layer (PML) is used to avoid the reflected electromagnetic wave at the boundary of the micro-LEDs. A 9 × 9 dipole source array is placed in the MQWs as a light source, and the space between dipoles is set to 0.8 µm. The emission wavelengths of the dipole sources in the RGB micro-LEDs are 630 nm, 520 nm, and 465 nm, respectively. A cross-sectional discrete Fourier transform (DFT) monitor is used to obtain the electric field distribution and a transmission box monitor is used to calculate the LEEs. Figure 7a–f show the electric field distributions of RGB flip-chip micro-LEDs with the double-stack DBR and WRA-DBR. In Figure 7a–f, region I is the bottom external space of micro-LEDs and region II is the top external space of the micro-LEDs. According to the electric field distributions shown in Figure 7a–f, it can be observed that there exists a weaker electric field intensity in the WRA-DBR in comparison with the electric field intensity in the double-stack DBR, indicating that the WRA-DBR has a higher reflectivity than the double-stack DBR. Therefore, more photons will be reflected upward to region II. Furthermore, compared to RGB flip-chip micro-LEDs with a double-stack DBR, stronger electric field intensity exists in the air of region II outside the RGB flip-chip micro-LEDs with the WRA-DBR, implying that the photons are more likely to be extracted from RGB flip-chip micro-LEDs with the WRA-DBR. Moreover, the LEEs of each face of the RGB micro-LEDs with the double-stack DBR and WRA-DBR are shown in Figure 7g–i. In Figure 7g, the LEEs of blue micro-LEDs with the double-stack DBR and blue micro-LEDs with the WRA-DBR are almost same. The reason for the similar LEEs in the blue micro-LEDs with the double-stack DBR and WRA-DBR is that the double-stack DBR and WRA-DBR both perform well in the blue light region. From the results of Figures 7h and 7i, it can be observed that green and red micro-LEDs with the WRA-DBR exhibit higher LEEs than green and red micro-LEDs with the double-stack DBR.

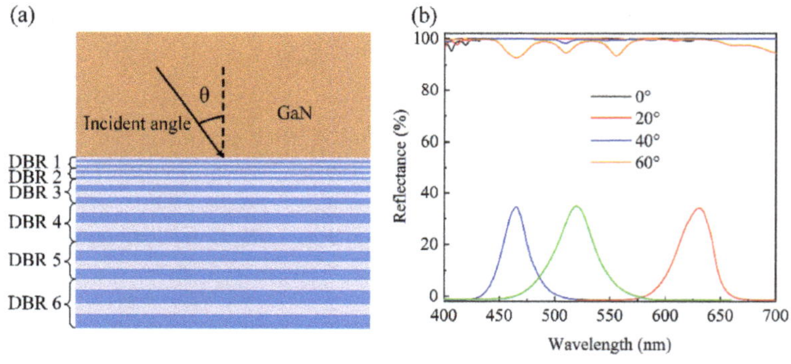


**Figure 5.** (**a**) Schematic illustration of the WRA-DBR structure. (**b**) Reflectance spectra of the WRA-DBR as a function of incident angle.

**Table 1.** Optical parameters for each layer of AlGaInP-based red micro-LED and GaN-based blue and green micro-LEDs used in the FDTD simulations.

| Red Micro-LED | Thickness (µm) | n | k | Green/Blue Micro-LED | Thickness (µm) | n | k |
|---|---|---|---|---|---|---|---|
| | | | | | | Green (Blue) | Green (Blue) |
| p-GaP | 2 | 3.322 | 0 | p-GaN | 0.5 | 2.38 (2.43) | $5 \times 10^{-3}$ ($6 \times 10^{-3}$) |
| p-AlInP | 0.42 | 3.014 | $4 \times 10^{-3}$ | n-GaN | 2.5 | 2.38 (2.43) | $5 \times 10^{-3}$ ($6 \times 10^{-3}$) |
| n-AlGaInP | 1.3 | 3.223 | $9 \times 10^{-3}$ | ITO | 0.28 | 1.89 (1.95) | $4 \times 10^{-3}$ ($6 \times 10^{-3}$) |
| n-GaAs | 0.08 | 3.856 | 0.196 | - | - | - | - |
| ITO | 0.28 | 1.79 | 0 | - | - | - | - |

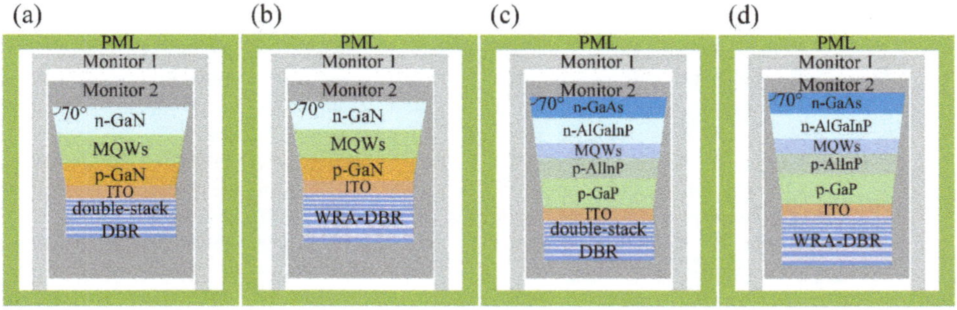


**Figure 6.** Simulation model of (**a**) blue/green micro-LED with double-stack DBR, (**b**) blue/green micro-LED with WRA-DBR, (**c**) red micro-LED with double-stack DBR, (**d**) red micro-LED with WRA-DBR.

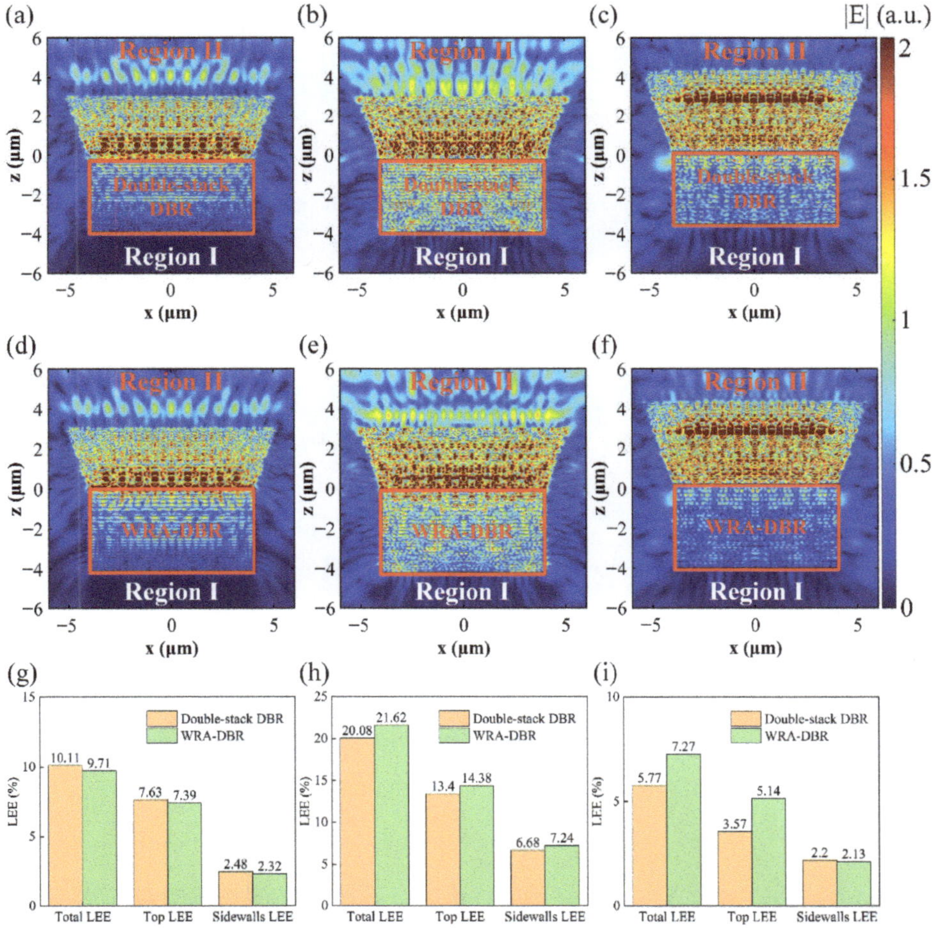


**Figure 7.** Simulated electric field distributions for a cross section of micro-LEDs: (**a**) blue, (**b**) green, (**c**) red micro-LEDs with double-stack DBR; (**d**) blue, (**e**) green, (**f**) red micro-LEDs with WRA-DBR. LEEs of each face (top and sidewalls) of (**g**) blue, (**h**) green, (**i**) red micro-LEDs with double-stack DBR and WRA-DBR.

## 4. Conclusions

In summary, we have designed a WRA-DBR consisting of six sub-DBRs with high reflectivity and less angular dependence within the visible light wavelength region. The average reflectivity of the WRA-DBR at a normal incident angle of light can reach up to 99.73% in the wavelength range of 400 nm to 700 nm. Moreover, the average reflectivity of the WRA-DBR is up to 97.93% when the incident angle of light is 60°, revealing that the WRA-DBR has less angular dependence. Furthermore, we investigate the influence of the double-stack DBR and WRA-DBR on the performance of RGB flip-chip micro-LEDs using the FDTD method. Compared to RGB flip-chip micro-LEDs with the double-stack DBR, stronger electric field intensity is found in the air outside the substrates of RGB flip-chip micro-LEDs with the WRA-DBR, indicating that the WRA-DBR can significantly improve the performance of RGB flip-chip micro-LEDs. Our work demonstrates the promising potential of the WRA-DBR for realization of high-efficiency RGB flip-chip micro-LEDs.

**Author Contributions:** Conceptualization, S.Z.; methodology, Y.S., X.Z. and S.Z.; validation, Y.S. and L.S.; formal analysis, Y.S., L.S. and P.D.; investigation, Y.S., L.S. and S.Z.; data curation, Y.S. and L.S.; writing—original draft preparation, Y.S.; writing—review and editing, S.Z.; visualization, Y.S.; supervision, S.Z.; project administration, S.Z.; funding acquisition, S.Z. All authors have read and agreed to the published version of the manuscript.

**Funding:** This research was funded by the National Natural Science Foundation of China (52075394) and the National Key R & D Program of China (2021YFB3600200). The authors also acknowledge valuable support from the National Youth Talent Support Program.

**Data Availability Statement:** The data that support the findings of this study are available from the corresponding author upon reasonable request.

**Acknowledgments:** The authors also acknowledge valuable support from the National Youth Talent Support Program.

**Conflicts of Interest:** The authors declare no conflict of interest.

## References

1. Huang, Y.M.; Chen, J.H.; Liou, Y.H.; Singh, K.J.; Tsai, W.C.; Han, J.; Lin, C.J.; Kao, T.S.; Lin, C.C.; Chen, S.C.; et al. High-Uniform and High-Efficient Color Conversion Nanoporous GaN-Based Micro-LED Display with Embedded Quantum Dots. *Nanomaterials* **2021**, *11*, 2696. [CrossRef] [PubMed]
2. Hsiang, E.L.; He, Z.; Huang, Y.; Gou, F.; Lan, Y.F.; Wu, S.T. Improving the Power Efficiency of Micro-LED Displays with Optimized LED Chip Sizes. *Crystals* **2020**, *10*, 494. [CrossRef]
3. Kishino, K.; Sakakibara, N.; Narita, K.; Oto, T. Two-dimensional multicolor (RGBY) integrated nanocolumn micro-LEDs as a fundamental technology of micro-LED display. *Appl. Phys. Express* **2019**, *13*, 014003. [CrossRef]
4. Gou, F.; Hsiang, E.L.; Tan, G.; Chou, P.T.; Li, Y.L.; Lan, Y.F.; Wu, S.T. Angular color shift of micro-LED displays. *Opt. Express* **2019**, *27*, A746–A757. [CrossRef]
5. Horng, R.H.; Chien, H.Y.; Tarntair, F.G.; Wuu, D.S. Fabrication and Study on Red Light Micro-LED Displays. *IEEE J. Electron Devices Soc.* **2018**, *6*, 1064–1069. [CrossRef]
6. Chang, K.P.; Tsai, Y.T.; Yen, C.C.; Horng, R.H.; Wuu, D.S. Structural design and performance improvement of flip-chip AlGaInP mini light-emitting diodes. *Semicond. Sci. Technol.* **2021**, *36*, 095008. [CrossRef]
7. Kim, T.K.; Islam, A.; Cha, Y.J.; Kwak, J.S. 32 × 32 Pixelated High-Power Flip-Chip Blue Micro-LED-on-HFET Arrays for Submarine Optical Communication. *Nanomaterials* **2021**, *11*, 3045. [CrossRef]
8. Huang, Y.; Hsiang, E.L.; Deng, M.Y.; Wu, S.T. Mini-LED, Micro-LED and OLED displays: Present status and future perspectives. *Light Sci. Appl.* **2020**, *9*, 105. [CrossRef]
9. Xiong, J.; Hsiang, E.L.; He, Z.; Zhan, T.; Wu, S.T. Augmented reality and virtual reality displays: Emerging technologies and future perspectives. *Light Sci. Appl.* **2021**, *10*, 216. [CrossRef]
10. Wu, Y.; Ma, J.; Su, P.; Zhang, L.; Xia, B. Full-Color Realization of Micro-LED Displays. *Nanomaterials* **2020**, *10*, 2482. [CrossRef]
11. Peng, D.; Zhang, K.; Chao, V.S.D.; Mo, W.; Lau, K.M.; Liu, Z. Full-Color Pixelated-Addressable Light Emitting Diode on Transparent Substrate (LEDoTS) Micro-Displays by CoB. *J. Display Technol.* **2016**, *12*, 742–746. [CrossRef]
12. Zhou, X.; Tian, P.; Sher, C.W.; Wu, J.; Liu, H.; Liu, R.; Kuo, H.C. Growth, transfer printing and colour conversion techniques towards full-colour micro-LED display. *Prog. Quantum Electron.* **2020**, *71*, 100263. [CrossRef]
13. Ding, K.; Avrutin, V.; Izyumskaya, N.; Özgür, Ü.; Morkoç, H. Micro-LEDs, a Manufacturability Perspective. *Appl. Sci.* **2019**, *9*, 1206. [CrossRef]

14. Zhou, S.; Liu, X.; Yan, H.; Chen, Z.; Liu, Y.; Liu, S. Highly efficient GaN-based high-power flip-chip light-emitting diodes. *Opt. Express* **2019**, *27*, A669–A692. [CrossRef] [PubMed]
15. Zhao, P.; Zhao, H. Analysis of light extraction efficiency enhancement for thin-film-flip-chip InGaN quantum wells light-emitting diodes with GaN micro-domes. *Opt. Express* **2012**, *20*, A765–A776. [CrossRef] [PubMed]
16. Jiang, P.; Peng, Y.; Mou, Y.; Cheng, H.; Chen, M.; Liu, S. Thermally stable multi-color phosphor-in-glass bonded on flip-chip UV-LEDs for chromaticity-tunable WLEDs. *Appl. Opt.* **2017**, *56*, 7921–7926. [CrossRef]
17. Konoplev, S.S.; Bulashevich, K.A.; Karpov, S.Y. From Large-Size to Micro-LEDs: Scaling Trends Revealed by Modeling. *Phys. Status Solidi Appl. Mater. Sci.* **2018**, *215*, 1700508. [CrossRef]
18. Liu, X.; Zhou, S.; Gao, Y.; Hu, H.; Liu, Y.; Gui, C.; Liu, S. Numerical simulation and experimental investigation of GaN-based flip-chip LEDs and top-emitting LEDs. *Appl. Opt.* **2017**, *56*, 9502–9509. [CrossRef]
19. Shchekin, O.B.; Epler, J.E.; Trottier, T.A.; Margalith, T.; Steigerwald, D.A.; Holcomb, M.O.; Martin, P.S.; Krames, M.R. High performance thin-film flip-chip InGaN–GaN light-emitting diodes. *Appl. Phys. Lett.* **2006**, *89*, 071109. [CrossRef]
20. Hu, H.; Tang, B.; Wan, H.; Sun, H.; Zhou, S.; Dai, J.; Chen, C.; Liu, S.; Guo, L.J. Boosted ultraviolet electroluminescence of InGaN/AlGaN quantum structures grown on high-index contrast patterned sapphire with silica array. *Nano Energy* **2020**, *69*, 104427. [CrossRef]
21. Tang, B.; Miao, J.; Liu, Y.; Zhou, S.; Xu, H.; Wan, H. Insights into the Influence of Sidewall Morphology on the Light Extraction Efficiency of Mini-LEDs. *IEEE Photon. J.* **2019**, *11*, 8200907. [CrossRef]
22. Guo, Y.; Zhang, Y.; Yan, J.; Xie, H.; Liu, L.; Chen, X.; Hou, M.; Qin, Z.; Wang, J.; Li, J. Light extraction enhancement of AlGaN-based ultraviolet light-emitting diodes by substrate sidewall roughening. *Appl. Phys. Lett.* **2017**, *111*, 011102. [CrossRef]
23. Zhuang, Z.; Iida, D.; Ohkawa, K. Effects of size on the electrical and optical properties of InGaN-based red light-emitting diodes. *Appl. Phys. Lett.* **2020**, *116*, 173501. [CrossRef]
24. Smith, J.M.; Ley, R.; Wong, M.S.; Baek, Y.H.; Kang, J.H.; Kim, C.H.; Gordon, M.J.; Nakamura, S.; Speck, J.S.; DenBaars, S.P. Comparison of size-dependent characteristics of blue and green InGaN microLEDs down to 1 μm in diameter. *Appl. Phys. Lett.* **2020**, *116*, 071102. [CrossRef]
25. Lu, S.; Zhang, Y.; Zhang, Z.H.; Zhu, B.; Zheng, H.; Tan, S.T.; Demir, H.V. High-Performance Triangular Miniaturized-LEDs for High Current and Power Density Applications. *ACS Photon.* **2021**, *8*, 2304–2310. [CrossRef]
26. Chao, S.H.; Yeh, L.H.; Wu, R.T.; Kawagishi, K.; Hsu, S.C. Novel patterned sapphire substrates for enhancing the efficiency of GaN-based light-emitting diodes. *RSC Adv.* **2020**, *10*, 16284–16290. [CrossRef]
27. Manley, P.; Walde, S.; Hagedorn, S.; Hammerschmidt, M.; Burger, S.; Becker, C. Nanopatterned sapphire substrates in deep-UV LEDs: Is there an optical benefit? *Opt. Express* **2020**, *28*, 3619–3635. [CrossRef]
28. Ooi, Y.K.; Zhang, J. Light Extraction Efficiency Analysis of Flip-Chip Ultraviolet Light-Emitting Diodes with Patterned Sapphire Substrate. *IEEE Photon. J.* **2018**, *10*, 8200913. [CrossRef]
29. Zhou, S.; Liu, S. *III-Nitride LEDs: From UV to Green*; Springer Nature: Singapore, 2022; pp. 91–95.
30. Wang, D.X.; Ferguson, I.T.; Buck, J.A. GaN-based distributed Bragg reflector for high-brightness LED and solid-state lighting. *Appl. Opt.* **2007**, *46*, 4763–4767. [CrossRef]
31. Huang, S.Y.; Horng, R.H.; Wang, W.K.; Wuu, D.S. GaN-Based Green Resonant Cavity Light-Emitting Diodes. *Jpn. J. Appl. Phys.* **2006**, *45*, 3433–3435. [CrossRef]
32. Zhi, T.; Tao, T.; Liu, B.; Yan, Y.; Xie, Z.; Zhao, H.; Chen, D. High Performance Wide Angle DBR Design for Optoelectronic Devices. *IEEE Photon. J.* **2021**, *13*, 8200206. [CrossRef]
33. Ding, X.; Gui, C.; Hu, H.; Liu, M.; Liu, X.; Lv, J.; Zhou, S. Reflectance bandwidth and efficiency improvement of light-emitting diodes with double-distributed Bragg reflector. *Appl. Opt.* **2017**, *56*, 4375–4380. [CrossRef] [PubMed]
34. Shi, L.; Zhao, X.; Du, P.; Liu, Y.; Lv, Q.; Zhou, S. Enhanced performance of GaN-based visible flip-chip mini-LEDs with highly reflective full-angle distributed Bragg reflectors. *Opt. Express* **2021**, *29*, 42276–42286. [CrossRef]
35. Kats, M.A.; Blanchard, R.; Genevet, P.; Capasso, F. Nanometre optical coatings based on strong interference effects in highly absorbing media. *Nat. Mater.* **2013**, *12*, 20–24. [CrossRef]
36. Wu, C.J.; Chu, B.H.; Weng, M.T. Analysis of Optical Reflection in a Chirped Distributed Bragg Reflector. *J. Electromagn. Waves Appl.* **2009**, *23*, 129–138. [CrossRef]
37. Pérez, E.X. Design, Fabrication and Characterization of Porous Silicon Multilayer Optical Devices. Ph.D. Thesis, Universitat Rovira i Virgili, Tarragona, Spain, 2008.
38. Hu, X.; Cai, J.; Liu, Y.; Zhao, M.; Chen, E.; Sun, J.; Yan, Q.; Guo, T. Design of inclined omni-directional reflector for sidewall-emission-free micro-scale light-emitting diodes. *Opt. Laser Technol.* **2022**, *154*, 108335. [CrossRef]

*electronics*

MDPI

*Review*

# Hybrid Quantum Dot as Promising Tools for Theranostic Application in Cancer

**Javed Ahmad [1,*], Anuj Garg [2], Gulam Mustafa [3], Mohammad Zaki Ahmad [1], Mohammed Aslam [4] and Awanish Mishra [5]**

[1] Department of Pharmaceutics, College of Pharmacy, Najran University, Najran 11001, Saudi Arabia
[2] Institute of Pharmaceutical Research, GLA University, Mathura 281406, India
[3] College of Pharmacy, Al-Dawadmi Campus, Shaqra University, Shaqra 11961, Saudi Arabia
[4] BBS Institute of Pharmaceutical & Allied Science, Greater Noida 203201, India
[5] Department of Pharmacology and Toxicology, National Institute of Pharmaceutical Education and Research (NIPER), Guwahati 781101, India
* Correspondence: jahmad18@gmail.com or jaahmed@nu.edu.sa

**Abstract:** Cancer is one of the leading causes of death worldwide. In the last few decades, cancer treatment has come a long way, but multidrug resistance (MDR) in cancer still has low survival rates. It means that much research is required for an accurate diagnosis and effective therapy. The new era of cancer research could include theranostic approaches and targeted delivery of chemotherapeutic agents utilizing the nanoparticulate system. Recently, there has been much interest gained among researchers for carbon-based and graphene-based quantum dots due to their higher biocompatibility and ease of biofunctionalization compared to conventional heavy metal quantum dots. Moreover, these quantum dots have various interesting utilities, including bioimaging, biosensing, quantum dots-mediated drug delivery, and their role in photodynamic therapy (PDT) and photothermal therapy (PTT). The current review highlighted the utility of hybrid quantum dots as a theranostic system in different cancers and discussed the various bio-molecules conjugated hybrid quantum dots investigated for diagnostic/therapeutic applications in cancer. The influence of conjugation of different biomolecules, such as folic acid, PEG, etc., with hybrid quantum dots on their biopharmaceutical attributes (such as aqueous solubility, tumor penetrability, stability of loaded therapeutics in the tumor microenvironment), delivery of drugs specifically to tumor tissues, and its therapeutic outcome in different cancer has also been discussed.

**Keywords:** hybrid quantum dots; cancer; bioimaging; theranostic; photodynamic therapy; photothermal therapy

**Citation:** Ahmad, J.; Garg, A.; Mustafa, G.; Ahmad, M.Z.; Aslam, M.; Mishra, A. Hybrid Quantum Dot as Promising Tools for Theranostic Application in Cancer. *Electronics* **2023**, *12*, 972. https://doi.org/10.3390/electronics12040972

Academic Editors: Lucas Lamata and Enzo Pasquale Scilingo

Received: 31 December 2022
Revised: 7 February 2023
Accepted: 8 February 2023
Published: 15 February 2023



## 1. Introduction

Cancer is the leading cause of death worldwide, accounting for nearly 10 million deaths in 2020 [1]. The correct diagnosis is crucial for accurate and effective treatment because all tumors need specific treatments such as surgery, radiotherapy, and chemotherapy. Novel therapeutic interventions and early diagnosis of cancer are major concerns for scientists, physicians, and healthcare professionals. In the last few decades, nanotechnology has been at the forefront of most cutting-edge research in many fields, such as medicine, diagnostics, bioimaging, and other biomedical applications [2]. Among the delivery approach of medicines in different cancers, various nanoparticles (NPs) are the most exploited carrier system for drug delivery in different cancer [3,4]. These carrier systems have different structures and dimensions, which may vary in size range from 1 to 100 nm, particularly for drug delivery in cancer. This gives them different physical and chemical properties that can be exploited for various purposes in cancer. The specific physicochemical properties of nanomaterials are exploited for focused ultrasonic heating therapy [5–7], radiofrequency (RF) ablation [8–10], magnetic fluid hyperthermia [11–13], and magnetic particle imaging [14,15] in cancer. NPs are widely exploited for bioimaging [16], drug delivery systems [17], therapeutic agents for photodynamic therapy (PDT) [18], photothermal therapy (PTT) [19], regenerative medicine [20], and smart biomaterials [21]. Furthermore, metallic NPs, especially gold [22], silver [23], platinum [24], and palladium [25], are the most investigated biocompatible NPs, for the diagnosis and treatment of cancer. Along with imaging agents to treat a wide range of diseases, including different carcinomas, the delivery of synthetic drugs, therapeutic peptides, and genes has raised interest in the theranostic approach, which can be simultaneously utilized for both diagnosis and treatment through a single system. Different nanomaterials have high penetrating efficiency to biological membranes, good biocompatibility, and can perform multiple functions due to their small size and ability to functionalization. This makes the utilization of nanomaterials a good candidate for theranostic application in different cancers [26].

Quantum dots (QDs) are a very small nanoparticulate system of organic (such as carbon-QDs, graphene-QDs) and inorganic nature (such as zinc sulfide–QDs, cadmium telluride-QDs, cadmium selenide-QDs) and vary in size ranges from 2 to 10 nm [27]. The small nano dimension is helpful to impart specific optical (high brightness, high quantum yield, high extinction coefficient, intermittent fluorescence signals, high stability against photobleaching) [28] and electronic properties that are exploited in different biomedical applications [27–29]. Its nanocrystal is distinguished by an energy band gap, required to excite an electron from one electronic band, i.e., a lower energy level, into another band, i.e., a higher energy level. Because they are so small, these nanocarrier systems of semiconductor origin can easily move electrons to a higher energy state, even when they are exposed to UV light. These properties of QDs are used in the diagnosis and treatment of various diseases, including cancer. This excitation scheme ultimately creates an electron-hole pair known as an exciton. The exciton gives off energy in the form of a fluorescent photon when it goes back to its ground state [30]. The 2–10 nm sized semiconducting nanocrystal started to act like the bulk Bohr exciton radius, and the particle's electrical and optical properties changed [31]. The inverse relationship between nanocrystal size and energy band gap is well-documented and understood. The inverse property says that as the size of the nanocrystal decreases, the energy band gap increases, and the corresponding excitation/emission wavelengths decrease. The size of the particle can be changed to change the color of the light that the QDs give off when they are exposed to UV light. By adjusting the particle size and size distribution of the QDs, a wide absorbance range with highly symmetric and narrow emission spectra can be achieved [32]. The different types of QDs are prepared by the bottom-up approach, which involves the assembling of their precursor in the molecular state into nanocrystals [33]. The promising techniques for the preparation of QDs are categorized into four basic approaches, which include biotemplate-based synthesis, colloidal synthesis, biogenic synthesis, and electrochemical assembly [27,33]. These methods for the preparation of QDs are illustrated in Figure 1.

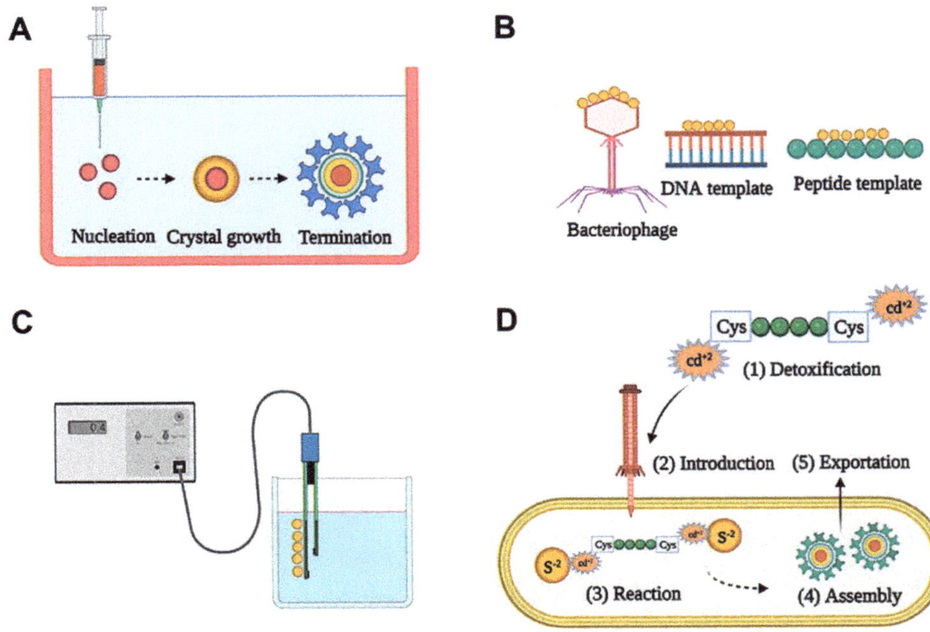


**Figure 1.** Schematic illustration of commonly used preparation approaches for QDs: (**A**) Colloidal synthesis approach. (**B**) Biotemplate-based synthesis approach. (**C**) Electrochemical assembly approach. (**D**) Biogenic synthesis approach. Reproduced from Abdellatif et al. [27], Dove Medical Press, 2022.

Carbon nanomaterials (specifically carbon nanotubes, carbon dots, graphene, and graphene derivatives) and other nanomaterials of organic and inorganic origin are popular for their extraordinary composition and excellent inherent properties for diverse applications such as fluorescent, fingerprinting, photocatalysis, electromagnetic shielding, and electric applications [3,34]. These nanomaterials have also made a pragmatic intervention in the field of biomedical engineering, contributing to research in tissue engineering, drug delivery, biosensing, bioimaging, and cancer theranostic [35]. It is functionalized with QDs as a hybrid nanoparticulate system called hybrid QDs, which have very small dimensions and theranostic utility in cancer. These hybrid systems are utilized to deliver drugs of synthetic/natural/biological origin and act as an imaging agent simultaneously, which are likely to increase their accumulation, specifically in the cancerous or tumor tissue. Thus, these versatile nanoparticulate systems as hybrid QDs have very wide utility in cancer detection/imaging and site-specific delivery of different types of therapeutics to the cancerous tissues (Illustrated in Figure 2).

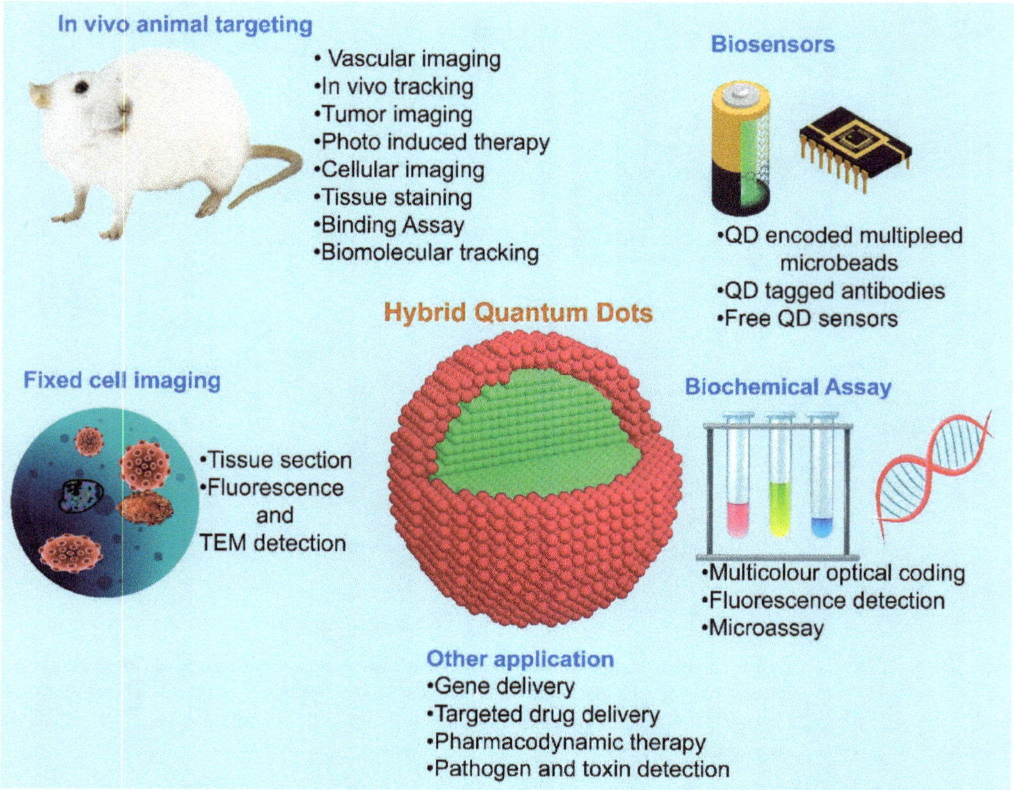


**Figure 2.** Illustration highlighting the utility of hybrid QDs as cancer theranostics for various applications.

The low drug efficacy in cancer may increase by improving the EPR (enhanced permeability and retention) effect and overcoming the tumor heterogeneity challenges through designing targeted hybrid QDs of a stealth nature [36]. RBC-camouflaged, light-responsive, carbon-based porous particles may be helpful in targeting and penetrating tumor tissues. Protein- and RBC membrane-targeted nanosponges improve targeting and circulation half-life. Porous carbon/silica and graphene QDs as hybrid systems are photoresponsive and tumor-penetrating drug carrier systems for theranostic application [37]. Cyclodextrins (CDs) are natural, water-soluble cyclic oligosaccharides with hydrophilic exteriors and hydrophobic interiors that are known for their utility in drug delivery applications. Their primary and secondary hydroxyl groups on the outside are easy to modify, and their lipophilic inner cavities can be filled with lipophilic moiety by the formation of an inclusion complex. These types of carriers are utilized to improve the aqueous solubility of water-insoluble therapeutics/imaging agents and serve as a promising carrier or drug delivery system in cancer management through hybridization with QDs for theranostic application due to the presence of numerous hydroxyl groups on their surface [38].

The present manuscript provides a detailed discussion and recent advancements in QDs for their utilization to improve the efficacy of loaded therapeutics and imaging applications in the effective management of cancer. It provided a discussion on hybrid quantum dots as a cancer theranostics and emphasized the recent research development (mainly in the last 5 years) in the area of cancer theranostics utilizing hybrid quantum dots.

## 2. Significance of Hybrid Quantum Dot as Cancer Theranostics

The significance of the theranostic approach in cancer treatment may utilize the simultaneous delivery of chemotherapeutics and photolytic agents for deep tumor penetration, which effectively damages and inhibits the tumor when treated with single irradiation. It may be utilized for tracking of progress of cancer therapy using incorporated imaging agents [39]. Hybrid QDs have been widely explored for their theranostic application in different types of cancers. However, toxicity issues of QDs due to their composition (heavy or inorganic materials) and nature (ROS generation and strong surface responses) raised concern for their modification/functionalization for biomedical applications. Hence, strategies have been conceptualized to minimize their toxicity and improve their biocompatibility through hybridization/functionalization with other moieties (such as polymers, lipids, polysaccharides, proteins, etc.), providing efficient accumulation in tumor tissues in addition to preventing their accumulation in healthy tissues [40,41]. Biological molecules are attached to QDs using hydrophilic surfactant shells with reactive groups such as COOH, NH2, or SH. Attachments are made using different methods, such as adsorption, covalent bonding, electrostatic interaction, etc. [39–41]. It has been reported to be conjugated with a wide range of biological molecules, including biotin [42], folic acid [43,44], antibodies [45], and peptides [46]. Silanization, which coats QDs with silica, is a good covalent coating method for hydroxyl-rich material surfaces. Silanization makes ligand molecules strongly cross-linked and chemically stable. The end terminal groups of the silane shell can expose thiol, phosphonate, or methyl terminal ends for subsequent QD coating and also make the material more biocompatible. Silanization is favored because it is less toxic than other ligands [47]. Silica shell thickness could control QD light responsiveness. The silica-coated QDs were modified with amino, carboxyl, and epoxy groups and stabilized with PEG segments to assess their applicability. These modified QDs efficiently conjugated with antibodies and were used as fluorescent labels in immunoassay detection [48]. An in vivo study has shown that emissive Si-QDs biodegrade quickly and produce non-toxic silicic acid that may be eliminated by urine [49].

The perspectives of hybrid QDs for their utility in cancer diagnosis/imaging and delivery of payload specifically to tumor tissues are discussed in the subsequent section.

### 2.1. *Perspectives of QDs for Diagnostic/Imaging Utility*

QDs in drug delivery may be utilized as therapeutic/imaging cargo that has photothermal and photodynamic features, making them excellent for bioimaging. Many clinically used photosensitizers (PSs) are not tumor-targeted; hence, they are treated with spatially controlled irradiation. After phototherapies, PS can increase reactive oxygen species (ROS) formation in healthy cells; hence, light exposure should be avoided to reduce skin photosensitivity. The hybrid QDs possess low toxicity and good biocompatibility, coupled with stable photoluminescence (PL), and therefore these are ideal candidates for both in vitro and in vivo bioimaging [40,41]. QDs by themselves are not as efficient as molecular PSs, but QDs can be used as antennae to improve light harvesting and energy transfer to molecular PSs because they absorb much light. NPs are commonly utilized for bioimaging, but their toxicity limits their utility. Because fluorescence imaging is very sensitive and has a good temporal and spatial resolution, hybrid QDs are a good choice for sensing and imaging cell targets. Hybrid QDs are chemically inert, dissolve well in water, are photostable, have a relationship between their optoelectronic properties and their shape and size, have fluorescence resonance energy transfer, high stability in physiological conditions, specific accumulation at target sites, are easy to modify on the surface [50], and have a high absorption coefficient because of hybridized C–C bonds. Therefore, these are good phototherapy chromophores.

Carbon QDs (C-QDs) synthesized and dispersed with excellent fluorescence, photostability, photobleaching resistance, and simple coupling with biological species [51]. C-QDs can carry $Ce_6$ and generate ROS. Using a 639 nm laser, water splitting produced oxygen and hydrogen in vivo. Increased oxygen yielded $^1O_2$ to improve PDT. C-QDs with specific cell targets can particularly detect malignant cells in different investigations. C-QDs conjugated with folic acid (FA) (C-dots-FA) to distinguish folate receptor (FR)-positive cancer cells from normal cells (FR-negative) by growing and analyzing NIH-3T3 and HeLa cells. Pheophytin (a natural, low-toxicity Mg-free chlorophyll derivative) was employed as a raw carbon source to synthesize C-QDs using a microwave technique [52]. QDs containing sulfur and nitrogen are used as PTT (photothermal therapy causes ease of cell death by protein denaturation and loosening of the cellular membrane by heating the tumor tissue exploiting irradiation of radiofrequency, ultrasound, microwaves, and magnetic fields, etc.) and PDT (photodynamic therapy utilizes a photosensitizer that absorbs light of a particular wavelength and produces oxygen-based molecular species to induce a cytotoxic effect) for cancers in animals [53,54]. PL and photoacoustic imaging [55–58] benefited from high photon conversion efficiencies. Passive targeting of QDs around cancer cells destroyed the tumor. Co-doped C-QDs had a strong photothermal conversion, optical and photoacoustic performance, and renal excretion [53,59]. N–O-CQDs with significant NIR absorption. Combining the biocompatible N-doped carbon dots (N-CDs) with folic acid, which possess a wide range of high-targeting capabilities (26 types of tumor cell lines) and alters the cellular metabolism leading to autophagy, results in a new targeted tumor therapy based on autophagy regulation [60]. Similarly, maleimide-terminated TTA1 aptamers complexed with CDs (TTA1–CDs), which is substantially expressed in HeLa and C6 (rat glioma cell line) but not in normal healthy CHO cells, exhibit a strong fluorescence along cancer cell membranes and minimal uptake in healthy cells [61].

Graphene quantum dots (GQDs) are one type of nanocarrier that has been seen in physics and chemical research due to their ultrasmall size, varied photoluminescence, and mechanical features [62]. Ultra-tiny GQDs exploiting imaging agents and labeling cell membranes are promising agents for drug transportation in cancer therapy because of their outstanding optical properties and transmembrane capabilities. The innate immune system and tumor heterogeneity continue to pose challenges to efficient tumor targeting and penetration; however, NIR irradiation, the energy created by photothermal conversion, can not only release therapeutic cargo but also burst the vesicle to suppress the tumor [63]. When NPs first enter the circulatory system, the innate immune system quickly recognizes them as foreign bodies and gets rid of them through the reticuloendothelial system and the mononuclear phagocyte system. This results in poor delivery efficiency. Because the tumor is a strong physiological barrier, only a small part of the dose injected gets to the deep tumor through the increased EPR effect, which helps particle accumulation. The high interstitial fluid pressure (IFP) and cancer-associated fibroblasts in tumors make it hard for therapeutic drugs to reach the perivascular cells of tumors [64,65]. Thus, in order to improve tumor therapy, it is crucial to create stealthy and permeable drug delivery systems for the efficient transportation of therapeutic agents.

For imaging or diagnostic applications, theranostic nanoplatforms must be robust enough to support them, have a superior cargo-loading and -releasing profile, and be able to do so. Hybridization between distinct NPs is a promising strategy because it can result in the accumulation of a wide range of chemical, physical, and biological properties inside a single complex. Because of their exceptional physical and chemical properties [32,66], GQDs have been put to use in a variety of biomedical settings. If GQD fluorescence could be made stable, it would greatly improve the efficiency of imaging in the life sciences. Due to their unique chemical, physical, and biological properties, graphene quantum dots (GQDs) and magnetic nanoparticles (MNPs) are two promising choices for use in these hybrids. Both magnetic resonance imaging (MRI) and computed tomography (CT) use contrast agents made of magnetic nanoparticles [67]. In addition to its use in biosensing and magnetic separation, this nanoparticle may also be put to use in hyperthermia therapy, thermo-ablation, targeted drug administration, and even bio-sensing. For example, magnetic nanoparticles could be added to GQDs to make even more complexes that could be used in medicine. The most commonly used magnetic nanoparticles are iron oxide NPs (usually $Fe_2O_3$ or $Fe_3O_4$), which can be classified as a pure metal, metal oxide, or magnetic nanocomposites [68]. Combining GQDs with other NPs, such as magnetic nanoparticles, could make them even better for use in biology.

Carbon quantum dots (CQDs), a novel kind of fluorescent carbon nanomaterial possessing the unique advantages of high stability, remarkable biocompatibility, easy synthesis and surface functionalization, and comparable optical characteristics, have been extensively studied, especially for bioimaging applications due to their tunable strong fluorescence emission property [69]. In light of the drawbacks of conventional chemotherapy, PTT has emerged as a viable option for treating cancer. Tumors can be heated from the inside out by injecting photothermal substances into the affected area or by targeting the tumor with other agents. These photothermal agents are designed to stimulate near-infrared (NIR) radiation and generate heat upon relaxation, killing cancer cells. Researchers have been investigating several facets of theranostic nanosystems [66] since it has been postulated that these systems, which meet both diagnostic and therapeutic needs, could be utilized to effectively eradicate cancer cells. It has been found that a wide range of organic, inorganic, organo-inorganic, and combinations can act as photothermal agents. Carbon nanomaterials, including carbon nanotubes, graphene, and graphene derivatives, have garnered significant attention due to their potential applications in fields as diverse as fingerprinting, photocatalysis, electromagnetic shielding, and electrics [70,71]. Due to their luminescence, versatile surface chemistry, easy cellular internalization, and high biocompatibility, CQDs are particularly promising in drug delivery. Additionally, the nano-formulation system also offers the possibility to increase drug solubility, bioavailability, and half-life. Although doxorubicin (DOX) is widely used for cancer treatment, it has many disadvantages, including a low EPR effect, low cellular internalization, and cytotoxicity to normal cells [72]. One approach to bypassing these problems is to use a multifunctional nanocarrier system for tumor-targeted drug delivery, which has the advantage of accumulating at the tumor site due to the increased EPR effect.

### 2.2. Perspectives of QDs for Therapeutic Utility

Different QD-based therapeutic systems for anticancer application have been widely explored in recent years [73,74]. Various studies have been conducted to investigate the potential of QDs for targeted drug delivery, PDT, PTT, and gene delivery in cancer treatment [75,76]. The QDs for cancer therapy have been investigated both at in vitro and in vivo levels. PDT is one of the most promising non-invasive cancer treatment approaches with limited side effects. It can be used alone or in combination with surgery, chemotherapy, or ionizing radiation to destroy undetected cancerous cells at the margins of resection. PDT uses photosensitizing drugs that are pharmacologically inactive until a particular light wavelength irradiates them in the presence of oxygen, which generates reactive oxygen species and induces cell death and tissue necrosis [77,78]. Graphene oxide (GO), an oxidized version of graphite, has received considerable attention during the past decade. GO, on the other hand, can be dispersed in water, which makes it a good candidate to investigate in a biological system. On the other hand, graphenes need to be surface functionalized to make them dispersible in water and safer for the biological environment [79,80]. Their synthesis and surface tailoring entail hazardous and toxic reagents, traces of which may remain with the material to demonstrate further toxicity in vitro/in vivo systems due to GO sheets having intrinsic toxicity. By delivering medications and energy in two different locations at the same time, a hybrid system of nano dimension may be able to lessen the adverse effects of cancer treatment and improve the distinctive properties required for precision medicine. The hybrid carrier systems are frequently eliminated from blood circulation very quickly, and piled-up tumors at the periphery close to the blood arteries. It has a short elimination half-life in blood and high tumor penetration. The membrane of a red blood cell (RBC) was given the appearance of a sponge by being composed of carbon composite. When it is exposed to light, it functions as both a "stealth agent" and a "photolytic carrier". It means that it transfers "tumor-penetrating agents" (such as graphene QDs and docetaxel) as well as heat. When compared to the nanosponge, the RBC-membrane-enveloped nanosponge demonstrates an eight-fold increase in accumulation in tumor tissue. This is because the RBC-membrane-enveloped nanosponge will be integrated with a specific protein that accumulates in tumor spheroids through high lateral bilayer fluidity [81]. The delivery of graphene QDs to tumor areas is accomplished by passing near-infrared light through a structure that is only one atom thick. This makes it much simpler for its therapeutic utilization to penetrate the cancerous tissue and improves the prognosis of cancer therapy utilizing the theranostic approach.

The pathway of accumulation and removal of QDs and hybrid QDs in in vivo systems are depicted in Figure 3.

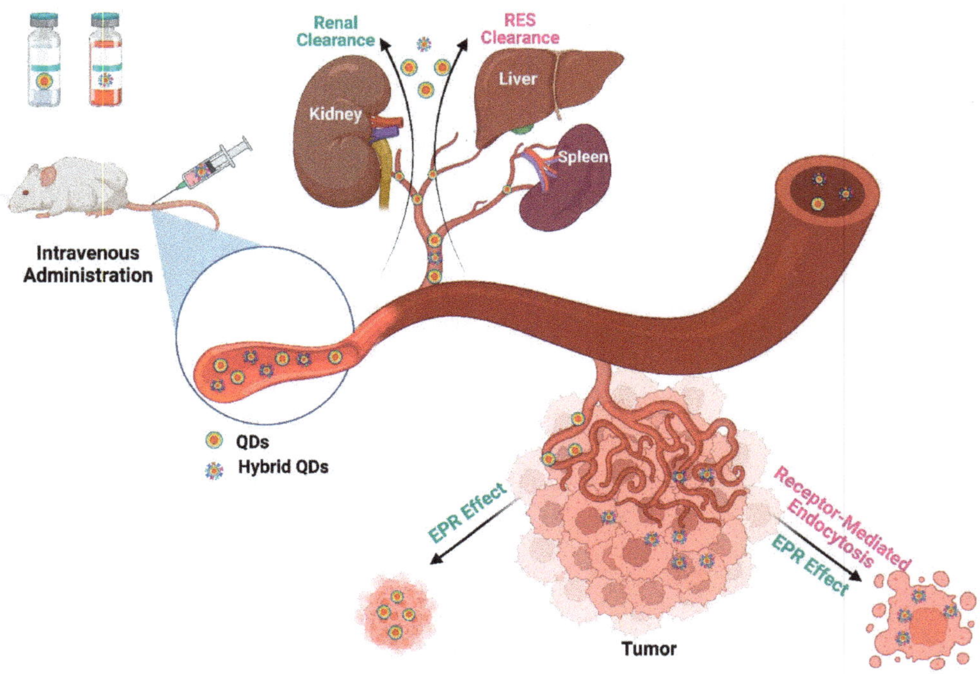


**Figure 3.** Schematic illustration highlights the accumulation and removal of QDs versus hybrid QDs. QDs are more prone to RES clearance and renal clearance compared to hybrid QDs of a stealth nature. The more targeted delivery of hybrid QDs compared to QDs resulted in major accumulation in tumor tissues due to receptor-mediated endocytosis and the EPR effect, leading to improved therapeutic outcomes. "Image created with BioRender.com".

Different types of hybrid QD-based theranostic systems were explored for cancer applications to improve the biopharmaceutical attributes of this nanoparticulate system (such as aqueous solubility, tumor penetrability, and stability of loaded therapeutics in the tumor microenvironment) and delivery of drug specifically to tumor tissues. Recent contemporary research conducted in this field is discussed in the subsequent section.

## 3. Hybrid Quantum Dot as Cancer Theranostics: Contemporary Research

### *3.1. Diagnostic Application*

Pei et al. developed fluorescent hyper-cross-linked-cyclodextrin–carbon quantum dot (CD-CQD) hybrid nanosponges with outstanding biocompatibility and intense bright blue fluorescence excited at 365 nm with a PLQY of 38.0% [82]. These hybrid QDs systems were generated by simple condensation polymerization of carbon quantum dots (CQDs) with cyclodextrin (CD) at a 1:5 feeding ratio for theranostic applications, specifically in malignancies. In another investigation, Fateh et al. made a hybrid nanostructure of graphene quantum dots (GQDs) and magnetic nanoparticles (MNP) by using hydrophobic interactions between long carbon chains on the surface of GQDs around the edges and MNP in the middle [83]. Pyrolysis was used to create GQDs, which were then changed using cetyl alcohol (CA) to produce surfactant-modified GQDs (CA-GQDs). Moreover, an oleate-iron complex has been utilized to make iron-oxide nanoparticles (IONP) as MNP. After that, CA-GQDs and IONP are mixed to make a structure with IONP in the middle and CA-GQDs all around it (IONP@CA-GQD). IONP@CA-GQD possesses both fluorescence and magnetic characteristics. At room temperature, IONP and IONP@CA-GQD have been tested for magnetization hysteresis loops in a moving magnetic field. There have been no observations of coercivity or remanence, indicating super-paramagnetism. The computed MS values for IONP and IONP@CA-GQD are 34.1 emu/g and 37.8 emu/g, respectively. Because of GQDs are fluorescent in nature, this hybrid structure could also be used for bioimaging [84,85].

A stable compound of graphene oxide (GO) and graphene quantum dots (GQD) was created by Kumavat et al. by electrostatic layer-by-layer assembly using a polyethylene imine bridge (GO-PEI-GQDs) [86]. In addition, various applications of the mono-equivalents of the GO-PEI-GQDs complex were compared, including cell imaging (diagnostics), photothermal, and oxidative stress responses in MDA-MB-231 breast cancer cells. When exposed to an 808 nm laser for 5 min at a concentration of up to 50 μg/mL, GO-PEI-GQDs displayed an outstanding photothermal response (44–49 °C). According to the study, GO-PEI-GQDs had synergistic effects on cancer cells. It has stable fluorescence imaging, improved photothermal effects, and cytotoxic actions. Composite materials made of GO and GQDs combine many different properties, which makes it possible to improve certain therapeutic systems, such as cancer theranostics [86].

Hyaluronic acid and QDs together have been proven to be useful tools for improving intracellular transport into liver cells. This is accomplished by interacting with CD44-receptors, which allows for in vivo real-time imaging [87,88]. The anionic polysaccharide chondroitin sulfate was employed to coat the positively charged oily core of the cadmium telluride (CdTe) QDs as cancer theranostic nanocapsules [89], which were also loaded with rapamycin and celecoxib as anticancer therapeutics [90]. Chondroitin sulfate nanocapsules have an exterior coating of cationic gelatin-coupled QDs placed on them to prevent non-specific uptake by healthy cells. Matrix metalloproteinase (MMPs) dissolved the gelatin at the tumor location, releasing therapeutic nanocapsules and QDs into cancer cells for therapeutic and imaging action as a cancer therapeutics. An ON–OFF effect, where the fluorescence of QDs was first quenched by energy transfer and then restored after bond cleavage in tumor cells, was seen in a study that substituted lactoferrin for gelatin [90]. Thus, using QDs fluorescence, the in vitro and in vivo localization of nanocapsules into breast tumors was observed.

Recent research related to hybrid QDs utilized for their diagnostic/imaging applicability in different types of cancers is summarized in Table 1.

**Table 1.** Summary of contemporary research carried out utilizing hybrid quantum dots for diagnostic/imaging applications in cancer.

| Type of QDs | Type of Cancer | Diagnostic/Imaging Technique | Outcome | Refs. |
|---|---|---|---|---|
| Lactoferrin QDs | Breast cancer | Fluorescence imaging | Intracellular uptake of QDs showed fluorescence fluorescent due to mercaptopropionic acid-capped cadmium telluride and was successfully used as theranostic | [89] (P:2018) |
| Gelatin/chondroitin QDs | Breast cancer | Fluorescence imaging | Matrix metalloproteinase layer enabled tracing their internalization into cancer cells and strong non-immunogenic response used as diagnostic | [90] (P:2018) |
| Magnetic graphene-QDs | Cancer cells | Electrochemical detection imaging | Images show high fluorescence in HeLa cells | [91] (P:2018) |

**Table 1.** *Cont.*

| Type of QDs | Type of Cancer | Diagnostic/Imaging Technique | Outcome | Refs. |
|---|---|---|---|---|
| Graphene-QDs | Cancer cells | MRI and fluorescence imaging | MRI and fluorescence imaging of living Hela cells and monitored intracellular drug release | [92] (P:2017) |
| Carbon-QDs | Tumor cells | Photoluminescence and photoacoustic imaging | Accumulation of C-QDs around the cancer cells via passive targeting with no active targeting species with fluorescence imaging | [93] (P:2018) |
| Carbon-QDs | Cervical cancer | Fluorescence imaging | TAT functionalization enhanced cell labeling and uptake, and that folate selectively tagged tumor cells | [94] (P:2013) |
| Carbon-QDs doped with Fluorine and Nitrogen | Squamous cell carcinoma | Near-infrared fluorescence (NIRF) and PET imaging | Carbon-QDs rapidly uptake by the tumor when administered subcutaneously as compared to intramuscular and intravenous | [95] (P:2013) |
| Carbon QDs doped with polyethyleneimine | Hepatocellular carcinoma | Bioimaging | Internalized QDs exhibit fluorescent emission authenticating their potential application for gene delivery and bioimaging | [96] (P:2012) |
| Magneton-fluorescence carbon-QDs conjugated with cDNA aptamer | Cervical cancer | Fluorescence and magnetic resonance (MR) imaging | DNA aptamer, which specifically recognizes the receptor tyrosine-protein kinase-like 7 (also known as colon carcinoma kinase 4, CCK4) for targeted dual mode fluorescence/magnetic resonance (MR) imaging | [97] (P:2018) |
| QDs-conjugated streptavidin probe | Breast cancer | Diagnosis | QDs-based immunohistochemistry demonstrates the prognostic value of EGFR area in the HER2-positive and lymph node-positive subtype of invasive breast cancer | [98] (P:2011) |
| Carboxyl-modified CdTe-QDs | HeLa and MCF-7 cells | Bioimaging | Sensing probes for cancer- biosensors was the detection of miRNA-21 on lysates of HeLa and MCF-7 cells and other biomarkers. | [99] (P:2022) |
| CdTe-QDs functionalized with single-stranded DNA | Non-specific cells | Fluorescence Diagnosis | QDs detect miRNA-122 within 40 min with enhanced intensity in proportion with miRNA-122 concentrations range 0.16–4.80 nM and has a low detection limit of 9.4 pM | [100] (P:2017) |
| Au-$SiO_2$-QDs | Breast cancer | Imaging | Photothermal effect provides real-time imaging capability, which makes it appealing as a potential theranostic tool for cancer treatment. | [101] (P:2018) |
| Graphene-QDs doped nitrogen | Skin cancer | Imaging and diagnosis | Fluorescence intensity of N-GQDs was quenched by the static quenching of UV-damaged DNA through the formation of an N-GQD/UV-damaged DNA complex | [102] (P:2022) |
| Iron selenide-QDs | Skin Cancer | Bioimaging | Synthesized QDs exhibit two bands of photon excitation property and high quantum yield which are suitable for second-window imaging | [103] (P:2019) |

### 3.2. Therapeutic Application

Pei et al. formulated doxorubicin (DOX) loaded-fluorescent hyper-cross-linked-cyclodextrin–carbon quantum dot (CD-CQD) hybrid nanosponges (DOX-β-CD-CQD) with a size of around 300 nm with a DOX loading capacity of 39.5% through host−guest complexation [82]. This is because of the supramolecular complexation of DOX with the CD units in the CD-CQD nanosponges. The developed DOX-CD-CQD nanosponges demonstrated pH-responsive controlled release in the simulated tumor microenvironment. The loaded DOX molecules in the surface layer of the DOX-CD-CQD were released in the first 30 h, similar to the pH 7.4 medium. Due to the higher solubility of DOX in acidic media attributed to its protonation, the supramolecular complexation of DOX with β-CD units had a lower inclusion constant and a greater release ratio than in pH 7.4 conditions. Due to the high formation constant, they took longer to get out of the loaded DOX inner layer. Protonated DOX diffusion was prevented by hydrophobic DOX-complexed CD. After 12 h of DOX release, with an accumulative release of approximately 50%, hydrophilic outer shells formed, facilitating protonated DOX diffusion out of the theranostic system. After 24 h of incubation, the DOX concentration gradient climbed to 1.7 μg/mL with the DOX-β-CD-CQD theranostic system concentration of 20 μg/mL. Cell viability (29%) was comparable to free DOX at 10 μg/mL. In terms of antitumor efficacy, the DOX-CD-CQD outperformed free DOX. The DOX-β-CD-CQD had an $IC_{50}$ of 5.00 μg/mL (equivalent to 0.425 μg/mL), compared to 2.26 μg/mL for free DOX. DOX-CD-CQD was internalized by HepG2 cells and accumulated in their nuclei, exhibiting better anticancer activity than the free drug [82]. In another investigation, Fateh et al. developed a hybrid nanostructure of cetyl alcohol-modified graphene quantum dots (CA-GQDs) and conjugated them with iron-oxide nanoparticles (IONP@CA-GQD) [44]. The study indicated no effect on the normal architecture of the liver and cardiac tissues after administration of these hybrid QDs at a dose of 3 mg/kg for 7 days in mice. Hence, IONP@CA-GQD can be offered as a potential drug delivery system for cancer theranostics. Moreover, IONP@CA-GQD was found to be more toxic for the tumor cells as compared to normal cells in the study [39,83]. It is due to the hydrophobic nature of the carbon chains of cetyl alcohol and oleic acid in the middle of IONP@CA-GQD, hydrophobic drugs can be loaded in this space [83,104]. Furthermore, the magnetic properties of IONP@CA-GQD would make cancer targeting feasible through this theranostic system. Similarly, Kim et al. examined QD-labeled hyaluronic acid (HA) derivatives for liver-targeted intracellular drug delivery. EDC activation of HA's carboxyl group and conjugation to ADH's amine group produced HA-ADH conjugates. After EDC and sufo-NHS activation of QD carboxyl groups, HA-ADH conjugates were tagged with QDs via amide bond formation. HA binds CD44 via its three carboxyl groups [88]. HA-QD conjugates were endocytosed via HA receptor-mediated endocytosis, as seen in the confocal microscopic images of B16F1 cells. HA receptors such as CD44 are significantly expressed in B16F1 cells. In the case of HEK293 cells without HA receptors, the cellular uptake of HA conjugates and QDs conjugates was noticeably reduced (as shown in Figure 4).

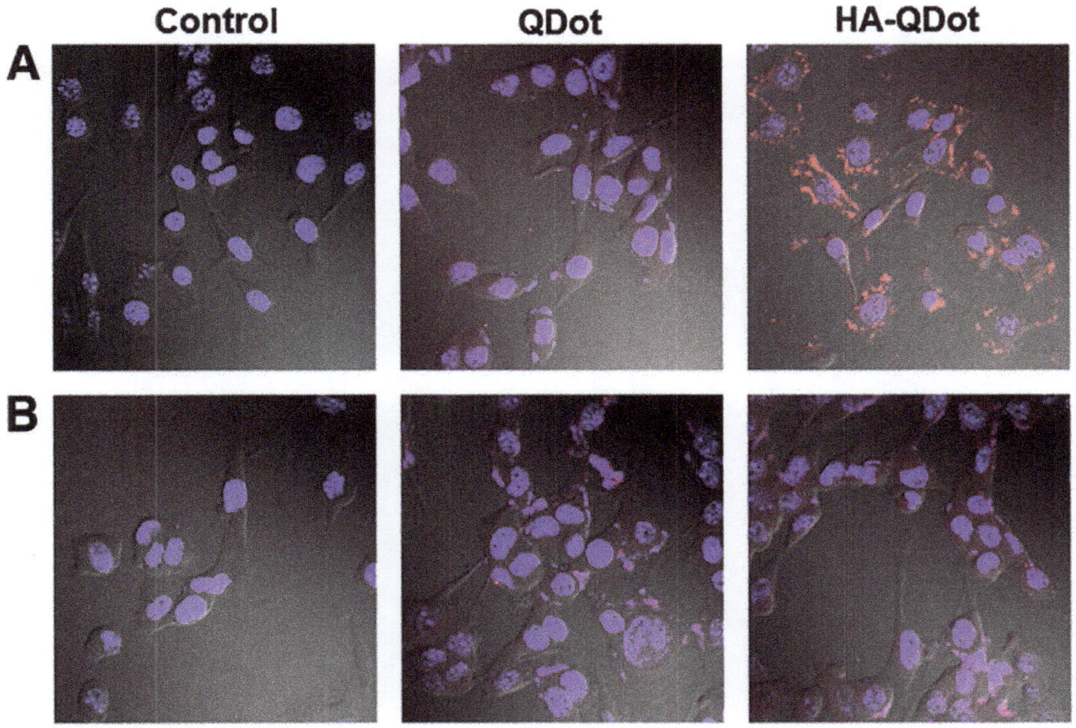


**Figure 4.** Illustration shows the cellular uptake of hyaluronic acid (HA) conjugated hybrid QDs after 2 h incubation in B16F1 cells—a high expression of HA receptors (**A**); HEK293 cells—without HA receptors (**B**). Confocal microscopic images reveal low cellular uptake of developed hybrid QDs system in HEK293 cells. Reproduced from Kim et al. [88], Elsevier, 2012.

In order to deliver targeted anticancer drugs, Chen et al. created a core-shell structured multifunctional nanocarrier system (ZnO-Au-PLA-GPPS-FA) consisting of ZnO-quantum dots-conjugated AuNPs as the core and folic acid (FA)-conjugated amphiphilic hyperbranched block copolymers as the shell. ZnO-quantum dots-conjugated AuNPs could be employed for photothermal therapy to kill tumor cells and fluorescent labeling, respectively. The outer hydrophilic block (GPPS-FA) and the inner hydrophobic block (PLA) were both biocompatible and biodegradable in in vivo system. The cancer cells may be targeted by an FA-conjugated multifunctional nanocarrier system, which may then be absorbed by the target cell through receptor-mediated endocytosis. Additionally, the presence of GPPS on the surface of multifunctional nanocarrier systems resulted in some of their anticancer effects [105]. In another investigation, Sung et al. made a targeted RBC-membrane-encased nanosponge (RBC-NS) that combines stealth and huge payloads of functional molecules to avoid the low EPR effect and the different types of tumors. This biocompatible, light-sensitive, carbon-based, porous particle looks like red blood cells (RBCs) and targets and gets into tumors well [39]. The targeted nanosponge, made of protein/RBC membranes (targeting and stealth properties), porous carbon/silica (hydrophobic, therapeutic agent transport), and graphene QDs (GQDs)/drug (photoresponsive, tumor-penetrating), were injected into a mouse model to deliver docetaxel (DTX) and GQDs to tumors (Figure 5). Moreover, Cetuximab (Ct), which can target tumors, is attached to the RBC layer to make it easier for particles to gather around tumors. The nanosponge delivers high amounts of GQD/DTX to the tumor as a photo-penetrative and photolytic agent. The Ct-RBC-GQD/NS-treated tumor can be heated to 68 °C for thermal tumor ablation. Ct-RBC-NS and Ct-NS elevated tumor temperatures to 62 and 53 °C, respectively. Irradiated saline-treated mice showed no temperature increase. Ct-RBC-GQD/stronger NS's improved photothermal conversion may be explained by the accumulation and photothermal combination effects. The localized heat of the NS releases GQD/DTX during NIR exposure, damaging the tumor and improving therapeutic drug penetration (as shown in Figure 5).

Contemporary research related to hybrid QDs utilized to improve the therapeutic performance of loaded drugs for anticancer activity in different types of cancers are summarized in Table 2.

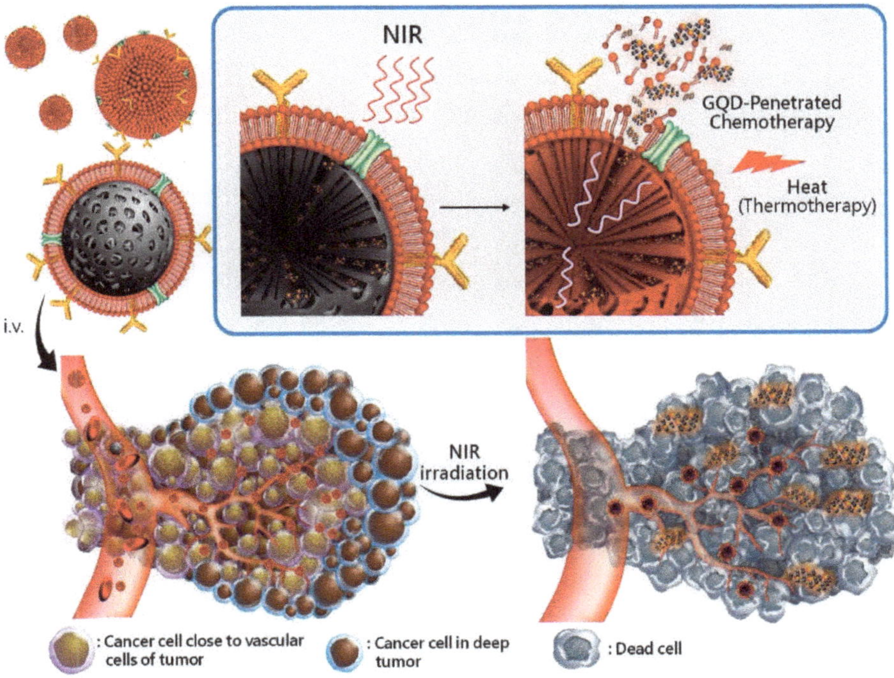


(A)

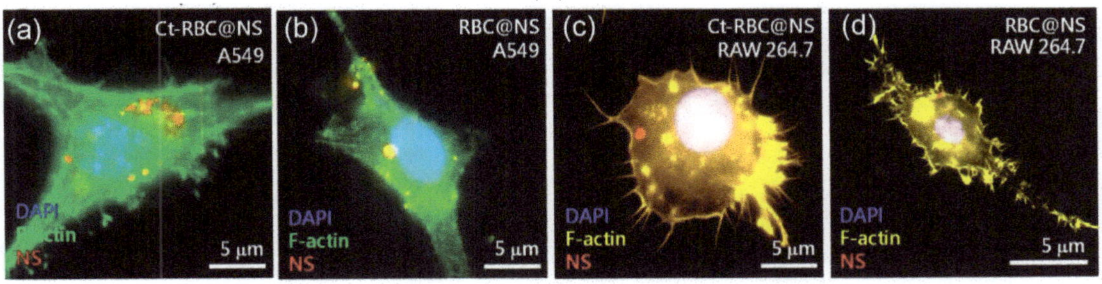


(B)

**Figure 5.** Schematic illustration highlights the penetration and accumulation of hybrid QDs (graphene quantum dots—GQDs) for RBC-membrane enveloped nanosponge-mediated targeted delivery in a tumor: (**A**) After the application of near-infrared (NIR) irradiation, generated heat leads to the penetration and accumulation of developed theranostic systems (GQDs with DTX) to deep tumors and the release of drug (DTX) into tumor cells ultimately causes cancer cell death. (**B**) Cellular uptake of RBC-membrane enveloped nanosponge with cetuximab (Ct-RBC@NS) and without cetuximab conjugation (RBC@NS) upon incubation for 2 h in A549 cancer cells and control RAW 264.7 cells. (a) Developed system conjugated with cetuximab (Ct-RBC@NS) in A549 cancer cells. It is monitored in the cytoplasm (green) and nuclei (blue). (b) Developed system without conjugation of cetuximab (RBC@NS) in A549 cancer cells. It is monitored in the cytoplasm (green) and nuclei (blue). (c) Developed system conjugated with cetuximab (Ct-RBC@NS) in control RAW 264.7 cells. (d) Developed system without conjugation of cetuximab (RBC@NS) in control RAW 264.7 cells. Reprinted with permission from Sung et al. [39], Copyright 2018, American Chemical Society.

**Table 2.** Summary of contemporary research carried out utilizing hybrid quantum dots for therapeutic/drug delivery applications in cancer.

| Type of QDs | Type of Cancer | In Vitro/In Vivo Model | Outcome | Refs. |
|---|---|---|---|---|
| Lactoferrin-QDs | Breast cancer | In vitro cancer cell line and in vivo tumor model | Enhanced cytotoxicity of breast cancer cells and in vivo antitumor efficacy | [89] (P:2018) |
| Gelatin/chondroitin-QDs | Breast cancer | In vitro cell line and in vivo model | Targeted internalization into cancer cells and enhanced cytotoxicity against breast cancer cells were demonstrated | [90] (P:2018) |
| Magnetic graphene-QDs | Cancerous cells | In vitro Hela cell line | G-QD susceptibility of cancerous HeLa cells to DOX is 13% higher and a promising material for cancer cell detection and targeted Dox | [91] (P:2018) |
| Graphene-QDs | Cancerous cells | In vitro Hela cell line | Cell viability study demonstrated the high cytotoxicity | [92] (P:2017) |
| Carbon-QDs doped with nitrogen and oxygen | Tumor cells | In vivo antitumor model | Nitrogen and oxygen co-doped C-QDs (N-O-CQDs) with strong absorbance in the NIR region leading to photothermal-based destruction of cancerous cells | [93] (P:2018) |
| Carbon QDs doped with polyethyleneimine | Hepatocellular carcinoma | In vitro COS-7 cells and HepG2 cells | Facilitate gene transfection in COS-7 and HepG2 cells with lower cytotoxicity | [96] (P:2012) |
| Magneton-fluorescence carbon-QDs conjugated with cDNA aptamer | Cervical cancer | In vitro cell line and In vivo tumor model | Targeted synergistic killing of lung cancer cells via PDT, PTT, and rapid release of DOX under simultaneous NIR laser irradiation | [97] (P:2018) |
| Au-$SiO_2$-QDs | Breast cancer | MCF-7 human breast cancer cells | A targeted synergistic anticancer effect that induced by DOX delivery and efficient heat generation by exploiting the photothermal effect of QDs-gold NPs. | [101] (P:2018) |
| ZnO-QDs | Cancerous cells | Hela cells | Studies showed that cytotoxicity by both blank and drug-loaded QDs provided high anticancer activity against Hela cells with folate targeting | [105] (P:2018) |
| Graphene-QDs on the surface of hollow $Cu_2S$ NPs | Breast cancer | MDA-MB-231 cells line | Flow cytometry showed a significant level of NIR-triggered Dox release inside MDA-MB-231 cells | [106] (P:2020) |
| Carbon-QDs with nuclear localization signal peptide | Lung cancer | Human lung carcinoma cells | Nucleus-targeted drug delivery of therapeutics functionalized with nuclear signal peptide to improve its antitumor activity | [107] (P:2016) |
| ZnO-QDs | Liver cancer | In vitro HepG2 cells | QDs significantly upregulated mRNA expressions, whereas the anti-apoptotic gene (Bcl-2) was down-regulated | [108] (P:2015) |
| CdSe-QDs | Hepatocellular carcinoma | In vitro HepG2 cancer cell | QDs successfully induced shrinkage and rupture of the membrane, and expression of an apoptotic gene (Bcl2) was positively comparing the untreated HepG2 cell line. | [109] (P:2021) |
| $Fe_3O_4$-$Ag_2O$ QDs/Cellulose fibers nanocomposites | Skin Cancer | In vitro cell line study | Magnetic QDs showed that the targeted cytotoxicity of the drug was increased when loaded on nanocomposites, compared to pure $Fe_3O4$-$Ag_2O$ quantum dots/cellulose fibers nanocomposites | [110] (P:2017) |
| CdTe-QDs and CdSe-QDs | Melanoma tumors | In vivo antitumor model | Result indicated CdTe and CdSe QDs irradiation-induced photothermal therapy shared great potential in the treatment of cancer | [111] (P:2012) |
| Graphene quantum dot mesoporous silica nanohybrids | Breast cancer | 4T1 cancer cell line; 4T1 tumor in Balb/c mice | Results indicate that developed hybrid QDs as powerful cancer theranostic for deep tumor localization and regression | [112] (P:2021) |
| Peptide-based graphene QDs | Breast cancer | HUVEC Cell line; 4T1 tumor-bearing Balb/c mice | Successfully demonstrated multifunctional theranostic peptideticles for targeted drug delivery and tracking in αv integrin overexpressed tumor model | [113] (P:2022) |
| Tryptophan–sorbitol-based carbon QDs | Liver cancer | Huh7 cell line; Huh7 cells bearing Balb/c mice | Promising cancer nanotheranostic system utilized for diagnosis, targeting, and PDT therapy in hepatocellular carcinoma | [114] (P:2022) |
| Mn-doped ZnS QDs | Breast cancer | 4T1 cancer cell line; 4T1 tumor in Balb/c mice | Theranostic system for image-guided therapy in breast tumor utilizing NIR-II fluorescence and magnetic resonance imaging | [115] (P:2022) |

## 4. Conclusions

The review concludes that hybrid QDs could be multi-modeled to treat different cancers, and therapeutic progress could be monitored in real-time. These QDs combined with different types of nanoparticulate systems (such as NPs of polymeric, lipid, and inorganic origin) to develop a theranostic system for cancer, particularly to improve the therapy outcome in MDR cancer. Further, this review concluded that carbon-based and graphene-based QDs had been extensively explored to conjugate them with different biomolecules to overcome the challenges associated with conventional QDs. Several preclinical studies showed that hybrid QDs could be successfully used as a theranostic system in cancer, bringing them closer to investigating its clinical utility. However, the literature review reveals that the clinical performance of hybrid QDs as cancer theranostics has not been addressed in detail as yet. Furthermore, the safety perspectives of the hybrid QDs in cancer should also be addressed systematically in future investigations as they may be accumulated in the healthy tissues due to failure of tumor-specific delivery that may increase the risks of untoward events. Although numerous in vivo studies have examined the distribution, accumulation, excretion, and toxic consequences of QDs, no consensus has been established. Moreover, due to the complexity of in vivo models, the replication of pharmacokinetics is difficult. However, certain in-vitro studies eased our basic understanding of mechanisms and possible adverse effects of various QDs. The type of QDs has an impact on their distribution within cells and clearance rate, which is directly related to their cytotoxicity. Based on the local accumulation and biological half-life, the possible cytotoxic potential of QDs can be anticipated. Thus, systematic investigation of the safety and efficacy of hybrid QDs should be of prime concern.

**Author Contributions:** J.A.: conceptualization, investigation, formal analysis, and validation; writing—original draft preparation, writing—review and editing, supervision, funding acquisition, and project administration; A.G.: investigation, formal analysis, validation, writing—original draft preparation; G.M.: investigation, formal analysis, validation, writing—original draft preparation; M.Z.A.: formal analysis, writing—review and editing, funding acquisition, project administration; M.A.: investigation, formal analysis, validation, writing—review and editing; A.M.: formal analysis, validation, writing—review and editing; All authors have read and agreed to the published version of the manuscript.

**Funding:** This work was funded by the Deanship of Scientific Research at Najran University, Saudi Arabia, under the National Research Priorities funding program grant code (NU/NRP/MRC/11/5).

**Institutional Review Board Statement:** Not applicable.

**Data Availability Statement:** Not applicable.

**Acknowledgments:** The authors are thankful to the Deanship of Scientific Research at Najran University for funding this work under the National Research Priorities funding program grant code (NU/NRP/MRC/11/5).

**Conflicts of Interest:** The authors declare no conflict of interest.

## References


1. Available online: https://www.who.int/news-room/fact-sheets/detail/cancer (accessed on 5 December 2022).
2. Kemp, J.A.; Kwon, Y.J. Cancer nanotechnology: Current status and perspectives. *Nano Converg.* **2021**, *8*, 34. [CrossRef]
3. Singh, R.; Deshmukh, R. Carbon nanotube as an emerging theranostic tool for oncology. *J. Drug Deliv. Sci. Technol.* **2022**, *74*, 103586. [CrossRef]
4. Siddique, S.; Chow, J.C. Recent advances in functionalized nanoparticles in cancer theranostics. *Nanomaterials* **2022**, *12*, 2826. [CrossRef]
5. Bernard, V.; Zobač, O.; Sopoušek, J.; Mornstein, V. AgCu bimetallic nanoparticles under effect of low intensity ultrasound: The cell viability study in vitro. *J. Cancer Res.* **2014**, *2014*, 971769. [CrossRef]
6. Zhao, Y.; Zhu, Y.; Fu, J.; Wang, L. Effective Cancer Cell Killing by Hydrophobic Nanovoid-Enhanced Cavitation under Safe Low-Energy Ultrasound. *Chem. Asian J.* **2014**, *9*, 790–796. [CrossRef] [PubMed]
7. Miller, D.L.; Smith, N.B.; Bailey, M.R.; Czarnota, G.J.; Hynynen, K.; Makin, I.R.S. Bioeffects Committee of the American Institute of Ultrasound in Medicine. (2012). Overview of therapeutic ultrasound applications and safety considerations. *J. Ultrasound Med.* **2012**, *31*, 623–634. [CrossRef] [PubMed]
8. Ashikbayeva, Z.; Aitkulov, A.; Atabaev, T.S.; Blanc, W.; Inglezakis, V.J.; Tosi, D. Green-Synthesized Silver Nanoparticle–Assisted Radiofrequency Ablation for Improved Thermal Treatment Distribution. *Nanomaterials* **2022**, *12*, 426. [CrossRef]

9. Glazer, E.S.; Curley, S.A. Non-invasive radiofrequency ablation of malignancies mediated by quantum dots, gold nanoparticles and carbon nanotubes. *Ther. Deliv.* **2011**, *2*, 1325–1330. [CrossRef]
10. Nguyen, D.T.; Tzou, W.S.; Zheng, L.; Barham, W.; Schuller, J.L.; Shillinglaw, B.; Sauer, W.H. Enhanced radiofrequency ablation with magnetically directed metallic nanoparticles. *Circ. Arrhythm. Electrophysiol.* **2016**, *9*, e003820. [CrossRef]
11. Ranoo, S.; Lahiri, B.B.; Nandy, M.; Philip, J. Enhanced magnetic heating efficiency at acidic pH for magnetic nanoemulsions stabilized with a weak polyelectrolyte. *J. Colloid Interface Sci.* **2020**, *579*, 582–597. [CrossRef]
12. Ranoo, S.; Lahiri, B.B.; Damodaran, S.P.; Philip, J. Tuning magnetic heating efficiency of colloidal dispersions of iron oxide nano-clusters by varying the surfactant concentration during solvothermal synthesis. *J. Mol. Liq.* **2022**, *360*, 119444. [CrossRef]
13. Dutz, S.; Buske, N.; Landers, J.; Gräfe, C.; Wende, H.; Clement, J.H. Biocompatible magnetic fluids of co-doped iron oxide nanoparticles with tunable magnetic properties. *Nanomaterials* **2020**, *10*, 1019. [CrossRef]
14. Du, Y.; Liu, X.; Liang, Q.; Liang, X.J.; Tian, J. Optimization and design of magnetic ferrite nanoparticles with uniform tumor distribution for highly sensitive MRI/MPI performance and improved magnetic hyperthermia therapy. *Nano Lett.* **2019**, *19*, 3618–3626. [CrossRef] [PubMed]
15. Dadfar, S.M.; Camozzi, D.; Darguzyte, M.; Roemhild, K.; Varvarà, P.; Metselaar, J.; Lammers, T. Size-isolation of superparamagnetic iron oxide nanoparticles improves MRI, MPI and hyperthermia performance. *J. Nanobiotechnol.* **2020**, *18*, 22. [CrossRef]
16. Sharma, P.; Brown, S.; Walter, G.; Santra, S.; Moudgil, B. Nanoparticles for bioimaging. *Adv. Colloid Interface Sci.* **2006**, *123*, 471–485. [CrossRef]
17. Couvreur, P. Nanoparticles in drug delivery: Past, present and future. *Adv. Drug Deliv. Rev.* **2013**, *65*, 21–23. [CrossRef]
18. Lucky, S.S.; Soo, K.C.; Zhang, Y. Nanoparticles in photodynamic therapy. *Chem. Rev.* **2015**, *115*, 1990–2042. [CrossRef]
19. Jaque, D.; Maestro, L.M.; del Rosal, B.; Haro-Gonzalez, P.; Benayas, A.; Plaza, J.L.; Martin Rodriguez, E.; Solé, J.G. Nanoparticles for photothermal therapies. *Nanoscale* **2014**, *6*, 9494–9530. [CrossRef]
20. Calabrese, G.; Petralia, S.; Franco, D.; Nocito, G.; Fabbi, C.; Forte, L.; Guglielmino, S.; Squarzoni, S.; Traina, F.; Conoci, S. A new Ag-nanostructured hydroxyapatite porous scaffold: Antibacterial effect and cytotoxicity study. *Mater. Sci. Eng. C* **2021**, *118*, 111394. [CrossRef] [PubMed]
21. Mele, E. Introduction: Smart Materials in Biomedicine. In *Smart Nanoparticles for Biomedicine*; Elsevier: Amsterdam, The Netherlands, 2008; pp. 1–13.
22. Dreaden, E.C.; Alkilany, A.M.; Huang, X.; Murphy, C.J.; El-Sayed, M.A. The golden age: Gold nanoparticles for biomedicine. *Chem. Soc. Rev.* **2012**, *41*, 2740–2779. [CrossRef]
23. Eckhardt, S.; Brunetto, P.S.; Gagnon, J.; Priebe, M.; Giese, B.; Fromm, K.M. Nanobio Silver: Its Interactions with Peptides and Bacteria, and Its Uses in Medicine. *Chem. Rev.* **2013**, *113*, 4708–4754. [CrossRef]
24. Jeyaraj, M.; Gurunathan, S.; Qasim, M.; Kang, M.H.; Kim, J.H. A Comprehensive Review on the Synthesis, Characterization, and Biomedical Application of Platinum Nanoparticles. *Nanomaterials* **2019**, *9*, 1719. [CrossRef] [PubMed]
25. Phan, T.T.V.; Huynh, T.C.; Manivasagan, P.; Mondal, S.; Oh, J. An Up-To-Date Review on Biomedical Applications of Palladium Nanoparticles. *Nanomaterials* **2020**, *10*, 66. [CrossRef]
26. Jain, A.; Tiwari, A.; Verma, A.; Saraf, S.; Jain, S.K. Combination cancer therapy using multifunctional liposomes. *Crit. Rev.™ Ther. Drug Carr. Syst.* **2020**, *37*, 105–134. [CrossRef] [PubMed]
27. Abdellatif, A.A.; Younis, M.A.; Alsharidah, M.; Al Rugaie, O.; Tawfeek, H.M. Biomedical applications of quantum dots: Overview, challenges, and clinical potential. *Int. J. Nanomed.* **2022**, *17*, 1951–1970. [CrossRef]
28. Wagner, A.M.; Knipe, J.M.; Orive, G.; Peppas, N.A. Quantum dots in biomedical applications. *Acta Biomater.* **2019**, *94*, 44–63. [CrossRef]
29. Bajwa, N.; Mehra, N.K.; Jain, K.; Jain, N.K. Pharmaceutical and biomedical applications of quantum dots. *Artif. Cells Nanomed. Biotechnol.* **2016**, *44*, 758–768. [CrossRef]
30. Liu, M.; Yazdani, N.; Yarema, M.; Jansen, M.; Wood, V.; Sargent, E.H. Colloidal quantum dot electronics. *Nat. Electron.* **2021**, *4*, 548–558. [CrossRef]
31. Smith, A.M.; Duan, H.; Mohs, A.M.; Nie, S. Bioconjugated quantum dots for in vivo molecular and cellular imaging. *Adv. Drug Deliv. Rev.* **2008**, *60*, 1226–1240. [CrossRef]
32. Xu, Q.; Gao, J.; Wang, S.; Wang, Y.; Liu, D.; Wang, J. Quantum dots in cell imaging and their safety issues. *J. Mater. Chem. B* **2021**, *9*, 5765–5779. [CrossRef]
33. Devi, S.; Kumar, M.; Tiwari, A.; Tiwari, V.; Kaushik, D.; Verma, R.; Batiha, G.E.S. Quantum dots: An emerging approach for cancer therapy. *Front. Mater.* **2022**, *8*, 585. [CrossRef]
34. Rahman, M.A.; Abul Barkat, H.; Harwansh, R.K.; Deshmukh, R. Carbon-based Nanomaterials: Carbon Nanotubes, Graphene, and Fullerenes for the Control of Burn Infections and Wound Healing. *Curr. Pharm. Biotechnol.* **2022**, *23*, 1483–1496. [CrossRef] [PubMed]
35. Kanthi Gudimella, K.; Gedda, G.; Kumar, P.S.; Babu, B.K.; Yamajala, B.; Rao, B.V.; Singh, P.P.; Kumar, D.; Sharma, A. Novel synthesis of fluorescent carbon dots from bio-based Carica Papaya Leaves: Optical and structural properties with antioxidant and anti-inflammatory activities. *Environ. Res.* **2022**, *204*, 111854. [CrossRef]
36. Zhang, H.; Jin, Y.; Chi, C.; Han, G.; Jiang, W.; Wang, Z.; Cheng, H.; Zhang, C.; Wang, G.; Sun, C.; et al. Sponge particulates for biomedical applications: Biofunctionalization, multi-drug shielding, and theranostic applications. *Biomaterials* **2021**, *273*, 120824. [CrossRef]

37. Ghaffarkhah, A.; Hosseini, E.; Kamkar, M.; Sehat, A.A.; Dordanihaghighi, S.; Allahbakhsh, A.; van der Kuur, C.I.; Arjmand, M. Synthesis, applications, and prospects of graphene quantum dots: A comprehensive review. *Small* **2022**, *18*, 2102683. [CrossRef]
38. Paduraru, D.N.; Niculescu, A.G.; Bolocan, A.; Andronic, O.; Grumezescu, A.M.; Bîrlă, R. An Updated overview of cyclodextrin-based drug delivery systems for cancer therapy. *Pharmaceutics* **2022**, *14*, 1748. [CrossRef] [PubMed]
39. Sung, S.Y.; Su, Y.L.; Cheng, W.; Hu, P.F.; Chiang, C.S.; Chen, W.T.; Hu, S.H. Graphene quantum dots-mediated theranostic penetrative delivery of drug and photolytics in deep tumors by targeted biomimetic nanosponges. *Nano Lett.* **2019**, *19*, 69–81. [CrossRef]
40. Yang, K.; Wan, J.; Zhang, S.; Zhang, Y.; Lee, S.-T.; Liu, Z. In Vivo Pharmacokinetics, Long-Term Biodistribution, and Toxicology of PEGylated Graphene in Mice. *ACS Nano* **2011**, *5*, 516–522. [CrossRef] [PubMed]
41. Zhang, S.; Yang, K.; Feng, L.; Liu, Z. In vitro and in vivo behaviors of dextran functionalized graphene. *Carbon* **2011**, *49*, 4040–4049. [CrossRef]
42. Lv, C.; Lin, Y.; Liu, A.A.; Hong, Z.Y.; Wen, L.; Zhang, Z.; Pang, D.W. Labeling viral envelope lipids with quantum dots by harnessing the biotinylated lipid-self-inserted cellular membrane. *Biomaterials* **2016**, *106*, 69–77. [CrossRef]
43. Hashemkhani, M.; Muti, A.; Sennaroğlu, A.; Acar, H.Y. Multimodal image-guided folic acid targeted Ag-based quantum dots for the combination of selective methotrexate delivery and photothermal therapy. *J. Photochem. Photobiol. B Biol.* **2020**, *213*, 112082. [CrossRef] [PubMed]
44. Singh, G.; Kumar, M.; Soni, U.; Arora, V.; Bansal, V.; Gupta, D.; Singh, H. Cancer cell targeting using folic acid/anti-HER2 antibody conjugated fluorescent CdSe/CdS/ZnS-Mercaptopropionic acid and CdTe-Mercaptosuccinic acid quantum dots. *J. Nanosci. Nanotechnol.* **2016**, *16*, 130–143. [CrossRef] [PubMed]
45. Sahoo, S.L.; Liu, C.H.; Kumari, M.; Wu, W.C.; Wang, C.C. Biocompatible quantum dot-antibody conjugate for cell imaging, targeting and fluorometric immunoassay: Crosslinking, characterization and applications. *RSC Adv.* **2019**, *9*, 32791–32803. [CrossRef]
46. Huang, H.; Bai, Y.L.; Yang, K.; Tang, H.; Wang, Y.W. Optical imaging of head and neck squamous cell carcinoma in vivo using arginine-glycine-aspartic acid peptide conjugated near-infrared quantum dots. *OncoTargets Ther.* **2013**, *6*, 1779–1787.
47. Lu, J.; Liong, M.; Li, Z.; Zink, J.I.; Tamanoi, F. Biocompatibility, biodistribution, and drug-delivery efficiency of mesoporous silica nanoparticles for cancer therapy in animals. *Small* **2010**, *6*, 1794–1805. [CrossRef] [PubMed]
48. Li, C.; Zou, Z.; Liu, H.; Jin, Y.; Li, G.; Yuan, C.; Jin, M. Synthesis of polystyrene-based fluorescent quantum dots nanolabel and its performance in H5N1 virus and SARS-CoV-2 antibody sensing. *Talanta* **2021**, *225*, 122064. [CrossRef]
49. Park, J.H.; Gu, L.; Von Maltzahn, G.; Ruoslahti, E.; Bhatia, S.N.; Sailor, M.J. Biodegradable luminescent porous silicon nanoparticles for in vivo applications. *Nat. Mater.* **2009**, *8*, 331–336. [CrossRef]
50. Zayed, D.G.; AbdElhamid, A.S.; Freag, M.S.; Elzoghby, A.O. Hybrid quantum dot-based theranostic nanomedicines for tumor-targeted drug delivery and cancer imaging. *Nanomedicine* **2019**, *14*, 225–228. [CrossRef] [PubMed]
51. Liu, W.; Li, C.; Ren, Y.; Sun, X.; Pan, W.; Li, Y.; Wang, J.; Wang, W. Carbon dots: Surface engineering and applications. *J. Mater. Chem. B* **2016**, *4*, 5772–5788. [CrossRef] [PubMed]
52. Li, B.; Lin, L.; Lin, H.; Wilson, B.C. Photosensitized singlet oxygen generation and detection: Recent advances and future perspectives in cancer photodynamic therapy. *J. Biophotonics* **2016**, *9*, 1314–1325. [CrossRef] [PubMed]
53. Das, R.K.; Panda, S.; Bhol, C.S.; Bhutia, S.K.; Mohapatra, S. N-doped carbon quantum dot (NCQD)-Deposited carbon capsules for synergistic fluorescence imaging and photothermal therapy of oral cancer. *Langmuir* **2019**, *35*, 15320–15329. [CrossRef] [PubMed]
54. Sekar, R.; Basavegowda, N.; Jena, S.; Jayakodi, S.; Elumalai, P.; Chaitanyakumar, A.; Baek, K.H. Recent Developments in Heteroatom/Metal-Doped Carbon Dot-Based Image-Guided Photodynamic Therapy for Cancer. *Pharmaceutics* **2022**, *14*, 1869. [CrossRef]
55. Proskurnin, M.A.; Khabibullin, V.R.; Usoltseva, L.O.; Vyrko, E.A.; Mikheev, I.V.; Volkov, D.S. Photothermal and optoacoustic spectroscopy: State of the art and prospects. *Phys.-Uspekhi* **2022**, *65*, 270. [CrossRef]
56. Walter, M.; Schubert, L.; Heberle, J.; Schlesinger, R.; Losi, A. Time-resolved photoacoustics of channelrhodopsins: Early energetics and light-driven volume changes. *Photochem. Photobiol. Sci.* **2022**, 1–10. [CrossRef]
57. He, Z.; Zhang, C.Y.; Lei, Y.; Song, G.; Yao, Y. Plasmonic nanomaterials: A versatile phototheranostic platform of cancers. *Materials Today* **2022**. [CrossRef]
58. Gellini, C.; Feis, A. Optothermal properties of plasmonic inorganic nanoparticles for photoacoustic applications. *Photoacoustics* **2021**, *23*, 100281. [CrossRef]
59. Song, Z.; Quan, F.; Xu, Y.; Liu, M.; Cui, L.; Liu, J. Multifunctional N, S co-doped carbon quantum dots with pH-and thermo-dependent switchable fluorescent properties and highly selective detection of glutathione. *Carbon* **2016**, *104*, 169–178. [CrossRef]
60. Liu, B.; Wei, S.; Liu, E.; Zhang, H.; Lu, P.; Wang, J.; Sun, G. Nitrogen-doped carbon dots as a fluorescent probe for folic acid detection and live cell imaging. *Spectrochim. Acta Part A Mol. Biomol. Spectrosc.* **2022**, *268*, 120661. [CrossRef]
61. Lee, C.H.; Rajendran, R.; Jeong, M.S.; Ko, H.Y.; Joo, J.Y.; Cho, S.; Chang, Y.W.; Kim, S. Bioimaging of targeting cancers using aptamer-conjugated carbon nanodots. *Chem. Commun.* **2013**, *49*, 6543–6545. [CrossRef] [PubMed]
62. Bacon, M.; Bradley, S.J.; Nann, T. Graphene quantum dots. *Part. Part. Syst. Charact.* **2014**, *31*, 415–428. [CrossRef]
63. Liu, H.; Li, C.; Qian, Y.; Hu, L.; Fang, J.; Tong, W.; Wang, H. Magnetic-induced graphene quantum dots for imaging-guided photothermal therapy in the second near-infrared window. *Biomaterials* **2020**, *232*, 119700. [CrossRef]

64. Sun, B.; Luo, C.; Yu, H.; Zhang, X.; Chen, Q.; Yang, W.; Wang, M.; Kan, Q.; Zhang, H.; Wang, Y.; et al. Disulfide bond-driven oxidation-and reduction-responsive prodrug nanoassemblies for cancer therapy. *Nano Lett.* **2018**, *18*, 3643–3650. [CrossRef]
65. Zhu, L.; Zhao, H.; Zhou, Z.; Xia, Y.; Wang, Z.; Ran, H.; Li, P.; Ren, J. Peptide-functionalized phase-transformation nanoparticles for low intensity focused ultrasound-assisted tumor imaging and therapy. *Nano Lett.* **2018**, *18*, 1831–1841. [CrossRef] [PubMed]
66. Chen, H.; Zhang, W.; Zhu, G.; Xie, J.; Chen, X. Rethinking cancer nanotheranostics. *Nat. Rev. Mater.* **2017**, *2*, 17024. [CrossRef] [PubMed]
67. Avasthi, A.; Caro, C.; Pozo-Torres, E.; Leal, M.P.; García-Martín, M.L. Magnetic nanoparticles as MRI contrast agents. *Top. Curr. Chem.* **2020**, *378*, 40. [CrossRef]
68. Pourmadadi, M.; Rahmani, E.; Shamsabadipour, A.; Mahtabian, S.; Ahmadi, M.; Rahdar, A.; Díez-Pascual, A.M. Role of Iron Oxide ($Fe_2O_3$) Nanocomposites in Advanced Biomedical Applications: A State-of-the-Art Review. *Nanomaterials* **2022**, *12*, 3873. [CrossRef] [PubMed]
69. Jana, P.; Dev, A. Carbon quantum dots: A promising nanocarrier for bioimaging and drug delivery in cancer. *Mater. Today Commun.* **2022**, *32*, 104068. [CrossRef]
70. Sasidharan, S.; Bahadur, D.; Srivastava, R. Protein-poly (amino acid) nanocore–shell mediated synthesis of branched gold nanostructures for computed tomographic imaging and photothermal therapy of cancer. *ACS Appl. Mater. Interfaces* **2016**, *8*, 15889–15903. [CrossRef]
71. Zhu, Y.; Murali, S.; Cai, W.; Li, X.; Suk, J.W.; Potts, J.R.; Ruoff, R.S. Graphene and graphene oxide: Synthesis, properties, and applications. *Adv. Mater.* **2010**, *22*, 3906–3924. [CrossRef] [PubMed]
72. Navya, P.N.; Kaphle, A.; Srinivas, S.P.; Bhargava, S.K.; Rotello, V.M.; Daima, H.K. Current trends and challenges in cancer management and therapy using designer nanomaterials. *Nano Converg.* **2019**, *6*, 23. [CrossRef]
73. Luo, G.; Long, J.; Zhang, B.; Liu, C.; Ji, S.; Xu, J.; Ni, Q. Quantum dots in cancer therapy. *Expert Opin. Drug Deliv.* **2012**, *9*, 47–58. [CrossRef] [PubMed]
74. Kumar, A.; Singh, K.R.; Ghate, M.D.; Lalhlenmawia, H.; Kumar, D.; Singh, J. Bioinspired quantum dots for cancer therapy: A mini-review. *Mater. Lett.* **2022**, *313*, 131742. [CrossRef]
75. Zhang, H.; Yee, D.; Wang, C. Quantum dots for cancer diagnosis and therapy: Biological and clinical perspectives. *Nanomedicine* **2008**, *3*, 83–91. [CrossRef] [PubMed]
76. Taghavi, S.; Abnous, K.; Taghdisi, S.M.; Ramezani, M.; Alibolandi, M. Hybrid carbon-based materials for gene delivery in cancer therapy. *J. Control Release* **2020**, *318*, 158–175. [CrossRef]
77. Robertson, C.A.; Evans, D.H.; Abrahamse, H. Photodynamic therapy (PDT): A short review on cellular mechanisms and cancer research applications for PDT. *J. Photochem. Photobiol. B Biol.* **2009**, *96*, 1–8. [CrossRef]
78. Thomsen, H.; Marino, N.; Conoci, S.; Sortino, S.; Ericson, M.B. Confined photo-release of nitric oxide with simultaneous two-photon fluorescence tracking in a cellular system. *Sci. Rep.* **2018**, *8*, 9753. [CrossRef]
79. Zhao, X.; Wei, Z.; Zhao, Z.; Miao, Y.; Qiu, Y.; Yang, W.; Jia, X.; Liu, Z.; Hou, H. Design and development of graphene oxide nanoparticle/chitosan hybrids showing pH-sensitive surface charge-reversible ability for efficient intracellular doxorubicin delivery. *ACS Appl. Mater. Interfaces* **2018**, *10*, 6608–6617. [CrossRef]
80. Zhuang, W.; He, L.; Wang, K.; Ma, B.; Ge, L.; Wang, Z.; Huang, J.; Wu, J.; Zhang, Q.; Ying, H. Combined adsorption and covalent linking of paclitaxel on functionalized nano-graphene oxide for inhibiting cancer cells. *ACS Omega* **2018**, *3*, 2396–2405. [CrossRef]
81. Zhao, C.; Song, X.; Liu, Y.; Fu, Y.; Ye, L.; Wang, N.; Wang, F.; Li, L.; Mohammadniaei, M.; Zhang, M.; et al. Synthesis of graphene quantum dots and their applications in drug delivery. *J. Nanobiotechnol.* **2020**, *18*, 1–32. [CrossRef]
82. Pei, M.; Pai, J.Y.; Du, P.; Liu, P. Facile synthesis of fluorescent hyper-cross-linked β-cyclodextrin-carbon quantum dot hybrid nanosponges for tumor theranostic application with enhanced antitumor efficacy. *Mol. Pharm.* **2018**, *15*, 4084–4091. [CrossRef]
83. Fateh, S.T.; Kamalabadi, M.A.; Aliakbarniya, A.; Jafarinejad-Farsangi, S.; Koohi, M.; Jafari, E.; Karam, Z.M.; Keyhanfar, F.; Dezfuli, A.S. Hydrophobic@ amphiphilic hybrid nanostructure of iron-oxide and graphene quantum dot surfactant as a theranostic platform. *OpenNano* **2022**, *6*, 100037. [CrossRef]
84. Schroeder, K.L.; Goreham, R.V.; Nann, T. Graphene quantum dots for theranostics and bioimaging. *Pharm. Res.* **2016**, *33*, 2337–2357. [CrossRef]
85. Dezfuli, A.S.; Kohan, E.; Fateh, S.T.; Alimirzaei, N.; Arzaghi, H.; Hamblin, M.R. Organic dots (O-dots) for theranostic applications: Preparation and surface engineering. *RSC Adv.* **2021**, *11*, 2253–2291. [CrossRef]
86. Kumawat, M.K.; Thakur, M.; Bahadur, R.; Kaku, T.; Prabhuraj, R.S.; Ninawe, A.; Srivastava, R. Preparation of gra-phene oxide-graphene quantum dots hybrid and its application in cancer theranostics. *Mater. Sci. Eng. C* **2019**, *103*, 109774. [CrossRef] [PubMed]
87. Kim, K.S.; Hur, W.; Park, S.J.; Hong, S.W.; Choi, J.E.; Goh, E.J.; Yoon, S.K.; Hahn, S.K. Bioimaging for targeted deliv-ery of hyaluronic acid derivatives to the livers in cirrhotic mice using quantum dots. *ACS Nano* **2010**, *4*, 3005–3014. [CrossRef]
88. Kim, K.S.; Kim, S.; Beack, S.; Yang, J.A.; Yun, S.H.; Hahn, S.K. In vivo real-time confocal microscopy for tar-get-specific delivery of hyaluronic acid-quantum dot conjugates. *Nanomed. Nanotechnol. Biol. Med.* **2012**, *8*, 1070–1073. [CrossRef]
89. AbdElhamid, A.S.; Helmy, M.W.; Ebrahim, S.M.; Bahey-El-Din, M.; Zayed, D.G.; Zein El Dein, E.A.; El-Gizawy, S.A.; El-zoghby, A.O. Layer-by-layer gelatin/chondroitin quantum dots-based nanotheranostics: Combined rapamy-cin/celecoxib delivery and cancer imaging. *Nanomedicine* **2018**, *13*, 1707–1730. [CrossRef] [PubMed]

90. AbdElhamid, A.S.; Zayed, D.G.; Helmy, M.W.; Ebrahim, S.M.; Bahey-El-Din, M.; Zein-El-Dein, E.A.; El-Gizawy, S.A.; El-zoghby, A.O. Lactoferrin-tagged quantum dots-based theranostic nanocapsules for combined COX-2 inhibi-tor/herbal therapy of breast cancer. *Nanomedicine* **2018**, *13*, 2637–2656. [CrossRef] [PubMed]
91. Chowdhury, A.D.; Ganganboina, A.B.; Tsai, Y.C.; Chiu, H.C.; Doong, R.A. Multifunctional GQDs-Concanavalin A@ $Fe_3O_4$ nanocomposites for cancer cells detection and targeted drug delivery. *Anal. Chim. Acta* **2018**, *1027*, 109–120. [CrossRef]
92. Su, X.; Chan, C.; Shi, J.; Tsang, M.K.; Pan, Y.; Cheng, C.; Gerile, O.; Yang, M. A graphene quantum dot@ $Fe_3O_4$@ $SiO_2$ based nanoprobe for drug delivery sensing and dual-modal fluorescence and MRI imaging in cancer cells. *Biosens. Bi-Oelectron.* **2017**, *92*, 489–495. [CrossRef]
93. Bao, X.; Yuan, Y.; Chen, J.; Zhang, B.; Li, D.; Zhou, D.; Jing, P.; Xu, G.; Wang, Y.; Holá, K.; et al. In vivo theranostics with near-infrared-emitting carbon dots—Highly efficient photothermal therapy based on passive targeting after intravenous administration. *Light Sci. Appl.* **2018**, *7*, 91. [CrossRef] [PubMed]
94. Bhunia, S.K.; Saha, A.; Maity, A.R.; Ray, S.C.; Jana, N.R. Carbon nanoparticle-based fluorescent bioimaging probes. *Sci. Rep.* **2013**, *3*, 1473. [CrossRef]
95. Huang, X.; Zhang, F.; Zhu, L.; Choi, K.Y.; Guo, N.; Guo, J.; Tackett, K.; Anilkumar, P.; Liu, G.; Quan, Q.; et al. Effect of injection routes on the biodistribution, clearance, and tumor uptake of carbon dots. *ACS Nano* **2013**, *7*, 5684–5693. [CrossRef]
96. Liu, C.; Zhang, P.; Zhai, X.; Tian, F.; Li, W.; Yang, J.; Liu, Y.; Wang, H.; Wang, W.; Liu, W. Nano-carrier for gene de-livery and bioimaging based on carbon dots with PEI-passivation enhanced fluorescence. *Biomaterials* **2012**, *33*, 3604–3613. [CrossRef]
97. Zhang, M.; Wang, W.; Cui, Y.; Chu, X.; Sun, B.; Zhou, N.; Shen, J. Magnetofluorescent Fe3O4/carbon quantum dots coated single-walled carbon nanotubes as dual-modal targeted imaging and chemo/photodynamic/photothermal tri-ple-modal therapeutic agents. *Chem. Eng. J.* **2018**, *338*, 526–538. [CrossRef]
98. Yang, X.Q.; Chen, C.; Peng, C.W.; Hou, J.X.; Liu, S.P.; Qi, C.B.; Gong, Y.P.; Zhu, X.B.; Pang, D.W.; Li, Y. Quantum dot-based quantitative immunofluorescence detection and spectrum analysis of epidermal growth factor receptor in breast cancer tissue arrays. *Int. J. Nanomed.* **2011**, *6*, 2265.
99. Martins, C.S.; LaGrow, A.P.; Prior, J.A. Quantum Dots for Cancer-Related miRNA Monitoring. *ACS Sens.* **2022**, *7*, 1269–1299. [CrossRef]
100. Lv, S.; Chen, F.; Chen, C.; Chen, X.; Gong, H.; Cai, C. A novel CdTe quantum dots probe amplified resonance light scattering signals to detect microRNA-122. *Talanta* **2017**, *165*, 659–663. [CrossRef] [PubMed]
101. Volsi, A.L.; Fiorica, C.; D'Amico, M.; Scialabba, C.; Palumbo, F.S.; Giammona, G.; Licciardi, M. Hybrid Gold/Silica/Quantum-Dots supramolecular-nanostructures encapsulated in polymeric micelles as potential theranostic tool for targeted cancer therapy. *Eur. Polym. J.* **2018**, *105*, 38–47. [CrossRef]
102. Jabeen, G.; Ahmad, M.H.; Aslam, M.; Riaz, S.; Hayat, A.; Nawaz, M.H. N-Doped graphene quantum dots (N-GQDs) as fluorescent probes for detection of UV induced DNA damage. *RSC Adv.* **2022**, *12*, 22458–22464. [CrossRef]
103. Kwon, J.; Jun, S.W.; Choi, S.I.; Mao, X.; Kim, J.; Koh, E.K.; Kim, Y.H.; Kim, S.K.; Hwang, D.Y.; Kim, C.S.; et al. FeSe quantum dots for in vivo multiphoton biomedical imaging. *Sci. Adv.* **2019**, *5*, eaay0044. [CrossRef] [PubMed]
104. Hsiao, M.H.; Mu, Q.; Stephen, Z.R.; Fang, C.; Zhang, M. Hexanoyl-chitosan-PEG copolymer coated iron oxide nano-particles for hydrophobic drug delivery. *ACS Macro Lett.* **2015**, *4*, 403–407. [CrossRef]
105. Chen, T.; Zhao, T.; Wei, D.; Wei, Y.; Li, Y.; Zhang, H. Core–shell nanocarriers with ZnO quantum dots-conjugated Au nanoparticle for tumor-targeted drug delivery. *Carbohydr. Polym.* **2013**, *92*, 1124–1132. [CrossRef] [PubMed]
106. Zheng, S.; Jin, Z.; Han, C.; Li, J.; Xu, H.; Park, S.; Park, J.O.; Choi, E.; Xu, K. Graphene quantum dots-decorated hol-low copper sulfide nanoparticles for controlled intracellular drug release and enhanced photothermal-chemotherapy. *J. Mater. Sci.* **2020**, *55*, 1184–1197. [CrossRef]
107. Yang, L.; Wang, Z.; Wang, J.; Jiang, W.; Jiang, X.; Bai, Z.; He, Y.; Jiang, J.; Wang, D.; Yang, L. Doxorubicin conjugated functionalizable carbon dots for nucleus targeted delivery and enhanced therapeutic efficacy. *Nanoscale* **2016**, *8*, 6801–6809. [CrossRef]
108. Ahmad, J.; Wahab, R.; Siddiqui, M.A.; Musarrat, J.; Al-Khedhairy, A.A. Zinc oxide quantum dots: A potential candi-date to detain liver cancer cells. *Bioprocess Biosyst. Eng.* **2015**, *38*, 155–163. [CrossRef] [PubMed]
109. Rahman, M.M.; Opo, F.A.; Asiri, A.M. Cytotoxicity Study of Cadmium-Selenium Quantum Dots (Cdse QDs) for De-stroying the Human HepG2 Liver Cancer Cell. *J. Biomed. Nanotechnol.* **2021**, *17*, 2153–2164. [CrossRef]
110. Fakhri, A.; Tahami, S.; Nejad, P.A. Preparation and characterization of Fe3O4-Ag2O quantum dots decorated cellu-lose nanofibers as a carrier of anticancer drugs for skin cancer. *J. Photochem. Photobiol. B Biol.* **2017**, *175*, 83–88. [CrossRef] [PubMed]
111. Chu, M.; Pan, X.; Zhang, D.; Wu, Q.; Peng, J.; Hai, W. The therapeutic efficacy of CdTe and CdSe quantum dots for photothermal cancer therapy. *Biomaterials* **2012**, *33*, 7071–7083. [CrossRef]
112. Prasad, R.; Jain, N.K.; Yadav, A.S.; Jadhav, M.; Radharani, N.N.V.; Gorain, M.; Srivastava, R. Ultrahigh Penetration and Retention of Graphene Quantum Dot Mesoporous Silica Nanohybrids for Image Guided Tumor Regression. *ACS Appl. Bio Mater.* **2021**, *4*, 1693–1703. [CrossRef]

113. Ghafary, S.M.; Rahimjazi, E.; Hamzehil, H.; Mousavi, S.M.M.; Nikkhah, M.; Hosseinkhani, S. Design and preparation of a theranostic peptideticle for targeted cancer therapy: Peptide-based codelivery of doxorubicin/curcumin and graphene quantum dots. *Nanomed. Nanotechnol. Biol. Med.* **2022**, *42*, 102544. [CrossRef] [PubMed]
114. Wang, Y.; Chen, J.; Tian, J.; Wang, G.; Luo, W.; Huang, Z.; Fan, X. Tryptophan-sorbitol based carbon quantum dots for theranostics against hepatocellular carcinoma. *J. Nanobiotechnol.* **2022**, *20*, 78. [CrossRef] [PubMed]
115. Li, Y.; Zhang, P.; Tang, W.; McHugh, K.J.; Kershaw, S.V.; Jiao, M.; Han, B. Bright, magnetic NIR-II quantum dot probe for sensitive dual-modality imaging and intensive combination therapy of cancer. *ACS Nano* **2022**, *16*, 8076–8094. [CrossRef] [PubMed]

*electronics*

MDPI

*Review*

# Recent Advances in the Characterized Identification of Mono-to-Multi-Layer Graphene and Its Biomedical Applications: A Review

Nargish Parvin †, Vineet Kumar †, Sang Woo Joo, Sang-Shin Park and Tapas Kumar Mandal *

School of Mechanical Engineering and IT, Yeungnam University, Gyeongsan-si 38541, Korea
* Correspondence: tpsmndl@yu.ac.kr
† These authors contributed equally to this work.

**Abstract:** The remarkable mechanical, electrical, and thermal capabilities of monolayer graphene make it a wonder substance. As the number of layers in graphene flakes increases to few-layer graphene (number of layers ≤ 5) and multi-layer graphene (number of layers ≤ 10), its properties are affected. In order to obtain the necessary qualities, it is crucial to manage the number of layers in the graphene flake. Therefore, in the current review, we discuss the various processes for producing mono- and few-/multi-layer graphene. The impact of mono-/few-/multi-layer graphene is then assessed with regard to its qualities (including mechanical, thermal, and optical properties). Graphene possesses unique electrical features, such as good carrier mobility, typical ambipolar behaviour, and a unique energy band structure, which might be employed in field effect transistors (FETs) and utilized in radio frequency (RF) circuits, sensors, memory, and other applications. In this review, we cover graphene's integration into devices for biomolecule detection as well as biomedical applications. The advantages of using graphene in each situation are explored, and samples of the most cutting-edge solutions for biomedical devices and other applications are documented and reviewed.

**Keywords:** 2D nanosheet; nanomaterial; layered material; monolayer; medical imaging

**Citation:** Parvin, N.; Kumar, V.; Joo, S.W.; Park, S.-S.; Mandal, T.K. Recent Advances in the Characterized Identification of Mono-to-Multi-Layer Graphene and Its Biomedical Applications: A Review. *Electronics* **2022**, *11*, 3345. https://doi.org/10.3390/electronics11203345

Academic Editor: Lucas Lamata

Received: 27 September 2022
Accepted: 14 October 2022
Published: 17 October 2022



## 1. Introduction

The sixth element is truly fascinating due to its allotropic forms [1,2]. Its allotropes are as soft as graphite [3,4] and as hard as diamond [5]. Graphite is three-dimensional structure that is made up of stacking multiple one-atom-thick layers, which are formed from strong $sp^2$ hybridized carbon atoms arranged in a hexagonal lattice [6,7]. This two-dimensional monatomic thick crystal structure consists of single atomic sheet of graphite, and is called "graphene" [8]. Recently, graphene, an allotrope of carbon, has become a hot topic of research due to its good physico-chemical properties [9,10]. When stacked, graphene forms different types of graphene, such as monolayer graphene [11], few-layer graphene [12] and multilayer graphene [13]. The properties of graphene are dependent on the number of stacked layers; for example, monolayer graphene has better properties than few-layer graphene and multi-layer graphene [14,15].

Scientists described the theoretical existence of single layer graphene more than 80 years ago [16]. Then, the practical existence of two-dimensional graphene was considered physically impossible [17]. However, in 2004, Geim et al. isolated single-sheet graphene via the scotch tape method and demonstrated its properties experimentally [18]. This was the first time researchers came to know about the remarkable properties of graphene [19]. Since then, there has been an exponential rise in studies of graphene-based materials to determine the various applications of their properties, such as in biomedical applications [20,21].

Due to the expert attention paid to graphene-based materials, high-performance materials were successfully produced. However, achieving these high performance levels involves various challenges, especially in relation to monolayer graphene [22,23]. One of these challenges is the synthesis of monolayer or few-layer graphene in bulk with high purity, which is an is extremely difficult process [24]. Other challenges include the restacking of monolayer graphene in few-layer and multi-layer graphene [25,26]. This restacking results in the decreased performance of the graphene-based devices. Therefore, the control of lateral size and aggregation states, in addition to the process of the oxidation of graphene in graphene oxide, is essential for developing graphene-based high-performance devices [27,28].

After achieving the synthesis of monolayer graphene, which exhibits good characteristics for a roadmap of graphene-based devices, the characterization of the order of stacking in graphene became a subject of interest for researchers [29,30]. These characterization techniques include atomic force microscopy (AFM) [31], Raman spectra [32], Raman mapping [33], and Transmission electron microscopy (TEM) [34]. Among these techniques, Raman spectra and TEM are methods frequently used to determine whether the synthesized material is monolayer graphene, few-layer graphene, or multi-layer graphene [32,34]. After characterization, the graphene material can be used for different applications, such as biosensors [35], tissue engineering [36], drug carriers [37], and other biomedical applications. In this review, we discuss the characterized differentiations of graphene and their particular advantages in biomedical applications based on the number of layers.

## 2. Production of Mono-, Few-, and Multi-Layer Graphene

Graphene, especially monolayer graphene, has received a great deal of attention since 2004 due to its good mechanical, electrical, and thermal properties [18]. However, the synthesis of monolayer graphene is extremely difficult and expensive [38,39]. Therefore, synthesizing few-layer graphene or multi-layer graphene is also a subject of interest [40]. A general scheme represents monolayer, few-layer, and multi-layer graphene (Figure 1). Mono-, few-, and multi-layer graphene can be synthesized using various methods, including micromechanical exfoliation [41], chemical vapour deposition [42], and chemical methods such as oxidizing graphene into graphene oxide [43] and then reducing it chemically [44] or thermally [45]. A few methods provide high-quality, large-scale few-layer and multi-layer graphene but small amounts of monolayer graphene, such as the chemical method; however, the purity and defect density remain matters of concern [46,47]. Similarly, large-scale monolayer graphene can be synthesized by chemical vapour deposition, but purity still is a topic of concern [48]. The various methods for producing mono-, few-, and multi-layer graphene are as follows.

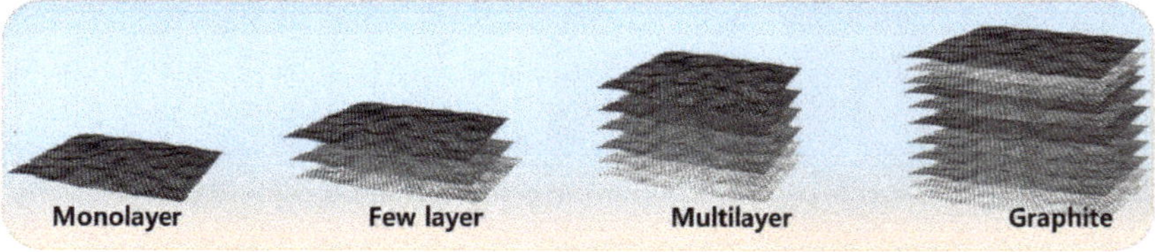


**Figure 1.** Schematic overview of monolayer, few-layer and multi-layer graphene.

### 2.1. *Synthesis of Few- to Multi-Layer Graphene*

#### 2.1.1. Exfoliation of Graphite

Graphite is an abundant material and well known for its lubricating properties [49,50]. However, it has poor mechanical, electrical, and thermal properties. Graphite can be exfoliated into different types of graphene, but we must first overcome the weak van der Waals forces which hold the different types of graphene into graphite [51]. A general scheme of the process for obtaining different layers of graphene is shown in Figure 2.

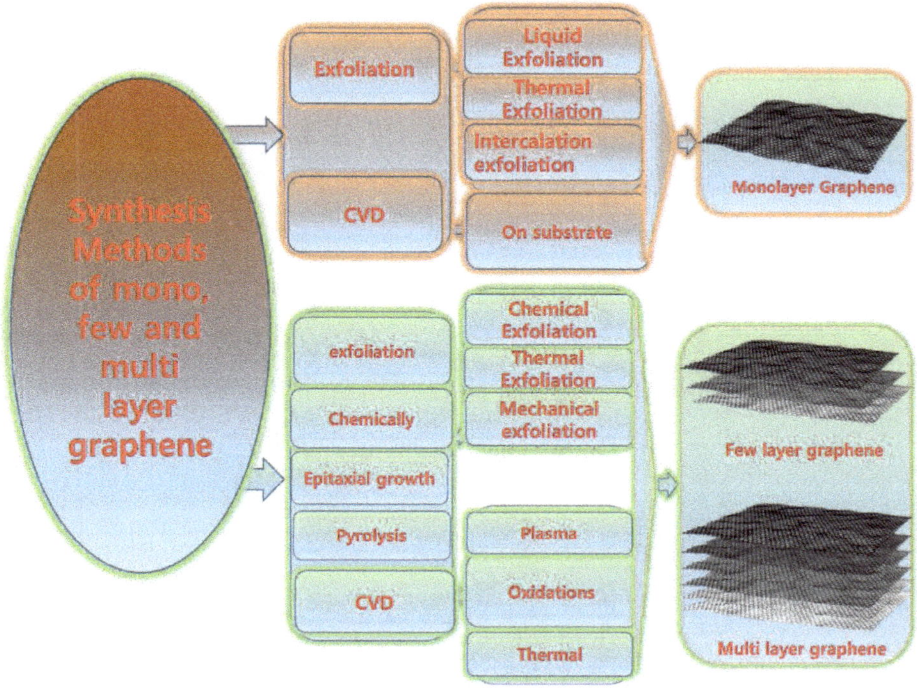


**Figure 2.** Schematic overview of the process for obtaining different layers of graphene.

(a) There are several ways to overcome these weak interactions, and the most promising among them is the sonication of graphite in different solvents; however, the yields of multi-layer and monolayer graphene [52] are very poor in this process.

(b) Another promising strategy to obtain graphene from graphite involves oxidizing graphite by various methods, such as Hummer's method, and reducing it chemically or thermally to obtain a large-scale yield [53]. However, the redox results of graphite into graphene mostly provide high yields of few-layer graphene or multi-layer graphene.

(c) Another method of producing few- and multi-layer graphene involves the exfoliation of graphite via graphite intercalation [54,55]. Different types of chemicals can be inserted to graphite interlayer space, thereby increasing the interlayer distance of adjacent graphene sheets in graphite. This phenomenon also changes the properties of graphene, since the increase in interlayer spacing affects electronic coupling between adjacent graphene sheets in graphite [56].

(d) Another way to exfoliate graphite into few- and multi-layer graphene is via ball milling [57]. This is a way of exfoliating graphite via mechanical exfoliation. Ball milling has been extensively used in the past to reduce the particle size of a material [58]. Scientists thus propose ball milling as a way to mechanically exfoliate graphite in small-size nano-graphite, increasing the mixing time to obtain few- or multi-layer graphene. Thus, ball milling is a promising technique for exfoliating graphite into graphene. The advantage of using ball milling to produce graphene is its low production cost, its easy handling, and its ability to produce graphene at large scale.

(e) The plasma synthesis method is another significant way to produce graphene with few-to-multiple layers. Microwave plasmas produced by surface waves at a stimulation frequency of 2.45 GHz and under atmospheric pressure conditions were successfully used to produce highly structured and stable self-standing graphene sheets [16]. There were also investigations into how the addition of hydrogen affects the density of the carbon precursor (C2, C) and the structural soundness of synthetic graphene sheets. Changes in the sp3/sp2 ratio and the C2 and C number densities were shown to be correlated [59]. Microwave-driven plasmas were used to control oxygen functions and the sp2/sp3 carbon ratio (~15) to a high degree [60].

#### 2.1.2. Synthesis of Monolayer Graphene

(a) In 2004, for the first time, Geim and Novoselov developed a method of synthesizing graphene using micromechanical cleavage as "scotch tape" via mechanical exfoliation [18]. This was for the first time in history that any scientist experimentally synthesized monolayer graphene. After synthesizing the monolayer graphene, these scientists further demonstrated its outstanding properties [61]. However, due to the uneven thickness of the graphene flakes and its high production costs, the mechanical exfoliation method was not suitable for the mass production of graphene that might be used to study graphene-based devices.
(b) Another method of producing monolayer graphene is the chemical vapour deposition method [62]. Monolayer graphene can be grown epitaxially on a silicon carbide substrate, and can be used for various applications, such as transistors. The size of the monolayer graphene grown depends on the size of the silicon wafer. The surface of the silicon wafer also influences the properties of the synthesized monolayer graphene.

## 3. Properties of Mono-, Few-, and Multi-Layer Graphene

The properties of graphene depend on a number of internal factors (such as the number of graphene layers stacked in a graphene flake) and external factors (such as temperature); these factors affect the final performance of devices based on graphene flakes. It is well known that monolayer graphene exhibits extremely good mechanical, thermal, and electrical properties, and that these properties decrease as the number of layers increases—such as from monolayer graphene flakes to few-layer and multi-layer graphene flakes. The sections below describe the effects of these factors on the mechanical, thermal, and optical properties of mono-, few-, and multi-layer graphene.

### *3.1. Mechanical Properties of Mono-, Few-, and Multi-Layer Graphene*

Figure 3 shows the effect of the number of layers in graphene flake on mechanical properties (such as fracture stress and fracture strain). As expected, it was found that the mechanical properties decrease as the number of layer increases from monolayer to few-layer and then to multi-layer graphene. However, these changes in mechanical properties are not significant, as shown by Zhang et al. [14]. It is clear that the fracture strain and tensile strength are not significantly affected by the transition from few-layer graphene to multi-layer graphene. This is due to the dominance of the "nano effect" of graphite as the number of layers increases from few-layer to multi-layer graphene. Zhang et al. [14] further report that the mechanical properties are significantly affected by temperature, which is an external factor. The studies show that, as the temperature increases from 300 K to 2000 K, the tensile strength decreases significantly from 125 GPa to around 43 GPa. This fall in tensile strength is due to the softening of the structures. In addition, it is well known, and was demonstrated by Zhang et at [14], that the atoms of graphene undergo severe movement at higher temperatures, leading to such reductions in the mechanical properties. Similarly, fracture strain (67%) and Young's modulus (23%) drop significantly as the temperature increases from 300 K to 2000 K [14].

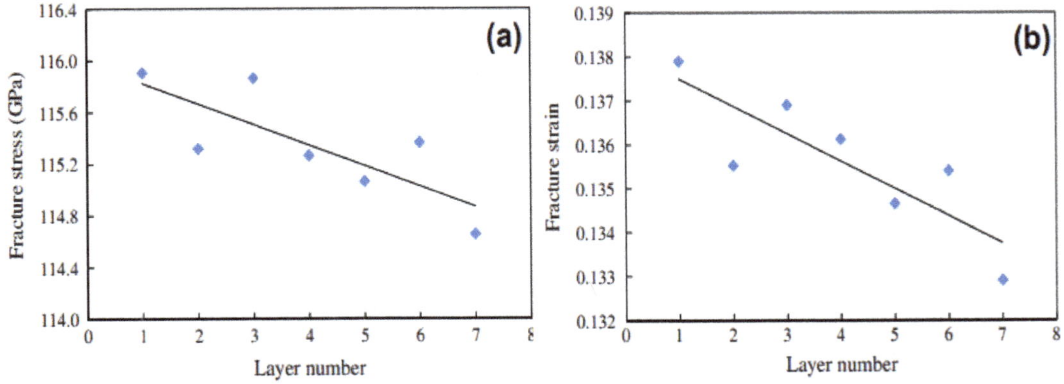


**Figure 3.** (**a**) The fracture stress and (**b**) fracture strain of mono- to few- to multi-layer graphene. Reproduced from [14] with the permission of Elsevier.

### *3.2. Thermal Properties of Mono-, Few-, and Multi-Layer Graphene*

The coefficient of thermal expansion (CTE) was studied as a function of temperature and the number of layers stacked in graphene flakes. It was found and demonstrated by Mag-isa et al. [15] that thermal properties such as CTE are highly dependent on internal factors (the number of layers in the graphene flake) and external factors (temperature) (Figure 4). For example, it was noted that, as the temperature increases from 20 °C to 140 °C, the COE is affected. This was expected, due to increase in the thermal kinetics of the processes as the temperature increases. Similarly, the thermal properties are affected by increasing the number of layers, as the thermal properties of monolayer graphene differ from those of few- and multi-layer graphene [12,15]. It is interesting to note that the results for few-layer graphene are placed between those of monolayer graphene and multi-layer graphene, as expected. Moreover, all of the values of CTE studies reported by Mag-isa et al. [15] and other researchers are negative. A negative CTE is generally regarded as a non-close-packed system with directional interactions, such as ice or graphene.

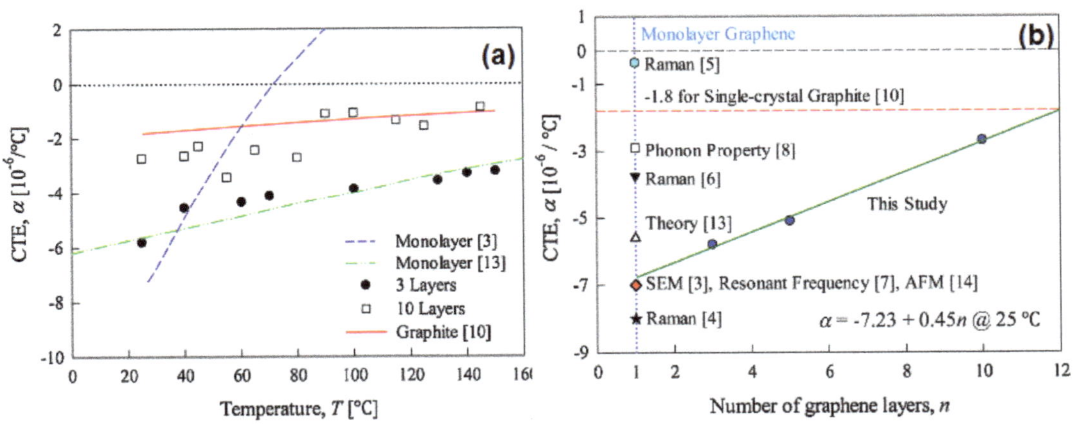


**Figure 4.** The thermal properties of the graphene-based flakes: (**a**) coefficient of thermal expansion against temperature; (**b**) coefficient of thermal expansion against mono-, few-, and multi-layer graphene. Reproduced from [15] with the permission of Elsevier.

### 3.3. Optical Properties of Mono-, Few-, and Multi-Layer Graphene

By taking into account the surface conductivity ($\sigma$), the refractive index ($n$) and the extinction coefficient ($k$) of graphene can be determined. Knowledge of the frequency-dependent surface conductivity sG, which can be determined from either a microscopic model or from experiments, is necessary for understanding the optical characteristics of graphene [63]. The surface conductivity ($\sigma_G$) of graphene is dictated by the high-frequency expression acquired from the Kubo model [64].

## 4. Determination of Mono-, Few-, and Multi-Layer Graphene

The determination of the number of graphene layers and defects is critical for tailoring its properties for intended applications, and involves microscopic as well as spectroscopic measurements.

### 4.1. Atomic Force Microscopy (AFM)

AFM is defined as high resolution microscopy with a resolution at the atomic scale. AFM can determine various features of a material, such as thickness, grain height, topological features, phase diagrams, and roughness. These features are then correlated with the properties and target applications of the material, such as graphene. Xu et al. [65] studied and employed the AFM technique to determine thickness of graphene flake. The corresponding histogram shows that the flake thickness ranges from 1.1 to 1.6 nm. By considering an interlayer distance 0.33 nm, a total of 4–5 layers can be estimated to be present in the graphene flake. It therefore falls into the category of few-layer graphene [65]. Chen et al. [66] demonstrated the use of AFM for determining the roughness of graphene flakes. The authors found that the pristine polymer forms a homogenous film with a mean roughness in the range of 2.2 nm. As the graphene is added to the polymer, the composites made of polymer-graphene also exhibit good film-formation properties, with a mean roughness of 3.1 nm [66]. Zhang et al. [67] also studied the early stages of atomic layer deposition (ALD) on epitaxial graphene (EG) via AFM (Figure 5). It can be seen that the EG-based sample shows flat topological features. The small surface roughness and very narrow distribution of height values can also be noticed. Fitted with a single, the Gaussian peak shows a full-width half-maximum (FWHM) of 0.2 nm. Similarly, different topological images of the specimen were shown as a function of different ALD cycles, increasing from 10 to 80 [68]. No significant differences can be seen in the topology images; the height values and Gaussian fit of FWHM remained practically the same as the number of ALD cycles increases. Gao et al. [69] also studied the thickness of graphene flakes with the help of the AFM technique. The authors state that the thickness of the graphene flake used in the work was in the range of 0.4–1 nm. This shows that the graphene flake was monolayer-to-few-layer graphene.

Yuan et al. [70] investigated the thickness of the graphene flake with the help of the AFM tool. The authors demonstrated the sheet-like topological feature of the graphene flake. They further investigated the flake thickness and its lateral dimension. It was found from the histogram that the graphene flake consisted mostly of monolayer graphene, while a low percentage of few-layer graphene and multi-layer graphene was witnessed. The lateral dimension was in the range of 2 to 10 μm, and therefore had very high shape anisotropy [70]. Finally, Bhuiyan et al. [71] investigated the graphene flake thickness with the help of AFM microscopy. From this study, the authors demonstrate the smooth, sheet-like topology of the graphene flake, which had a thickness of around 1 nm; this shows that the graphene flake used in the work is few-layer graphene [71].

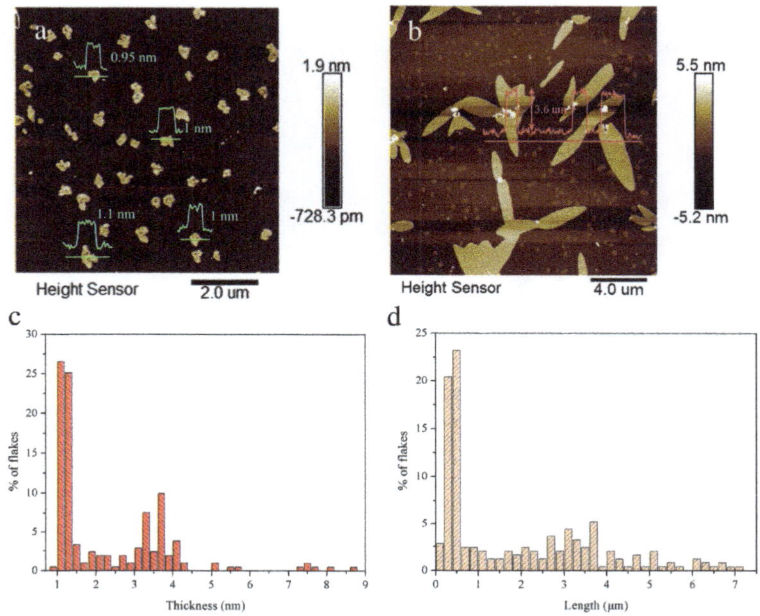


**Figure 5.** (**a**) AFM images of monolayer graphene flakes with adjacent height profiles; (**b**) different types of few-layer graphene flakes with adjacent height profiles; (**c**) the adjacent thickness of the flakes; (**d**) the length distribution of different flakes analyzed through AFM microscopy. Reproduced from [67]. [CC-BY].

### 4.2. *Transmission Electron Microscopy (TEM)*

The TEM technique involves the transmission of a beam of electrons through a sample that is usually 100 nm thick and suspended on a grid. This technique gives an image of atomic-scale resolution and can be used in determining the characteristics of mono-, few-, and multi-layer graphene.

Araby et al. [72] demonstrated the use of TEM in determining the graphene thickness via high resolution TEM images (Figure 6). The images show that the graphene flake is few-layer graphene, the type of graphene obtained using an indium catalyst at 150 °C. On the other hand, Long et al. [73] examined the morphology of graphene flakes via TEM and HRTEM. The TEM images show the wrinkled morphology of the graphene flakes, indicating a stable structure. The HRTEM images of the graphene flakes show the stacking morphology of borophene on the reduced graphene oxide. Moreover, the investigation shows a good interaction between borophene and the reduced graphene oxide [73]. Navik et al. [74] investigated the morphology of few-to-multi-layer graphene via HRTEM. This technique shows that, in the graphene flakes, the distribution of few-layer graphene is around 70% and the remaining 30% belongs to multi-layer graphene. The HRTEM images provided by the authors show exfoliated graphene sheets in which the exfoliation process produces large transparent graphene flakes, which mostly belong to few-layer graphene [74]. Ding et al. [75] studied the number of graphene layers per graphene flake via HRTEM. The authors demonstrated that the graphene flake used in the work contains mono- and bi-layer graphene. Thus, a high degree of exfoliation was achieved in the work and the remarkable properties of the related materials were studied.

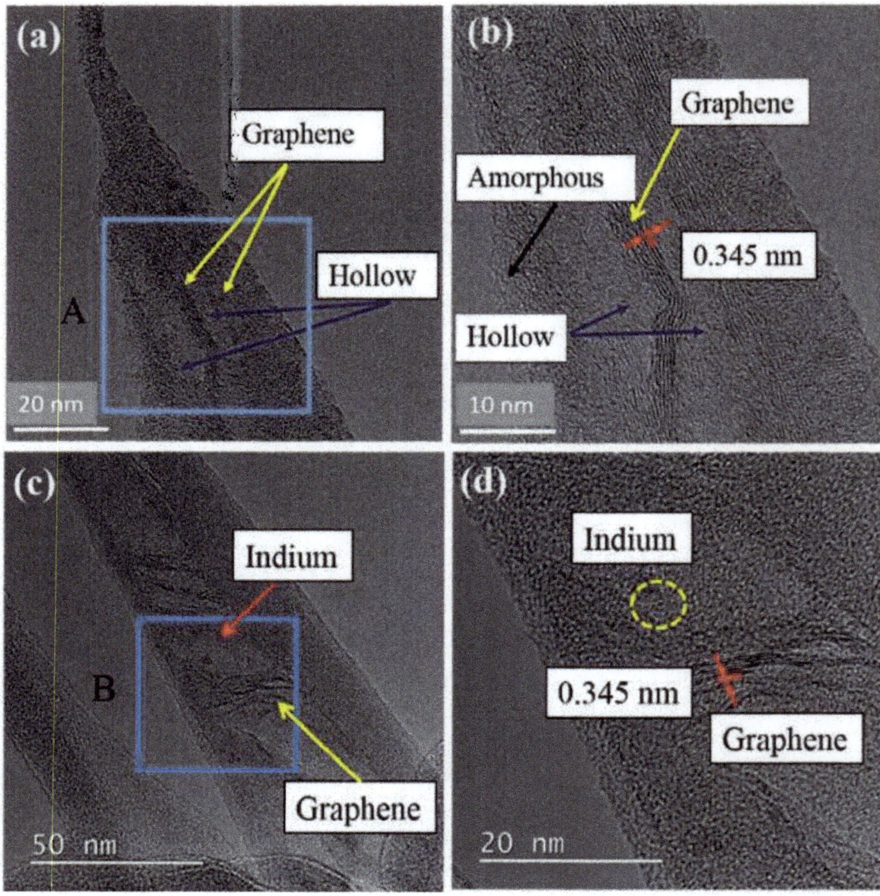


**Figure 6.** (**a–d**) Typical TEMs of graphene nano flakes in different forms, synthesized by an indium catalyst at 150 °C. Reproduced from [72] (CC BY).

### 4.3. Raman Spectra

Raman spectra (named after Sir C.V. Raman from India) is a famous spectroscopy technique that provides information about the vibrational modes of molecules, etc. It is most prominently used in determining the number of graphene layers in a graphene flake. The graphene flake generally shows three characteristic bands, namely, the D-band, G-band, and 2D band [76]. The position of the 2D band is different for different types of graphene flake. It is around 2702 $cm^{-1}$ for tri-layer graphene, and increases with the increase in the number of layers to as much as 2720 $cm^{-1}$ for multi-layer graphene and 2725 $cm^{-1}$ for bulk graphite [14]. Campanelli et al. [77] used Raman spectra to demonstrate the interaction of $Yb^{3+}$ with monolayer and bi-layer graphene. Their paper demonstrates that there is a good interaction between $Yb^{3+}$ and graphene flakes, as confirmed using Raman spectra. A shift of ~5 $cm^{-1}$ to ~7 $cm^{-1}$ was observed at different Raman bands in the presence of $Yb^{3+}$, indicating a good interaction [77]. Wang et al. [78] demonstrated the use of the Raman spectra technique for identifying monolayer and bi-layer graphene. The authors reported on systematic and statistical tests of the signatures of the so-called buffer layer and its coupling with the epitaxial graphene layer via Raman spectroscopy. The findings show a coupling between graphene and buffer layer [78]. Silva et al. [79] demonstrated the different number of graphene flakes with mono-, few-, and multi-layer graphene via

Raman spectra (Figure 7). A statistical analysis of the different types of graphene flake with mono-, few-, and multi-layer graphene was presented. The protocol is based on the position of the 2D band, which is different for different type of graphene layers, as discussed above. The authors demonstrated that the Raman spectra is suitable for the statistical analysis of mono-, few-, and multi-layer graphene with relatively low costs [79].

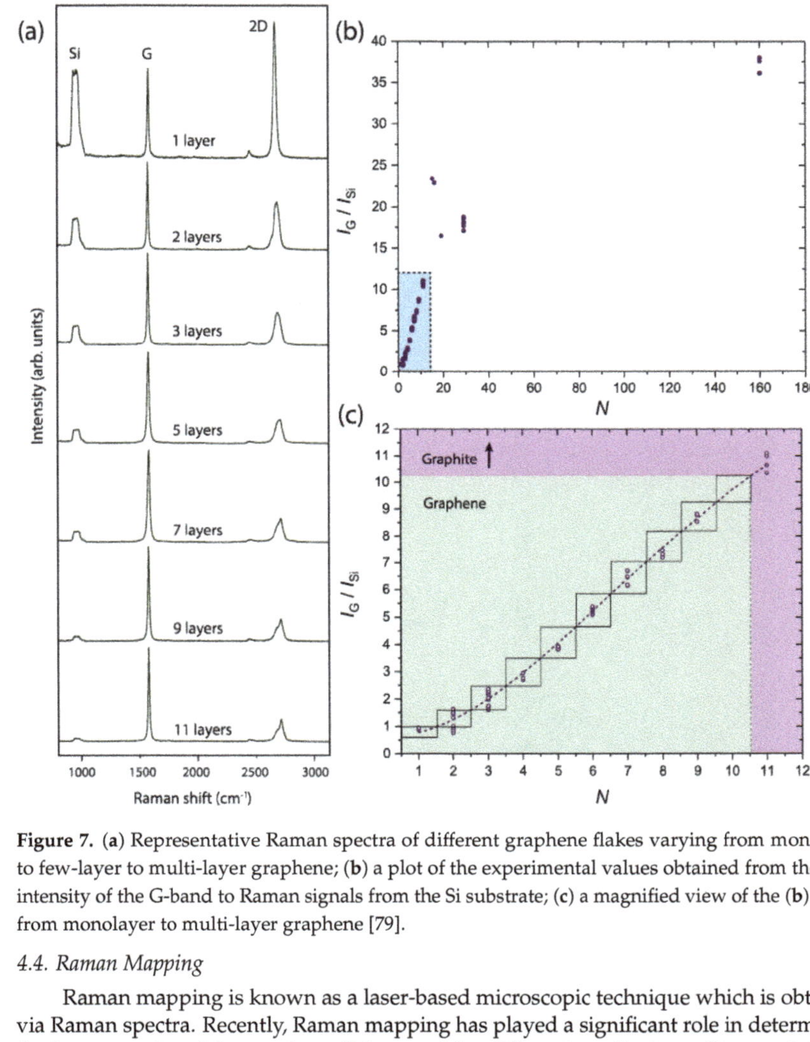


**Figure 7.** (**a**) Representative Raman spectra of different graphene flakes varying from monolayer to few-layer to multi-layer graphene; (**b**) a plot of the experimental values obtained from the ratio intensity of the G-band to Raman signals from the Si substrate; (**c**) a magnified view of the (**b**) graph from monolayer to multi-layer graphene [79].

### *4.4. Raman Mapping*

Raman mapping is known as a laser-based microscopic technique which is obtained via Raman spectra. Recently, Raman mapping has played a significant role in determining the homogeneity of the graphene flakes used for different applications. Huang et al. [80] studied monolayer graphene through Raman mapping at different temperatures of over 150 °C and 250 °C. The Raman signatures indicate the formation of defects in graphene, even at low temperatures. The Raman maps show that the ratio of intensity of the D to G bands was 1.85 and 4.01 for 150 °C and 250 °C, respectively, indicating higher defect density for the 250 °C sample [80]. Bouhafs et al. showed [81] the synthesis and characterization of large-area few-layer graphene by chemical vapor deposition. The Raman map used in this study shows clear ABA and ABC stacking in the studied few-layer graphene. Bleu et al. [82] studied few-layer graphene via Raman spectra and Raman mapping. The most significant peaks for the few-layer graphene studied through Raman spectra are the D, G, and 2D bands that are located around 1350, 1580, and 2700 $cm^{-1}$, respectively (Figure 8).

In addition, the intensity ratio of the D to G band can be used to estimate the defect density, and the intensity ratio of 2D to G reveals the number of graphene layers stacked. The histogram was estimated via Raman maps and few-layer graphene was found to be dominant in the sample [82]. Moreover, the Raman maps show the good homogeneity of the synthesized few-layer graphene, with low defect density.

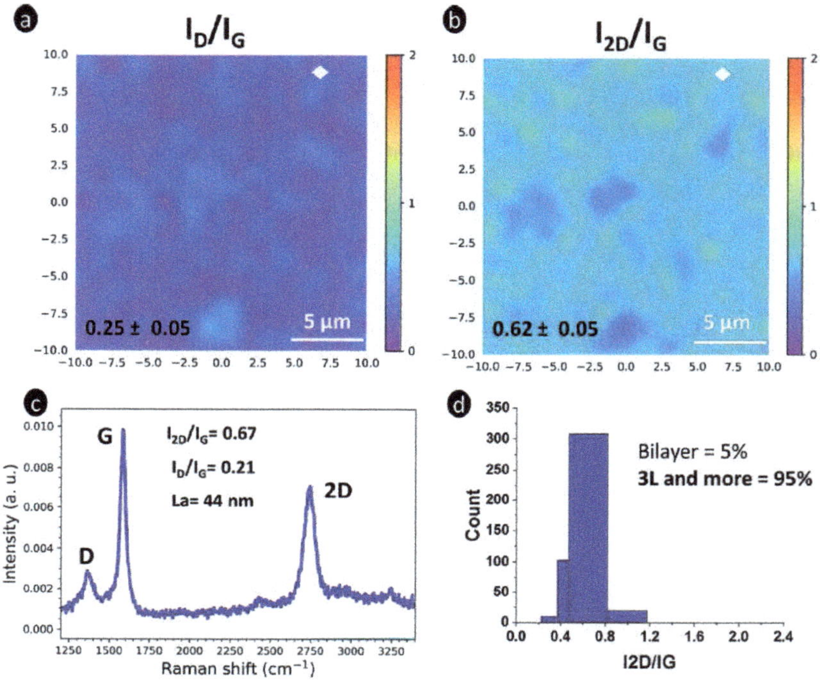


**Figure 8.** Raman analysis of the synthesized few-layer graphene; (**a**) a Raman map of the ratio of intensity of the D-band to the G-band; (**b**) a Raman map of ratio of the intensity of the 2D band to the G-band; (**c**) Raman spectra of few-layer graphene; (**d**) the distribution of the number of graphene layers in few-layer graphene. Reproduced from [82] with the permission of Elsevier.

## 5. Biomedical Devices and Other Biomedical Applications

### *5.1. Use as a Biomedical Device*

The anatomical characteristics of different layers of graphene are different. Mono-to-multi-layer graphene is highly sensitive and is used in sensor applications, including biomolecules, elements, gas, pressure, and electrochemical detection [83,84]. Due to its large surface area, electrical conductivity, biocompatibility, easy operation, tunable structural transformation, and excellent mechanical strength, layered graphene is used as a biomedical device for regenerative drugs [85,86], bio-mining [87] detection [88], and cancer cures [89].

On the other hand, there is a basic restriction on the use of monolayer graphene in device applications: that is, the light–graphene interaction is weak due to the atomic thickness of graphene. In particular, graphene exhibits a wavelength-independent absorption of 2.3% in the visible and near-infrared regions, which limit its further application in areas such as sensors and photodetectors [90]. In recent years, several approaches have been proposed to solve these problems, such as the angle-insensitive broadband absorption enhancement of graphene [91], and electrically modulating and switching the infrared absorption of monolayer graphene [92].

## 5.2. *Use as a Sensor Device*

Additionally, biomolecules with hydrophobic domains or systems, such as DNA strands or proteins, have a tendency to naturally adsorb on graphene [93]. Another simple way to modify graphene for the targeted immobilization of bio-receptor units is to functionalize it [94].

### 5.2.1. Nucleic Acids Sensor Device

Compared with devices made using bilayer and few-layer graphene, sensor devices made with functionalized monolayer graphene exhibit superior sensitivity, better repeatability, and stronger construction [95]. Monolayer-graphene-based FET devices are used as nucleic acid sensors with high selectivity and low detection limits [96,97]. In this device, the fixed graphene layers are determined by using 1-pyrenebutanoic acid succinimidyl ester (PBASE) as a link chemical and acquired a 10 pM detection limit (Figure 9). To recognize pesticides, more GFET detectors based on microfluidics were created [98].

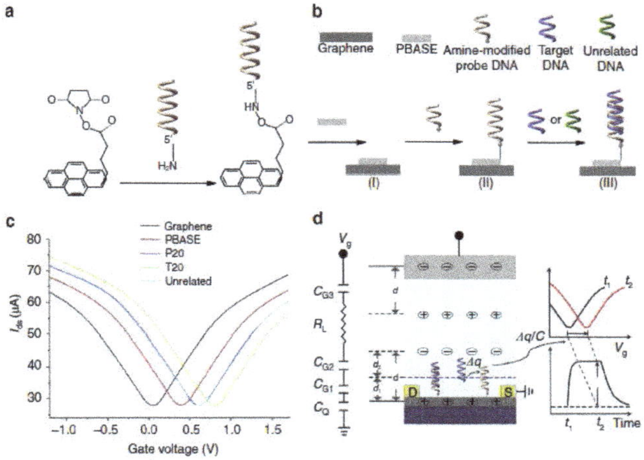


**Figure 9.** Characteristics of G-FETs. (**a**) The relationship between PBASE's molecular structure and the probe DNA. (**b**) PBASE is used to functionalize the graphene surface. A probe DNA (shown in orange) is immobilized by interacting with the PBASE. The probe DNA is then hybridized with the target DNA (shown in blue) and unrelated control DNAs (in green). (**c**) Graphs of the source drain current versus the constant source drain voltage (0.1 V) for different conditions of graphene. (**d**) A schematic showing the equivalent circuit made up of four parallel plate capacitors and a resistance (RL) linked in series and the sensing model of a G-FET. Reproduced from [97] (CC BY).

### 5.2.2. Mammalian Cell Sensor Device

For cell-sensing applications, layered graphene is used as an immobilizer for detecting cells or various pathogens. Recently, scientists developed few-layer graphene-based FETs with HER2-specific aptamers and the ability to detect very low concentrations of HER2-cells [98]. Moreover, the devices exhibited highly sensitive detection against breast cancer cells and SK-BK-3 cells at the single-cell level.

### 5.2.3. Monolayer Graphene and Immune Sensor Devices

The surface modifications of functionalized specific polymers improve the interfacial interactions between them when used as high-performing medical devices [99,100]. As evidence, when the monolayers of graphene are mixed with a bioactive immobilizer (biotinylated cholera toxin), they showed a remarkable response to the anti-CT test [101]. Additionally, pyrene-NTA-conjugated graphene-monolayer-based sensor devices can detect up to 4 $pL^{-1}$ of anti-CT, which is also a very significant result for anti-CT approaches [102].

A high-sensitivity surface plasmon resonance (SPR) biosensor made of a monolayer of graphene, a four-layer MoS2, and a layer of gold was studied [103]. The maximal sensitivity of the sensor, 282°/RIU, was shown by computational models using the transfer matrix method (TMM), which was about two times more sensitive than the typical Au-based SPR sensor. Additionally, the Kretschmann-configured manufactured sensors were employed to detect okadaic acid (OA). To create a competitive inhibitory immunoassay, the okadaic acid-bovine serum albumin bioconjugate (OA-BSA) was mounted on the graphene layer of the sensors. According to the findings, the sensor had a very low sensitivity limit (LOD) for OA of 1.18 ng/mL, which is roughly 22.6 times lower than that of a traditional Au immunoassay. Figure 10 depicts the schematic diagram of the OA detection process (Figure 10).

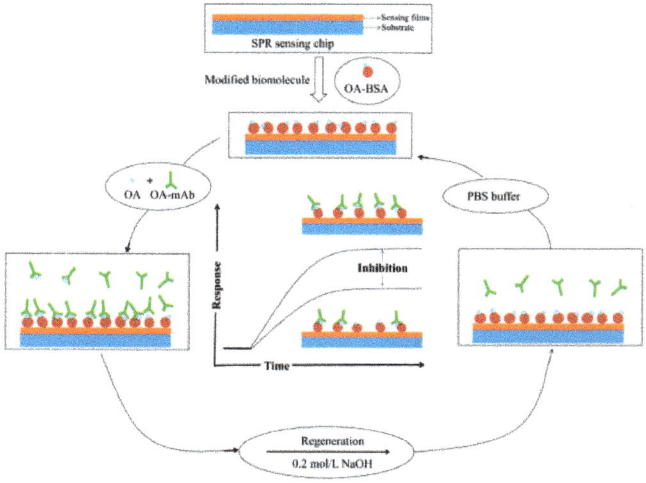


**Figure 10.** The flowchart for the method of detecting okadaic acid (OA). Reproduced from [103] (CC BY).

5.2.4. The Development of Imaging Devices for Biomedical Purposes

Bio-imaging is an important aspect of regenerative medicine [104]. Imaging techniques use different layers of graphene (a single layer to multiple layers) for imaging contrast agents such as fluorescence/confocal imaging [105,106], Forster resonance energy transfer (FRET) imaging [106], coherent anti-Stokes Raman scattering imaging (CARS) [107–109], magnetic resonance imaging (MRI) [110], surface-enhanced Raman scattering (SERS) [111], ultrasound imaging [112,113], photoacoustic imaging [114], electron paramagnetic resonance imaging (EPRI) [115,116], and positron-emission tomography (PET) [117,118]. Among these techniques, a few, such as computer tomography (CT), PET, MRI, and ultrasound imaging, are well-established and for use in humans. These imaging techniques used to better understand and monitor the affected site of the body for therapeutic purposes. Additionally, there are two fundamental necessities for bio-imaging: (1) rapid, selective, and sensitive instruments, and (2) good, non-toxic, biocompatible, biodegradable imaging contrast agents that have good bio clearance and are capable of crossing the physiological body membrane barrier.

Currently, the most developed EPRI technique is used to detect and quantify various physiological parameters in a cancer-effected microenvironment in in vivo models, such as pO2 [119–121], pH [116,122], and oxidation situations [116,123]. Normally, for this method, TAM (triphenylmethyl radical) [124] or other compound chemicals [125] are used. In EPRI for localized oxygenation sensing, carbon-based ink is used [126]. CARS imaging technology is highly sensitive in providing signals to process images in in vivo, ex vivo, and in cell cultures, up to micrometer-level penetrations [127,128].

Mono-to-multi-layered graphene-based materials are used in the biomedical field as contrast agents [129,130]. The key benefit of this is that the processes are biocompatible, less toxic, and metal-free [131]. Mono-to-multi-layered graphene-based materials are mainly used in fluorescence imaging [132] where they are used in modified form, display layer, and surface-chemistry-dependent fluorescence [133]. To enhance the biocompatibility of mono-to-multi-layered graphene, a few common polymers are used, such as polyethylene mine [134], PEG (polyethylene glycol) [135–138], polystyrene [139], and polypeptides [140]. Recently, scientists shortened the varied imaging techniques of different layers of graphene used, and determined that various imaging techniques can be performed on one platform [141]. Therefore, we assume that, in the coming decade, different layers of graphene-based materials and their diverse morphology will play a significant role in biomedical applications and in determining the future of therapeutic approaches.

### *5.3. Use in Other Biomedical Applications*

#### 5.3.1. As a Delivery Carrier and in Treatment

##### Ligand-Based Drug Delivery

Most interferences with drug-delivery systems are due to the presence of many proteolytic enzymes in the cytoplasm of a cell. Graphene and its derivatives are used as drug carriers or carriers that bypass those interfering enzymes. Its properties make it a promising carrier for biomedical applications; for example, COOH and OH are the main group ligands of GO that are acceptable for binding to various organic molecules, such as proteins, DNA, and many polymers. This ligand makes it biocompatible, cell selectable, and efficient. Some reported polymer binding ligands are used to deliver anti-cancer drugs such as doxorubicin (DOX) [142]. Other reports noted that the drug Ibuprofen has been delivered using chitosan-containing GO [138].

Graphene's unmodified basal plane site with open surface $\pi$ electrons is hydrophobic, creating $\pi$-$\pi$ interactions for loading drugs [143,144]. GO has been transformed into a carrier for the supply of water-soluble cancer drugs. The solubility of physiological and aqueous solutions can be increased by functional NGO with PEG [145]. Figure 11 shows a scheme of monolayer, few-layer, and multi-layer graphene applications.

**Figure 11.** Cartoon illustrating the applications of monolayer, few-layer, and multi-layer graphene.

Stimuli Response Delivery

Some drug delivery occurs due to changes in temperature, pH, light, and salt concentrations, whereby the polymer ligand can detect the particular physiological change and deliver the drug. A few reports show that the biopolymers bound with graphene materials are capable of pH-responsive drug delivery or of being used as transporters [146]. For example, DOX is an anticancer drug that is easily released from the GO-Dox complex in a low pH body environment due to its high solubility at a low pH or in acidic conditions [147]. On the other hand, pH-sensitive delivery has been shown to be effective in the treatment of tumors using folic acid with DOX and camptothecin [148]. A photochemically regulated gene-delivery carrier was also reported where the low molecular weight PEI and rGO were combined with polyethylene glycol (PEG) and delivered the plasmid DNA by physiochemical assays [149].

Gene Delivery

Gene delivery is a good alternative way to cure many genetic diseases, as genes can be delivered to a cell through a vehicle or carrier. Many scientific works have reported gene delivery through GO or GO-based materials. To prepare for gene delivery, polyehylenimine (PEI) was adhered to the surface of the GO sheet via covalent coupling and electrostatic contact with stacking plasmid DNA (PDNA) [150]. The long chain [138] and the branching chain are parallel to the GO [151]. In contrast to other complex compounds, PEI has high transfection effectiveness and low cytotoxicity. Additionally, chitosan complex GO has been used for the delivery of anti-cancer medication drugs and plasmid DNA [152].

5.3.2. Therapeutic Applications

Many results report that the different layers of graphene or graphene oxide can inhibit the growth of tumor spheres in various cell lines, including breast, lung, ovarian, pancreatic cancers, and glioblastoma [153]. Few-layer GO not only inhibits tumor development when injected into mice as an immunocompetent containing CT26 colon cancer cells [154], but also inhibits autophagy [155], immunity, and cell death [156]. In xenograft models, the conjugate PEG-GO exhibits malignancy and thus has a high therapeutic value [157]. An example of this is an rGO/iron oxide NP coated with PEG (rGO-IONP-PEG), which can be used for photothermal treatment (PTT) in vivo and as a triple-mode sensor for fluorescence, photoacoustic tomography (PAT), and magnetic resonance imaging (MRI) [158,159]. Another example is fluorinated GO, which can be used as a magnetically sensitive drug carrier and can be imaged via MRI and photoacoustic tomography [160].

5.3.3. Tissue Engineering for Cell Growth

Monolayer Graphene and Scaffold Formation for Fibroblasts

In a recent study, monolayer graphene's cytotoxicity and usefulness as a scaffold for murine fibroblast L929 cell lines were assessed. The authors demonstrated the impact of monolayer graphene toxicity through tests on cell viability, morphology, cytoskeleton architecture (microfilaments and microtubules), cell adhesion, and migration. They discovered that the fibroblasts grown on a monolayer of graphene showed modifications in cell attachment, motility, and cytoskeleton organization [161]. Within 24 h of culture, monolayer graphene was discovered to show no cytotoxicity toward L929 fibroblasts and to improve cell growth and adhesion. Additionally, as shown by the results, the monolayer graphene aided in the migration of cells. In the end, it was shown that monolayer graphene is not hazardous to murine subcutaneous connective tissue fibroblasts, and may even help injured tissues heal [161].

Few-Layer and Multi-Layer Graphene and Scaffold Formation

Graphene with different numbers of layers has different effects on different cell line cultures. A mammalian cell culture was tested using a few-layer graphene sheet platform. They described an investigation on the behavior of NIH-3T3 fibroblasts on several carbon nano-materials layers, including carbon nanotubes, RGO, and GEO. Layers treated with carbon nanomaterials demonstrated excellent compatibility and improved gene transfection efficiency [162]. When compared to conventional polystyrene tissue culture plates, graphene-chitosan hybrid films have demonstrated positive outcomes in tissue engineering to repair and enhance tissue function [163].

Differentiation of Stem Cells

Stem cells are important for living organisms due to their continuous regeneration powers and ability to convert into any kind of cell type where required. In regenerative medicine, stem cells are used as a therapeutic agent. Different layers of graphene are used in stem cell cultures for different purposes. The different layers of graphene showed various types of toxicity effects, electrical conductivity behaviors, biocompatibility, etc. A recent study showed that human mesenchymal stem cells (hMSCs) could differentiate into neurons very well on a surface made of few-layer graphene because the graphene served as an electrically coupled cell-adhesion layer for the hMSCs [164]. In another study, multi-layer graphene also improved the osteogenic differentiation to a greater extent than the general growth factor [165]. A comparative study found that osteogenic differentiation was stimulated but adipogenic differentiation was impeded on graphene as compared to the GO. This occurred due to the different surface characteristics of different layers of graphene or graphene derivatives which come from chemical functionalization [166,167].

#### 5.3.4. Anti-Microbial Effects

The different layers of graphene and their derivatives showed various diversified antimicrobial effects on both Gram-positive and Gram-negative bacteria [168,169]. Few-layer graphene oxide and reduced graphene oxide both showed antibacterial effects on their surfaces [168]. Meanwhile, some reports showed that the bacterial growth was intensified, rather than inhibited, on the monolayer graphene surface [170]. The monolayer GO is more effective than the few-layer and multi-layer GOs because monolayer GO is more efficient in charge transfer with bacterial cells [169]. This happens due to oxidative stress induced by membrane disruption [170,171].

#### 5.3.5. Digital and Analog Devices and Equipment

Graphene is frequently employed by researchers in the field of digital and analog device and equipment. A novel electro-optical encoder based on a graphene–$Al_2O_3$ multi-layer stack has been designed. It was demonstrated that employing various chemical opportunities for graphene yields the desired encoding operation for the intended structure [172]. In a different study, a switch made of graphene was proposed as a solution to power and footprint issues. The structure may be a good contender for PIC uses, since adding graphene monolayers improves the interaction between light and the material, and lowers the cutoff intensity for the switching function [173]. By altering the chemical prospects of graphene, a graphene-based plasmonic waveguide was created and studied for switching functions at terahertz frequencies [174]. A graphene layer and a silicon ridge were added above and below the channel, respectively, to create a new plasmonic channel. Low energy consumption and the opportunity for on-chip integration are benefits of the sub-wavelength scale [175]. The graphene waveguides are encased between two $SnO_2$ layers in a small, graphene-based plasmonic D flip-flop, which is shown. The flip-flop's compact area is a crucial component for use in optical integrated circuits [176].

## 6. Conclusions, Current Trends, and Future Prospects

Since 2004, when its remarkable properties were first demonstrated, graphene has attracted a good deal of attention from scientists around the world. These properties change as monolayer graphene is restacked to form few-layer graphene, and few-layer graphene is stacked to form multi-layer graphene. However, the outstanding properties are limited to monolayer graphene, while few-layer and multi-layer graphene only show the "nano effect" of graphite. However, the frequent use of monolayer graphene is difficult due to the challenges of synthesizing it in bulk and its processing costs. Scientists have come up with different approaches to synthesize graphene, such as the exfoliation of graphite into graphene or the synthesis of graphene from chemical vapor deposition. Chemical vapor deposition is the most suitable technique for synthesizing monolayer graphene, while the exfoliation of graphite can be used to synthesize few-layer and multi-layer graphene. A significant advancement in the characterization of graphene flakes has been witnessed, and various methods for characterizing graphene have been proposed by scientists. Among them, AFM, TEM, and Raman spectra have been found to be the most promising.

The future of graphene is bright. For the last two decades, researchers around the world have demonstrated a great interest in graphene. The increasing number of publications on graphene are evidence of the attention being paid to graphene. In particular, monolayer graphene is of key interest due to its outstanding mechanical, electrical, and thermal properties. Synthesizing monolayer graphene in bulk remains a key challenge, and new methods for synthesizing graphene will be explored in the coming decade.

Its remarkable and easy modification facilities, which include surface alterations using any type of required ligands, render differently layered 2D graphene materials very interesting for biomedical purposes. Monolayer graphene and few-layer graphene have no toxicity, higher biocompatibility, and are highly capable of strong UV absorption, SERS, and fluorescence and fluorescence quenching; this makes them some of the most powerful nanomaterials for biosensors, therapeutics, and tissue engineering, as well as for biomedical electronic devices. Still, a few challenges remain, even though the different layers of graphene show promising results concerning their functionalization and optimizations by fractionation based on the number of layers, size, morphology, and chemical functionalities. Moreover, there are many incomparable benefits and still many opportunities to explore the interesting characteristics of these materials and their possible uses. We presume that interdisciplinary efforts with chemistry, biology, and engineering will stimulate the automatic recognition of graphene-layer-based platforms for biomedical applications, and will lead to various successful and innovative applications. In addition, we conclude that different layers of graphene-based materials and their diverse morphology will perform a significant function in the subsequent decade in biomedical applications, and will shape the future of therapeutics.

**Author Contributions:** N.P., V.K. and T.K.M.: conceptualization, validation, writing—original draft preparation; T.K.M.: supervision; S.-S.P. and S.W.J.: validation. All authors have read and agreed to the published version of the manuscript.

**Funding:** This research received no external funding.

**Conflicts of Interest:** The authors declare no conflict of interest.

## References

1. Dong, H.; Zhang, Z.; Feng, Z.; Kang, J.; Wu, D.; Wang, Q.; Li, J.; Su, R. Origins of low lattice thermal conductivity in 2D carbon allotropes. *J. Mater. Res. Technol.* **2021**, *11*, 1982–1990. [CrossRef]
2. Liu, W.-D.; Yu, Y.; Dargusch, M.; Liu, Q.; Chen, Z.-G. Carbon allotrope hybrids advance thermoelectric development and applications. *Renew. Sustain. Energy Rev.* **2021**, *141*, 110800. [CrossRef]
3. Mailian, A.; Panosyan, Z.; Yengibaryan, Y.; Mailian, M. Identification of Carbon Allotropes in Tribolayers Obtained by Rubbing of Graphite. *Mater. Today Proc.* **2017**, *4*, 6842–6848. [CrossRef]
4. Kabir, H.; Zhu, H.; May, J.; Hamal, K.; Kan, Y.; Williams, T.; Echeverria, E.; McIlroy, D.N.; Estrada, D.; Davis, P.H.; et al. The sp2-sp3 carbon hybridization content of nanocrystalline graphite from pyrolyzed vegetable oil, comparison of electrochemistry and physical properties with other carbon forms and allotropes. *Carbon* **2019**, *144*, 831–840. [CrossRef]
5. Liu, Y.; Jiang, X.; Fu, J.; Zhao, J. New metallic carbon: Three dimensionally carbon allotropes comprising ultrathin diamond nanostripes. *Carbon* **2018**, *126*, 601–610. [CrossRef]
6. Zhang, X.; Schneider, R.; Müller, E.; Gerthsen, D. Practical aspects of the quantification of sp2-hybridized carbon atoms in diamond-like carbon by electron energy loss spectroscopy. *Carbon* **2016**, *102*, 198–207. [CrossRef]
7. Theye, M.-L.; Paret, V. Spatial organization of the sp2-hybridized carbon atoms and electronic density of states of hydrogenated amorphous carbon films. *Carbon* **2002**, *40*, 1153–1166. [CrossRef]

8. Mandal, T.K.; Hou, Y.; Gao, Z.Y.; Ning, H.; Yang, W.S.; Gao, M.Y. Graphene oxide-based sensor for ultrasensitive visual detection of fluoride. *Adv. Sci.* **2016**, *3*, 1600217. [CrossRef]
9. Bin Hamid, M.A.; Chan, K.T.; Raymond Ooi, C.H.; Zainuddin, H.; Mohd Shah, N.; Shahrol Nidzam, N.N. Structural stability and electronic properties of graphene/germanene heterobilayer. *Results Phys.* **2021**, *28*, 104545. [CrossRef]
10. Zhang, H.; Zhang, B.; Gao, Q.; Song, J.; Han, G. A review on microstructures and properties of graphene-reinforced aluminum matrix composites fabricated by friction stir processing. *J. Manuf. Process.* **2021**, *68*, 126–135. [CrossRef]
11. Wu, Y.-C.; Shao, J.-L.; Zhan, H. Damage and self-healing characteristics of monolayer graphene enhanced Cu under ballistic impact. *Mech. Mater.* **2021**, *155*, 103736. [CrossRef]
12. Graf, D.; Molitor, F.; Ensslin, K.; Stampfer, C.; Jungen, A.; Hierold, C.; Wirtz, L. Spatially Resolved Raman Spectroscopy of Single- and Few-Layer Graphene. *Nano Lett.* **2007**, *7*, 238–242. [CrossRef] [PubMed]
13. Liu, B.; Cao, S.; Gao, N.; Cheng, L.; Liu, Y.; Zhang, Y.; Feng, D. Thermosetting CFRP interlaminar toughening with multi-layers graphene and MWCNTs under mode I fracture. *Compos. Sci. Technol.* **2019**, *183*, 107829. [CrossRef]
14. Zhang, Y.Y.; Gu, Y.T. Mechanical properties of graphene: Effects of layer number, temperature and isotope. *Comput. Mater. Sci.* **2013**, *71*, 197–200. [CrossRef]
15. Mag-isa, A.E.; Kim, S.-M.; Kim, J.-H.; Oh, C.-S. Variation of thermal expansion coefficient of freestanding multilayer pristine graphene with temperature and number of layers. *Mater. Today Commun.* **2020**, *25*, 101387. [CrossRef]
16. Tsyganov, D.; Bundaleska, N.; Henriques, J.; Felizardo, E.; Dias, A.; Abrashev, M.; Kissovski, J.; Botelho do Rego, A.M.; Ferraria, A.M.; Tatarova, E. Simultaneous Synthesis and Nitrogen Doping of Free-Standing Graphene Applying Microwave Plasma. *Materials* **2020**, *13*, 4213. [CrossRef]
17. Mukri, M.-R.; Elias, M.S.; Aziz, M.; Tanemura, M.; Mohd Yusop, M.Z. Structural Modification of Pristine Graphene Network Towards Nanoporous Graphene Membrane: A Review. *J. Appl. Membr. Sci. Technol.* **2018**, *22*. [CrossRef]
18. Novoselov, K.S.; Geim, A.K.; Morozov, S.V.; Jiang, D.; Zhang, Y.; Dubonos, S.V.; Grigorieva, I.V.; Firsov, A.A. Electric field effect in atomically thin carbon films. *Science* **2004**, *306*, 666–669. [CrossRef]
19. Geim, A.K.; Novoselov, K.S. The rise of graphene. In *Nanoscience and Technology*; Macmillan Publishers Ltd.: London, UK, 2009; pp. 11–19.
20. Shende, P.; Pathan, N. Potential of carbohydrate-conjugated graphene assemblies in biomedical applications. *Carbohydr. Polym.* **2021**, *255*, 117385. [CrossRef]
21. Song, S.; Shen, H.; Wang, Y.; Chu, X.; Xie, J.; Zhou, N.; Shen, J. Biomedical application of graphene: From drug delivery, tumor therapy, to theranostics. *Colloids Surf. B Biointerfaces* **2020**, *185*, 110596. [CrossRef]
22. Lin, L.; Peng, H.; Liu, Z. Synthesis challenges for graphene industry. *Nat. Mater.* **2019**, *18*, 520–524. [CrossRef]
23. Urban, K.W. The challenges of graphene. *Nat. Mater.* **2011**, *10*, 165–166. [CrossRef]
24. Reina, G.; González-Domínguez, J.M.; Criado, A.; Vázquez, E.; Bianco, A.; Prato, M. Promises, facts and challenges for graphene in biomedical applications. *Chem. Soc. Rev.* **2017**, *46*, 4400–4416. [CrossRef] [PubMed]
25. Jeong, J.H.; Kim, Y.H.; Roh, K.C.; Kim, K.-B. Effect of thermally decomposable spacers on graphene microsphere structure and restacking of graphene sheets during electrode fabrication. *Carbon* **2019**, *150*, 128–135. [CrossRef]
26. Xie, Q.; Zhang, Y.; Zhao, P. Facile fabrication of honeycomb-like restacking-inhibited graphene architecture with superior electrochemical performance for energy storage. *Mater. Lett.* **2018**, *225*, 93–96. [CrossRef]
27. Kota, M.; Park, H.S. Restacking-inhibited nitrogen-incorporated mesoporous reduced graphene oxides for high energy supercapacitors. *Ceram. Int.* **2018**, *44*, 3195–3200. [CrossRef]
28. Xu, G.; Yuan, J.; Geng, X.; Dou, H.; Chen, L.; Yan, X.; Zhu, H. Caterpillar-like graphene confining sulfur by restacking effect for high performance lithium sulfur batteries. *Chem. Eng. J.* **2017**, *322*, 454–462. [CrossRef]
29. Tambe, P. Synthesis and characterization of acid treated reduced graphene oxide. *Mater. Today Proc.* **2022**, *49*, 1294–1297. [CrossRef]
30. Fu, H.; Gao, B.; Liu, Z.; Liu, W.; Wang, Z.; Wang, M.; Li, J.; Feng, Z.; Reza Kamali, A. Electrochemical performance of honeycomb graphene prepared from acidic graphene oxide via a chemical expansion method. *J. Electroanal. Chem.* **2022**, *920*, 116545. [CrossRef]
31. Hauquier, F.; Alamarguy, D.; Viel, P.; Noël, S.; Filoramo, A.; Huc, V.; Houzé, F.; Palacin, S. Conductive-probe AFM characterization of graphene sheets bonded to gold surfaces. *Appl. Surf. Sci.* **2012**, *258*, 2920–2926. [CrossRef]
32. Agarwal, P.B.; Paulchowdhury, P.; Mukherjee, A.; Lohani, P.; Thakur, N.K. Optimization of oxygen plasma based etching of single layered graphene through Raman and FESEM characterization. *Mater. Today Proc.* **2022**, *48*, 616–618. [CrossRef]
33. Shtepliuk, I.; Ivanov, I.G.; Pliatsikas, N.; Iakimov, T.; Jamnig, A.; Sarakinos, K.; Yakimova, R. Probing the uniformity of silver-doped epitaxial graphene by micro-Raman mapping. *Phys. B Condens. Matter* **2020**, *580*, 411751. [CrossRef]
34. Pelaez-Fernandez, M.; Bermejo, A.; Benito, A.M.; Maser, W.K.; Arenal, R. Detailed thermal reduction analyses of graphene oxide via in-situ TEM/EELS studies. *Carbon* **2021**, *178*, 477–487. [CrossRef]
35. Wang, Z.; Yu, H.; Zhao, Z. Silk fibroin hydrogel encapsulated graphene filed-effect transistors as enzyme-based biosensors. *Microchem. J.* **2021**, *169*, 106585. [CrossRef]
36. GV, Y.D.; Prabhu, A.; Anil, S.; Venkatesan, J. Preparation and characterization of dexamethasone loaded sodium alginate-graphene oxide microspheres for bone tissue engineering. *J. Drug Deliv. Sci. Technol.* **2021**, *64*, 102624. [CrossRef]

37. Karimi, S.; Namazi, H. $Fe_3O_4$@PEG-coated dendrimer modified graphene oxide nanocomposite as a pH-sensitive drug carrier for targeted delivery of doxorubicin. *J. Alloys Compd.* **2021**, *879*, 160426. [CrossRef]
38. Sharif, S.; Ahmad, K.S.; Rehman, F.; Bhatti, Z.; Thebo, K.H. Two-dimensional graphene oxide based membranes for ionic and molecular separation: Current status and challenges. *J. Environ. Chem. Eng.* **2021**, *9*, 105605. [CrossRef]
39. Ikram, R.; Jan, B.M.; Ahmad, W. Advances in synthesis of graphene derivatives using industrial wastes precursors; prospects and challenges. *J. Mater. Res. Technol.* **2020**, *9*, 15924–15951. [CrossRef]
40. Rotte, N.K.; Naresh, V.; Muduli, S.; Reddy, V.; Srikanth, V.V.S.; Martha, S.K. Microwave aided scalable synthesis of sulfur, nitrogen co-doped few-layered graphene material for high-performance supercapacitors. *Electrochim. Acta* **2020**, *363*, 137209. [CrossRef]
41. Salussolia, G.; Barbieri, E.; Pugno, N.M.; Botto, L. Micromechanics of liquid-phase exfoliation of a layered 2D material: A hydrodynamic peeling model. *J. Mech. Phys. Solids* **2020**, *134*, 103764. [CrossRef]
42. Hwang, G.; Kim, T.; Shin, J.; Shin, N.; Hwang, S. Machine learnings for CVD graphene analysis: From measurement to simulation of SEM images. *J. Ind. Eng. Chem.* **2021**, *101*, 430–444. [CrossRef]
43. Yildiz, G.; Bolton-Warberg, M.; Awaja, F. Graphene and graphene oxide for bio-sensing: General properties and the effects of graphene ripples. *Acta Biomater.* **2021**, *131*, 62–79. [CrossRef]
44. Guo, J.; Mao, B.; Li, J.; Wang, X.; Yang, X. Rethinking the reaction pathways of chemical reduction of graphene oxide. *Carbon* **2021**, *171*, 963–967. [CrossRef]
45. Barkauskas, J.; Gaidukevič, J.; Niaura, G. Thermal reduction of graphite oxide in the presence of nitrogen-containing dyes. *Carbon Lett.* **2021**, *31*, 1097–1110. [CrossRef]
46. Wadekar, P.H.; Ghosh, A.; Khose, R.V.; Pethsangave, D.A.; Mitra, S.; Some, S. A novel chemical reduction/co-precipitation method to prepare sulfur functionalized reduced graphene oxide for lithium-sulfur batteries. *Electrochim. Acta* **2020**, *344*, 136147. [CrossRef]
47. Vázquez-Sánchez, P.; Rodríguez-Escudero, M.A.; Burgos, F.J.; Llorente, I.; Caballero-Calero, O.; González, M.M.; Fernández, R.; García-Alonso, M.C. Synthesis of Cu/rGO composites by chemical and thermal reduction of graphene oxide. *J. Alloys Compd.* **2019**, *800*, 379–391. [CrossRef]
48. Demetriou, G.; Biancalana, F.; Abraham, E.; Ji, W.; Wang, Y.; Kar, A.K. Direct observation of an irradiance dependent nonlinear refraction in CVD single layer graphene. *Opt. Commun.* **2021**, *481*, 126535. [CrossRef]
49. Jiang, X.; Song, J.; Chen, S.; Su, Y.; Fan, H.; Zhang, Y.; Hu, L. In-situ fabricated bulk metallic glass/graphite composites with a 3D lubricating layer: Tribological properties under dry sliding and in seawater. *Tribol. Int.* **2020**, *148*, 106301. [CrossRef]
50. Liu, Q.; Pang, M.; Chen, J.; Liu, G.; Zhang, L. Microstructure and properties characterization of Ti-containing Ni60/Graphite self-lubricating composite coatings applied on 300 M ultra-high strength steel by laser cladding. *Mater. Chem. Phys.* **2021**, *266*, 124554. [CrossRef]
51. Ferreira, M.P.; da Nova Mussel, W.; Dutra, P.R.; Ângela de Barros Correia Menezes, M.; Pedrosa, T.A. Physical-chemical exfoliation of pristine graphite flakes. *Radiat. Phys. Chem.* **2021**, *188*, 109652. [CrossRef]
52. Tene, T.; Guevara, M.; Viteri, E.; Maldonado, A.; Pisarra, M.; Sindona, A.; Vacacela Gomez, C.; Bellucci, S. Calibration of Fermi Velocity to Explore the Plasmonic Character of Graphene Nanoribbon Arrays by a Semi-Analytical Model. *Nanomaterials* **2022**, *12*, 2028. [CrossRef] [PubMed]
53. Aixart, J.; Díaz, F.; Llorca, J.; Rosell-Llompart, J. Increasing reaction time in Hummers' method towards well exfoliated graphene oxide of low oxidation degree. *Ceram. Int.* **2021**, *47*, 22130–22137. [CrossRef]
54. Rozmanowski, T.; Krawczyk, P. Methanol electrooxidation at $NiCl_2$–$FeCl_3$–graphite intercalation compound affected by ozone treatment. *J. Phys. Chem. Solids* **2021**, *157*, 110223. [CrossRef]
55. Yamamoto, H.; Matsumoto, K.; Hagiwara, R. Stage-number dependence of intercalated species for fluorosilicate graphite intercalation compounds: Pentafluorosilicate vs. hexafluorosilicate. *J. Fluor. Chem.* **2021**, *242*, 109714. [CrossRef]
56. Cahen, S.; Vangelisti, R. Chemical vapor transport for intercalation reactions: Synthesis of a 1st stage DyCl3 graphite intercalation compound. *J. Solid State Chem.* **2021**, *299*, 122185. [CrossRef]
57. Suvarna, K.S.; Binitha, N.N. Graphene preparation by jaggery assisted ball-milling of graphite for the adsorption of Cr(VI). *Mater. Today Proc.* **2020**, *25*, 236–240. [CrossRef]
58. Tenorio Gonzalez, F.N.; Barajas Rosales, I.R.; Vera Serna, P.; Sánchez de Jesus, F.; Bolarin Miró, A.M.; Garrido Hernández, A.; Kusý, M. Reducing the crystallite and particle size of SrFe12O19 with PVA by high energy ball milling. *J. Alloys Compd.* **2019**, *771*, 464–470. [CrossRef]
59. Tsyganov, D.; Bundaleska, N.; Tatarova, E.; Dias, A.; Henriques, J.; Rego, A.; Ferraria, A.; Abrashev, M.V.; Dias, F.M.; Luhrs, C.C.; et al. On the plasma-based growth of 'flowing' graphene sheets at atmospheric pressure conditions. *Plasma Sources Sci. Technol.* **2016**, *25*, 015013. [CrossRef]
60. Tatarova, E.; Dias, A.; Henriques, J.; Abrashev, M.; Bundaleska, N.; Kovacevic, E.; Bundaleski, N.; Cvelbar, U.; Valcheva, E.; Arnaudov, B.; et al. Towards large-scale in free-standing graphene and N-graphene sheets. *Sci. Rep.* **2017**, *7*, 10175. [CrossRef]
61. Zhao, C.; Hong, Y.; Chu, X.; Dong, Y.; Hu, Z.; Sun, X.; Yan, S. Enhanced ferroelectric properties of P(VDF-TrFE) thin film on single-layer graphene simply adjusted by crystallization condition. *Mater. Today Energy* **2021**, *20*, 100678. [CrossRef]
62. Wu, Y.; Wang, S.; Komvopoulos, K. A review of graphene synthesis by indirect and direct deposition methods. *J. Mater. Res.* **2020**, *35*, 76–89. [CrossRef]
63. Depine, R.A. *Graphene Optics: Electromagnetic Solution of Canonical Problems*; IOP Publishing: Bristol, UK, 2016; ISBN 978-1-6817-4309-7.

64. Hanson, G.W. Dyadic Green's functions and guided surface waves for a surface conductivity model of graphene. *J. Appl. Phys.* **2008**, *103*, 064302. [CrossRef]
65. Xu, X.; Sun, S.; Luo, J.; Ma, R.; Lin, J.; Fang, L.; Zhang, P.; Chen, Y. Few-layer graphene prepared via microwave irradiation of black sesame for supercapacitor applications. *Chem. Eng. J.* **2021**, *425*, 130664. [CrossRef]
66. Chen, J.; Liu, R.; Zhu, L.; Chen, W.; Dong, C.; Wan, Z.; Cao, W.; Zhang, X.; Peng, R.; Wang, M. Sb2S3-based bulk/nano planar heterojunction film solar cells with graphene/polymer composite layer as hole extracting interface. *Mater. Lett.* **2021**, *300*, 130190. [CrossRef]
67. Zhang, Z.; Jin, H.; Wu, C.; Ji, J. Efficient Production of High-Quality Few-Layer Graphene Using a Simple Hydrodynamic-Assisted Exfoliation Method. *Nanoscale Res. Lett.* **2018**, *13*, 416. [CrossRef] [PubMed]
68. Schilirò, E.; Lo Nigro, R.; Panasci, S.E.; Gelardi, F.M.; Agnello, S.; Yakimova, R.; Roccaforte, F.; Giannazzo, F. Aluminum oxide nucleation in the early stages of atomic layer deposition on epitaxial graphene. *Carbon* **2020**, *169*, 172–181. [CrossRef]
69. Gao, B.; Hu, C.; Fu, H.; Sun, Y.; Li, K.; Hu, L. Preparation of single-layer graphene based on a wet chemical synthesis route and the effect on electrochemical properties by double layering surface functional groups to modify graphene oxide. *Electrochim. Acta* **2020**, *361*, 137053. [CrossRef]
70. Yuan, S.-J.; Dong, B.; Dai, X.-H. Facile and scalable synthesis of high-quality few-layer graphene from biomass by a universal solvent-free approach. *Appl. Surf. Sci.* **2021**, *562*, 150203. [CrossRef]
71. Bhuiyan, A.G.; Terai, T.; Katsuzaki, T.; Takeda, N.; Hashimoto, A. Growth of single crystalline Si on graphene using RF-MBE: Orientation control with an AlN interface layer. *Appl. Surf. Sci.* **2021**, *548*, 149295. [CrossRef]
72. Araby, M.I.; Rosmi, M.S.; Vishwakarma, R.; Sharma, S.; Wakamatsu, Y.; Takahashi, K.; Kalita, G.; Kitazawa, M.; Tanemura, M. Graphene formation at 150 °C using indium as catalyst. *RSC Adv.* **2017**, *7*, 47353–47356. [CrossRef]
73. Long, C.; Xie, X.; Fu, J.; Wang, Q.; Guo, H.; Zeng, W.; Wei, N.; Wang, S.; Xiong, Y. Supercapacitive brophene-graphene aerogel as elastic-electrochemical dielectric layer for sensitive pressure sensors. *J. Colloid Interface Sci.* **2021**, *601*, 355–364. [CrossRef] [PubMed]
74. Navik, R.; Gai, Y.; Wang, W.; Zhao, Y. Curcumin-assisted ultrasound exfoliation of graphite to graphene in ethanol. *Ultrason. Sonochem.* **2018**, *48*, 96–102. [CrossRef] [PubMed]
75. Ding, J.; Zhao, H.; Yu, H. Graphene nanofluids based on one-step exfoliation and edge-functionalization. *Carbon* **2021**, *171*, 29–35. [CrossRef]
76. Ferrari, A.C.; Meyer, J.C.; Scardaci, V.; Casiraghi, C.; Lazzeri, M.; Mauri, F.; Piscanec, S.; Jiang, D.; Novoselov, K.S.; Roth, S.; et al. Raman Spectrum of Graphene and Graphene Layers. *Phys. Rev. Lett.* **2006**, *97*, 187401. [CrossRef]
77. Campanelli, D.A.; Menezes, J.W.; Valsecchi, C.; Zegarra, L.B.R.; Jacinto, C.; Armas, L.E.G. Interaction between $Yb^{3+}$ doped glasses substrates and graphene layers by raman spectroscopy. *Thin Solid Films* **2020**, *712*, 138315. [CrossRef]
78. Wang, T.; Huntzinger, J.-R.; Bayle, M.; Roblin, C.; Decams, J.-M.; Zahab, A.-A.; Contreras, S.; Paillet, M.; Landois, P. Buffer layers inhomogeneity and coupling with epitaxial graphene unravelled by Raman scattering and graphene peeling. *Carbon* **2020**, *163*, 224–233. [CrossRef]
79. Silva, D.L.; Campos, J.L.E.; Fernandes, T.F.D.; Rocha, J.N.; Machado, L.R.P.; Soares, E.M.; Miquita, D.R.; Miranda, H.; Rabelo, C.; Vilela Neto, O.P.; et al. Raman spectroscopy analysis of number of layers in mass-produced graphene flakes. *Carbon* **2020**, *161*, 181–189. [CrossRef]
80. Huang, S.; Li, S.; Hsu, K.-J.; Villalobos, L.F.; Agrawal, K.V. Systematic design of millisecond gasification reactor for the incorporation of gas-sieving nanopores in single-layer graphene. *J. Membr. Sci.* **2021**, *637*, 119628. [CrossRef]
81. Bouhafs, C.; Pezzini, S.; Geisenhof, F.R.; Mishra, N.; Mišeikis, V.; Niu, Y.; Struzzi, C.; Weitz, R.T.; Zakharov, A.A.; Forti, S.; et al. Synthesis of large-area rhombohedral few-layer graphene by chemical vapor deposition on copper. *Carbon* **2021**, *177*, 282–290. [CrossRef]
82. Bleu, Y.; Bourquard, F.; Michalon, J.-Y.; Lefkir, Y.; Reynaud, S.; Loir, A.-S.; Barnier, V.; Garrelie, F.; Donnet, C. Transfer-free graphene synthesis by nickel catalyst dewetting using rapid thermal annealing. *Appl. Surf. Sci.* **2021**, *555*, 149492. [CrossRef]
83. Lebioda, M.; Pawlak, R.; Szymański, W.; Kaczorowski, W.; Jeziorna, A. Laser Patterning a Graphene Layer on a Ceramic Substrate for Sensor Applications. *Sensors* **2020**, *20*, 2134. [CrossRef]
84. Kaur, M.; Kaur, M.; Sharma, V.K. Nitrogen-doped graphene and graphene quantum dots: A review onsynthesis and applications in energy, sensors and environment. *Adv. Colloid Interface Sci.* **2018**, *259*, 44–64. [CrossRef]
85. Rodriguez-Losada, N.; Aguirre, J. The impact of graphene on neural regenerative medicine. *Neural Regen. Res.* **2017**, *12*, 1071. [CrossRef]
86. Jahanshahi, M.; Kowsari, E.; Haddadi-Asl, V.; Khoobi, M.; Bazri, B.; Aryafard, M.; Lee, J.H.; Kadumudi, F.B.; Talebian, S.; Kamaly, N.; et al. An innovative and eco-friendly modality for synthesis of highly fluorinated graphene by an acidic ionic liquid: Making of an efficacious vehicle for anti-cancer drug delivery. *Appl. Surf. Sci.* **2020**, *515*, 146071. [CrossRef]
87. Potsi, G.; Bourlinos, A.B.; Mouselimis, V.; Poláková, K.; Chalmpes, N.; Gournis, D.; Kalytchuk, S.; Tomanec, O.; Błoński, P.; Medved', M.; et al. Intrinsic photoluminescence of amine-functionalized graphene derivatives for bioimaging applications. *Appl. Mater. Today* **2019**, *17*, 112–122. [CrossRef]
88. Gu, H.; Tang, H.; Xiong, P.; Zhou, Z. Biomarkers-based Biosensing and Bioimaging with Graphene for Cancer Diagnosis. *Nanomaterials* **2019**, *9*, 130. [CrossRef] [PubMed]

89. Kovalska, E.; Lesongeur, P.; Hogan, B.T.; Baldycheva, A. Multi-layer graphene as a selective detector for future lung cancer biosensing platforms. *Nanoscale* **2019**, *11*, 2476–2483. [CrossRef] [PubMed]
90. Nair, R.R.; Blake, P.; Grigorenko, A.N.; Novoselov, K.S.; Booth, T.J.; Stauber, T.; Peres, N.M.R.; Geim, A.K. Fine Structure Constant Defines Visual Transparency of Graphene. *Science* **2008**, *320*, 1308. [CrossRef] [PubMed]
91. Sang, T.; Gao, J.; Yin, X.; Qi, H.; Wang, L.; Jiao, H. Angle-Insensitive Broadband Absorption Enhancement of Graphene Using a Multi-Grooved Metasurface. *Nanoscale Res. Lett.* **2019**, *14*, 105. [CrossRef]
92. Chen, J.; Chen, S.; Gu, P.; Yan, Z.; Tang, C.; Xu, Z.; Liu, B.; Liu, Z. Electrically modulating and switching infrared absorption of monolayer graphene in metamaterials. *Carbon* **2020**, *162*, 187–194. [CrossRef]
93. Divya, J.; Selvendran, S.; Raja, A.S.; Sivasubramanian, A. Graphene-Au-Coated Plasmonic Sensor Based on D-Shaped Bezier Polygonal Hollow Core Photonic Crystal Fiber. *Braz. J. Phys.* **2021**, *51*, 1314–1323. [CrossRef]
94. Georgakilas, V.; Otyepka, M.; Bourlinos, A.B.; Chandra, V.; Kim, N.; Kemp, K.C.; Hobza, P.; Zboril, R.; Kim, K.S. Functionalization of Graphene: Covalent and Non-Covalent Approaches, Derivatives and Applications. *Chem. Rev.* **2012**, *112*, 6156–6214. [CrossRef]
95. Serban, B.-C.; Cobianu, C.; Buiu, O.; Bumbac, M.; Dumbravescu, N.; Avramescu, V.; Nicolescu, C.M.; Brezeanu, M.; Radulescu, C.; Craciun, G.; et al. Quaternary Oxidized Carbon Nanohorns—Based Nanohybrid as Sensing Coating for Room Temperature Resistive Humidity Monitoring. *Coatings* **2021**, *11*, 530. [CrossRef]
96. Kumar Manoharan, A.; Jayavel, R.; Shanmugam, M.; Sengottaiyan, C.; Chinnathambi, S.; Mohankumar, N. A Compact Sensory Platform Based pH Sensor Using Graphene Field Effect Transistor. *J. Nanosci. Nanotechnol.* **2021**, *21*, 3299–3305. [CrossRef]
97. Xu, S.; Zhan, J.; Man, B.; Jiang, S.; Yue, W.; Gao, S.; Guo, C.; Liu, H.; Li, Z.; Wang, J.; et al. Real-time reliable determination of binding kinetics of DNA hybridization using a multi-channel graphene biosensor. *Nat. Commun.* **2017**, *8*, 14902. [CrossRef] [PubMed]
98. Ahmed, S.; Jiang, X.; Wang, C.; Kalsoom, U.E.; Wang, B.; Khan, J.; Muhammad, Y.; Duan, Y.; Zhu, H.; Ren, X.; et al. An Insightful Picture of Nonlinear Photonics in 2D Materials and their Applications: Recent Advances and Future Prospects. *Adv. Opt. Mater.* **2021**, *9*, 2001671. [CrossRef]
99. Hu, Z.; Shao, Q.; Moloney, M.G.; Xu, X.; Zhang, D.; Li, J.; Zhang, C.; Huang, Y. Nondestructive Functionalization of Graphene by Surface-Initiated Atom Transfer Radical Polymerization: An Ideal Nanofiller for Poly(p-phenylene benzobisoxazole) Fibers. *Macromolecules* **2017**, *50*, 1422–1429. [CrossRef]
100. Devi, N.A.; Swain, B.P. Investigation of Metal-Oxide/Reduced Graphene-Oxide Nanocomposites for Gas Sensor Applications. In *Nanostructured Materials and Their Applications*; Springer: Singapore, 2021; pp. 211–227.
101. Nurrohman, D.T.; Chiu, N.-F. A Review of Graphene-Based Surface Plasmon Resonance and Surface-Enhanced Raman Scattering Biosensors: Current Status and Future Prospects. *Nanomaterials* **2021**, *11*, 216. [CrossRef]
102. Wenjuan, Y.; Le Goff, A.; Spinelli, N.; Holzinger, M.; Diao, G.-W.; Shan, D.; Defrancq, E.; Cosnier, S. Electrogenerated trisbipyridyl Ru(II)-/nitrilotriacetic-polypyrene copolymer for the easy fabrication of label-free photoelectrochemical immunosensor and aptasensor: Application to the determination of thrombin and anti-cholera toxinantibody. *Biosens. Bioelectron.* **2013**, *42*, 556–562. [CrossRef]
103. Cai, H.; Wang, M.; Wu, Z.; Liu, J.; Wang, X. Performance Enhancement of SPR Biosensor Using Graphene–$MoS_2$ Hybrid Structure. *Nanomaterials* **2022**, *12*, 2219. [CrossRef] [PubMed]
104. Fatima, N.; Qazi, U.Y.; Mansha, A.; Bhatti, I.A.; Javaid, R.; Abbas, Q.; Nadeem, N.; Rehan, Z.A.; Noreen, S.; Zahid, M. Recent developments for antimicrobial applications of graphene-based polymeric composites: A review. *J. Ind. Eng. Chem.* **2021**, *100*, 40–58. [CrossRef]
105. Wychowaniec, J.K.; Iliut, M.; Zhou, M.; Moffat, J.; Elsawy, M.A.; Pinheiro, W.A.; Hoyland, J.A.; Miller, A.F.; Vijayaraghavan, A.; Saiani, A. Designing Peptide/Graphene Hybrid Hydrogels through Fine-Tuning of Molecular Interactions. *Biomacromolecules* **2018**, *19*, 2731–2741. [CrossRef] [PubMed]
106. Vafaei, S.; Allabush, F.; Tabaei, S.R.; Male, L.; Dafforn, T.R.; Tucker, J.H.R.; Mendes, P.M. Förster Resonance Energy Transfer Nanoplatform Based on Recognition-Induced Fusion/Fission of DNA Mixed Micelles for Nucleic Acid Sensing. *ACS Nano* **2021**, *15*, 8517–8524. [CrossRef] [PubMed]
107. Wychowaniec, J.K.; Litowczenko, J.; Tadyszak, K. Fabricating versatile cell supports from nano- and micro-sized graphene oxide flakes. *J. Mech. Behav. Biomed. Mater.* **2020**, *103*, 103594. [CrossRef] [PubMed]
108. Koren, K.; Zieger, S.E. Optode Based Chemical Imaging—Possibilities, Challenges, and New Avenues in Multidimensional Optical Sensing. *ACS Sens.* **2021**, *6*, 1671–1680. [CrossRef]
109. Parvin, N.; Mandal, T.; Roy, P. Soluble Nanoparticles Versatile Tools for Plant Tissue Imaging. *J. Bionanosci.* **2013**, *7*, 256–259. [CrossRef]
110. Kim, S.; Moriya, S.; Maruki, S.; Fukaminato, T.; Ogata, T.; Kurihara, S. Adsorption and release on three-dimensional graphene oxide network structures. *R. Soc. Open Sci.* **2021**, *8*, 201585. [CrossRef]
111. Driscoll, J.; Moirangthem, A.; Yan, I.K.; Patel, T. Fabrication and Characterization of a Biomaterial Based on Extracellular-Vesicle Functionalized Graphene Oxide. *Front. Bioeng. Biotechnol.* **2021**, *9*, 686510. [CrossRef]
112. Moody, A.S.; Dayton, P.A.; Zamboni, W.C. Imaging methods to evaluate tumor microenvironment factors affecting nanoparticle drug delivery and antitumor response. *Cancer Drug Resist.* **2021**, *4*, 382. [CrossRef]
113. Jakubovic, R.; Ramjist, J.; Gupta, S.; Guha, D.; Sahgal, A.; Foster, F.S.; Yang, V.X.D. High-Frequency Micro-Ultrasound Imaging and Optical Topographic Imaging for Spinal Surgery: Initial Experiences. *Ultrasound Med. Biol.* **2018**, *44*, 2379–2387. [CrossRef]

114. Phan, T.T.V.; Huynh, T.-C.; Manivasagan, P.; Mondal, S.; Oh, J. An Up-To-Date Review on Biomedical Applications of Palladium Nanoparticles. *Nanomaterials* **2019**, *10*, 66. [CrossRef] [PubMed]
115. Tadyszak, K.; Wychowaniec, J.; Litowczenko, J. Biomedical Applications of Graphene-Based Structures. *Nanomaterials* **2018**, *8*, 944. [CrossRef] [PubMed]
116. Parvin, N.; Jin, Q.; Wei, Y.; Yu, R.; Zheng, B.; Huang, L.; Zhang, Y.; Wang, L.; Zhang, H.; Gao, M.; et al. Few-Layer Graphdiyne Nanosheets Applied for Multiplexed Real-Time DNA Detection. *Adv. Mater.* **2017**, *29*, 201606755. [CrossRef]
117. Dolatkhah, M.; Hashemzadeh, N.; Barar, J.; Adibkia, K.; Aghanejad, A.; Barzegar-Jalali, M.; Omidi, Y. Graphene-based multifunctional nanosystems for simultaneous detection and treatment of breast cancer. *Colloids Surf. B Biointerfaces* **2020**, *193*, 111104. [CrossRef]
118. Jang, S.-C.; Kang, S.-M.; Lee, J.Y.; Oh, S.Y.; Vilian, A.E.; Lee, I.; Han, Y.-K.; Park, J.H.; Cho, W.-S.; Roh, C.; et al. Nano-graphene oxide composite for in vivo imaging. *Int. J. Nanomed.* **2018**, *13*, 221–234. [CrossRef]
119. Matsumoto, K.; Mitchell, J.B.; Krishna, M.C. Multimodal Functional Imaging for Cancer/Tumor Microenvironments Based on MRI, EPRI, and PET. *Molecules* **2021**, *26*, 1614. [CrossRef]
120. Sheth, D.B.; Gratzl, M. Electrochemical mapping of oxygenation in the three-dimensional multicellular tumour hemi-spheroid. *Proc. R. Soc. A Math. Phys. Eng. Sci.* **2019**, *475*, 20180647. [CrossRef]
121. Kuppusamy, P.; Shankar, R.A.; Zweier, J.L. In vivo measurement of arterial and venous oxygenation in the rat using 3D spectral-spatial electron paramagnetic resonance imaging. *Phys. Med. Biol.* **1998**, *43*, 1837–1844. [CrossRef]
122. Khramtsov, V.V.; Bobko, A.A.; Tseytlin, M.; Driesschaert, B. Exchange Phenomena in the Electron Paramagnetic Resonance Spectra of the Nitroxyl and Trityl Radicals: Multifunctional Spectroscopy and Imaging of Local Chemical Microenvironment. *Anal. Chem.* **2017**, *89*, 4758–4771. [CrossRef] [PubMed]
123. Nguyen, H.V.-T.; Chen, Q.; Paletta, J.T.; Harvey, P.; Jiang, Y.; Zhang, H.; Boska, M.D.; Ottaviani, M.F.; Jasanoff, A.; Rajca, A.; et al. Nitroxide-Based Macromolecular Contrast Agents with Unprecedented Transverse Relaxivity and Stability for Magnetic Resonance Imaging of Tumors. *ACS Cent. Sci.* **2017**, *3*, 800–811. [CrossRef]
124. Boś-Liedke, A.; Walawender, M.; Woźniak, A.; Flak, D.; Gapiński, J.; Jurga, S.; Kucińska, M.; Plewiński, A.; Murias, M.; Elewa, M.; et al. EPR Oximetry Sensor—Developing a TAM Derivative for In Vivo Studies. *Cell Biochem. Biophys.* **2018**, *76*, 19–28. [CrossRef]
125. Mrówczyński, R.; Coy, L.E.; Scheibe, B.; Czechowski, T.; Augustyniak-Jabłokow, M.; Jurga, S.; Tadyszak, K. Electron Paramagnetic Resonance Imaging and Spectroscopy of Polydopamine Radicals. *J. Phys. Chem. B* **2015**, *119*, 10341–10347. [CrossRef] [PubMed]
126. Ferini, G.; Valenti, V.; Tripoli, A.; Illari, S.I.; Molino, L.; Parisi, S.; Cacciola, A.; Lillo, S.; Giuffrida, D.; Pergolizzi, S. Lattice or Oxygen-Guided Radiotherapy: What If They Converge? Possible Future Directions in the Era of Immunotherapy. *Cancers* **2021**, *13*, 3290. [CrossRef]
127. Hirai, N.; Maeda, Y.; Hashimoto, K.; Andriana, B.B.; Matsuyoshi, H.; Sato, H. Coherent Anti-Stokes Raman Scattering Spectroscopy Using a Double-Wavelength-Emission Electronically Tuned Ti:Sapphire Laser. *Appl. Spectrosc.* **2021**, *75*, 988–993. [CrossRef] [PubMed]
128. Pence, I.J.; Goss, A.; Fast, A.; Brinkmann, M.; Hellwig, T.; Fallnich, C.; Evans, C.L. Development of a clinical coherent Raman imaging system for in vivo drug monitoring. In *Proceedings of the Biophotonics Congress 2021*; Optica Publishing Group: Washington, DC, USA; 2021; p. NTu3C.4. [CrossRef]
129. Yeh, P.-C.; Yoon, S.; Kurniawan, D.; Chung, Y.G.; Chiang, W.-H. Unraveling the Fluorescence Quenching of Colloidal Graphene Quantum Dots for Selective Metal Ion Detection. *ACS Appl. Nano Mater.* **2021**, *4*, 5636–5642. [CrossRef]
130. Lin, J.; Huang, Y.; Huang, P. Graphene-Based Nanomaterials in Bioimaging. In *Biomedical Applications of Functionalized Nanomaterials*; Elsevier: Amsterdam, The Netherlands, 2018; pp. 247–287.
131. Wen, W.; Song, Y.; Yan, X.; Zhu, C.; Du, D.; Wang, S.; Asiri, A.M.; Lin, Y. Recent advances in emerging 2D nanomaterials for biosensing and bioimaging applications. *Mater. Today* **2018**, *21*, 164–177. [CrossRef]
132. Liang, L.; Peng, X.; Sun, F.; Kong, Z.; Shen, J.-W. A review on the cytotoxicity of graphene quantum dots: From experiment to simulation. *Nanoscale Adv.* **2021**, *3*, 904–917. [CrossRef]
133. Samanta, S.; Banerjee, S.L.; Bhattacharya, K.; Singha, N.K. Graphene Quantum Dots-Ornamented Waterborne Epoxy-Based Fluorescent Adhesive via Reversible Addition–Fragmentation Chain Transfer-Mediated Miniemulsion Polymerization: A Potential Material for Art Conservation. *ACS Appl. Mater. Interfaces* **2021**, *13*, 36307–36319. [CrossRef]
134. Bapli, A.; Chatterjee, A.; Gautam, R.K.; Jana, R.; Seth, D. Modulation of the Protein–Ligand Interaction in the Presence of Graphene Oxide: A Detailed Spectroscopic Study. *Langmuir* **2021**, *37*, 5034–5048. [CrossRef]
135. Ghosh, S.; Chatterjee, K. Poly(Ethylene Glycol) Functionalized Graphene Oxide in Tissue Engineering: A Review on Recent Advances. *Int. J. Nanomed.* **2020**, *15*, 5991–6006. [CrossRef] [PubMed]
136. Anusuya, T.; Kumar, V.; Kumar, V. Hydrophilic graphene quantum dots as turn-off fluorescent nanoprobes for toxic heavy metal ions detection in aqueous media. *Chemosphere* **2021**, *282*, 131019. [CrossRef]
137. A Ocsoy, M.; Yusufbeyoglu, S.; Ildiz, N.; Ulgen, A.; Ocsoy, I. DNA Aptamer-Conjugated Magnetic Graphene Oxide for Pathogenic Bacteria Aggregation: Selective and Enhanced Photothermal Therapy for Effective and Rapid Killing. *ACS Omega* **2021**, *6*, 20637–20643. [CrossRef] [PubMed]
138. Huang, Q.; Cai, Y.; Zhang, X.; Liu, J.; Liu, Z.; Li, B.; Wong, H.; Xu, F.; Sheng, L.; Sun, D.; et al. Aligned Graphene Mesh-Supported Double Network Natural Hydrogel Conduit Loaded with Netrin-1 for Peripheral Nerve Regeneration. *ACS Appl. Mater. Interfaces* **2021**, *13*, 112–122. [CrossRef] [PubMed]

139. Oh, S.D.; Kim, J.; Lee, D.H.; Kim, J.H.; Jang, C.W.; Kim, S.; Choi, S.-H. Structural and optical characteristics of graphene quantum dots size-controlled and well-aligned on a large scale by polystyrene-nanosphere lithography. *J. Phys. D Appl. Phys.* **2016**, *49*, 025308. [CrossRef]
140. Borandeh, S.; Alimardani, V.; Abolmaali, S.S.; Seppälä, J. Graphene Family Nanomaterials in Ocular Applications: Physicochemical Properties and Toxicity. *Chem. Res. Toxicol.* **2021**, *34*, 1386–1402. [CrossRef] [PubMed]
141. Jampilek, J.; Kralova, K. Advances in Drug Delivery Nanosystems Using Graphene-Based Materials and Carbon Nanotubes. *Materials* **2021**, *14*, 1059. [CrossRef]
142. Wang, F.; Duo, T.; Wang, Y.; Xiao, Z.; Xu, A.; Liu, R. Novel Polyethyleneimine/κ-Carrageenan Composite from Facile One-Step Fabrication for the Removal of Copper Ion from Aqueous Solution. *J. Polym. Environ.* **2022**, *30*, 1001–1011. [CrossRef]
143. Oh, J.-S.; Park, J.-S.; Lee, E.-J. Enhanced Effect of Polyethyleneimine-Modified Graphene Oxide and Simvastatin on Osteogenic Differentiation of Murine Bone Marrow-Derived Mesenchymal Stem Cells. *Biomedicines* **2021**, *9*, 501. [CrossRef]
144. Jonoush, Z.A.; Farahani, M.; Bohlouli, M.; Niknam, Z.; Golchin, A.; Hatamie, S.; Rezaei-Tavirani, M.; Omidi, M.; Zali, H. Surface Modification of Graphene and its Derivatives for Drug Delivery Systems. *Mini-Rev. Org. Chem.* **2021**, *18*, 78–92. [CrossRef]
145. Olson, E.; Plaut, J.S.; Barnhill, S.A.; Sabuncu, S.; Dambacher, C.M.; Speese, S.D.; Ranganathan, S.V.; Branchaud, B.P.; Yildirim, A. Enzyme-Instructed Formation of β-Sheet-Rich Nanoplatelets for Label-Free Protease Sensing. *ACS Appl. Nano Mater.* **2021**, *4*, 7800–7810. [CrossRef]
146. Pontrelli, G.; Toniolo, G.; McGinty, S.; Peri, D.; Succi, S.; Chatgilialoglu, C. Mathematical modelling of drug delivery from pH-responsive nanocontainers. *Comput. Biol. Med.* **2021**, *131*, 104238. [CrossRef]
147. Ren, M.-X.; Wang, Y.-Q.; Lei, B.-Y.; Yang, X.-X.; Hou, Y.-L.; Meng, W.-J.; Zhao, D.-L. Magnetite nanoparticles anchored on graphene oxide loaded with doxorubicin hydrochloride for magnetic hyperthermia therapy. *Ceram. Int.* **2021**, *47*, 20686–20692. [CrossRef]
148. Yoon, J.; Lim, J.; Shin, M.; Lee, S.-N.; Choi, J.-W. Graphene/$MoS_2$ Nanohybrid for Biosensors. *Materials* **2021**, *14*, 518. [CrossRef]
149. Lemus, L.M.R.; Azamar-Barrios, J.A.; Ortiz-Vazquez, E.; Quintana-Owen, P.; Freile-Pelegrín, Y.; Perera, F.G.; Madera-Santana, T.J. Development and physical characterization of novel bio-nanocomposite films based on reduced graphene oxide, agar and melipona honey. *Carbohydr. Polym. Technol. Appl.* **2021**, *2*, 100133. [CrossRef]
150. Mathew, T.; Sree, R.A.; Aishwarya, S.; Kounaina, K.; Patil, A.G.; Satapathy, P.; Hudeda, S.P.; More, S.S.; Muthucheliyan, K.; Kumar, T.N.; et al. Graphene-based functional nanomaterials for biomedical and bioanalysis applications. *FlatChem* **2020**, *23*, 100184. [CrossRef]
151. Işıklan, N.; Hussien, N.A.; Türk, M. Synthesis and drug delivery performance of gelatin-decorated magnetic graphene oxide nanoplatform. *Colloids Surf. A Physicochem. Eng. Asp.* **2021**, *616*, 126256. [CrossRef]
152. Kesavan, S.; Meena, K.; Sharmili, S.A.; Govindarajan, M.; Alharbi, N.S.; Kadaikunnan, S.; Khaled, J.M.; Alobaidi, A.S.; Alanzi, K.F.; Vaseeharan, B. Ulvan loaded graphene oxide nanoparticle fabricated with chitosan and d-mannose for targeted anticancer drug delivery. *J. Drug Deliv. Sci. Technol.* **2021**, *65*, 102760. [CrossRef]
153. Ebrahimi, M.; Teimouri, M.; Pooladi, M. The Synergistic Anticancer Traits of Graphene Oxide Plus Doxorubicin Against BT474 and MCF7 Breast Cancer Stem Cells In Vitro. *Appl. Biochem. Biotechnol.* **2021**, *193*, 3586–3601. [CrossRef]
154. Zhou, F.; Wang, M.; Luo, T.; Qu, J.; Chen, W.R. Photo-activated chemo-immunotherapy for metastatic cancer using a synergistic graphene nanosystem. *Biomaterials* **2021**, *265*, 120421. [CrossRef]
155. Mudaliar, P.; Pradeep, P.; Abraham, R.; Sreekumar, E. Targeting cap-dependent translation to inhibit Chikungunya virus replication: Selectivity of p38 MAPK inhibitors to virus-infected cells due to autophagy-mediated down regulation of phospho-ERK. *J. Gen. Virol.* **2021**, *102*, 001629. [CrossRef]
156. Xie, L.; Chen, F.; Du, H.; Zhang, X.; Wang, X.; Yao, G.; Xu, B. Graphene oxide and indole-3-acetic acid cotreatment regulates the root growth of Brassica napus L. via multiple phytohormone pathways. *BMC Plant Biol.* **2020**, *20*, 101. [CrossRef] [PubMed]
157. Parvin, N.; Mandal, T.K. Dually emissive P,N-co-doped carbon dots for fluorescent and photoacoustic tissue imaging in living mice. *Microchim. Acta* **2017**, *184*, 1117–1125. [CrossRef]
158. Zhang, X.; Zhang, X.; Song, L.; Huang, X.; Li, Y.; Qiao, M.; Liu, W.; Zhang, T.; Qi, Y.; Wang, W.; et al. An ultrasensitive, homogeneous fluorescence quenching immunoassay integrating separation and detection of aflatoxin M1 based on magnetic graphene composites. *Microchim. Acta* **2021**, *188*, 59. [CrossRef]
159. Vasilopoulos, V.; Pitou, M.; Fekas, I.; Papi, R.; Ouranidis, A.; Pavlidou, E.; Patsalas, P.; Choli-Papadopoulou, T. Graphene-Wrapped Copper Nanoparticles: An Antimicrobial and Biocompatible Nanomaterial with Valuable Properties for Medical Uses. *ACS Omega* **2020**, *5*, 26329–26334. [CrossRef]
160. Liu, C.; Chen, F.; Tang, Y.; Huo, P. An environmentally friendly nanocomposite polypyrrole@silver/reduced graphene oxide with high catalytic activity for bacteria and antibiotics. *J. Mater. Sci. Mater. Electron.* **2021**, *32*, 15211–15225. [CrossRef]
161. Mandal, T.K.; Lee, Y.R.; Parvin, N. Red phosphorus Decorating Graphene oxide Nanosheets: Label-free DNA detection. *Biomater. Sci.* **2020**, *8*, 125–131. [CrossRef] [PubMed]
162. Ryoo, S.-R.; Kim, Y.-K.; Kim, M.-H.; Min, D.-H. Behaviors of NIH-3T3 Fibroblasts on Graphene/Carbon Nanotubes: Proliferation, Focal Adhesion, and Gene Transfection Studies. *ACS Nano* **2010**, *4*, 6587–6598. [CrossRef] [PubMed]
163. Tiwari, S.; Patil, R.; Dubey, S.K.; Bahadur, P. Graphene nanosheets as reinforcement and cell-instructive material in soft tissue scaffolds. *Adv. Colloid Interface Sci.* **2020**, *281*, 102167. [CrossRef] [PubMed]

164. Kang, E.-S.; Kim, H.; Han, Y.; Cho, Y.-W.; Son, H.; Luo, Z.; Kim, T.-H. Enhancing osteogenesis of adipose-derived mesenchymal stem cells using gold nanostructure/peptide-nanopatterned graphene oxide. *Colloids Surf. B Biointerfaces* **2021**, *204*, 111807. [CrossRef] [PubMed]
165. Long, X.; Duan, L.; Weng, W.; Cheng, K.; Wang, D.; Ouyang, H. Light-induced osteogenic differentiation of BMSCs with graphene/$TiO_2$ composite coating on Ti implant. *Colloids Surf. B Biointerfaces* **2021**, *207*, 111996. [CrossRef] [PubMed]
166. Lee, E.A.; Kwak, S.-Y.; Yang, J.-K.; Lee, Y.-S.; Kim, J.-H.; Kim, H.D.; Hwang, N.S. Graphene oxide film guided skeletal muscle differentiation. *Mater. Sci. Eng. C* **2021**, *126*, 112174. [CrossRef] [PubMed]
167. Yang, Y.; Jiang, X.; Li, Y.; Jin, W.; Huang, X. Construction of halogenated graphenes by halogenation of hydrogenated graphene. *Compos. Commun.* **2021**, *25*, 100771. [CrossRef]
168. Feng, X.; Wang, J.; Cai, P.; Yang, Z.; Shen, J.; Zhang, Y.; Zhang, X. Graphene/protamine assembled hybrid paper with antibacterial activity. *Colloids Surf. A Physicochem. Eng. Asp.* **2021**, *625*, 126977. [CrossRef]
169. Du, J.; Wang, T.; Zhou, Q.; Hu, X.; Wu, J.; Li, G.; Li, G.; Hou, F.; Wu, Y. Graphene oxide enters the rice roots and disturbs the endophytic bacterial communities. *Ecotoxicol. Environ. Saf.* **2020**, *192*, 110304. [CrossRef]
170. Ruiz, O.N.; Fernando, K.A.S.; Wang, B.; Brown, N.A.; Luo, P.G.; McNamara, N.D.; Vangsness, M.; Sun, Y.-P.; Bunker, C.E. Graphene Oxide: A Nonspecific Enhancer of Cellular Growth. *ACS Nano* **2011**, *5*, 8100–8107. [CrossRef]
171. Abdalla, O.; Wahab, M.A.; Abdala, A. Fabrication of Graphene Oxide-Based Membranes and their Applications in Water Treatment. *Curr. Pharm. Biotechnol.* **2021**, *22*, 1686–1704. [CrossRef]
172. Haddadan, F.; Soroosh, M.; Alaei-Sheini, N. Designing an electro-optical encoder based on photonic crystals using the graphene–$Al_2O_3$ stacks. *Appl. Opt.* **2020**, *59*, 2179. [CrossRef]
173. Jalali Azizpour, M.R.; Soroosh, M.; Dalvand, N.; Seifi-Kavian, Y. All-Optical Ultra-Fast Graphene-Photonic Crystal Switch. *Crystals* **2019**, *9*, 461. [CrossRef]
174. Haddadan, F.; Soroosh, M.; Alaei-Sheini, N. Cross-talk reduction in a graphene-based ultra-compact plasmonic encoder using an Au nano-ridge on a silicon substrate. *Appl. Opt.* **2022**, *61*, 3209. [CrossRef]
175. Haddadan, F.; Soroosh, M. Design and simulation of a subwavelength 4-to-2 graphene-based plasmonic priority encoder. *Opt. Laser Technol.* **2023**, *157*, 108680. [CrossRef]
176. Bagheri, F.; Soroosh, M.; Haddadan, F.; Seifi-Kavian, Y. Design and simulation of a compact graphene-based plasmonic D flip-flop. *Opt. Laser Technol.* **2022**, *155*, 108436. [CrossRef]

MDPI
St. Alban-Anlage 66
4052 Basel
Switzerland
Tel. +41 61 683 77 34
Fax +41 61 302 89 18
www.mdpi.com

*Electronics* Editorial Office
E-mail: electronics@mdpi.com
www.mdpi.com/journal/electronics

MDPI

www.ingramcontent.com/pod-product-compliance
Lightning Source LLC
Chambersburg PA
CBHW061452210926
51850CB00008B/1689
* 9 7 8 3 0 3 6 5 7 9 8 9 4 *